Springer Proceedings in Mathematics & Statistics

Volume 175

Springer Proceedings in Mathematics & Statistics

This book series features volumes composed of selected contributions from workshops and conferences in all areas of current research in mathematics and statistics, including operation research and optimization. In addition to an overall evaluation of the interest, scientific quality, and timeliness of each proposal at the hands of the publisher, individual contributions are all refereed to the high quality standards of leading journals in the field. Thus, this series provides the research community with well-edited, authoritative reports on developments in the most exciting areas of mathematical and statistical research today.

More information about this series at http://www.springer.com/series/10533

Ricardo Cao · Wenceslao González Manteiga
Juan Romo
Editors

Nonparametric Statistics

2nd ISNPS, Cádiz, June 2014

Springer

Editors
Ricardo Cao
Department of Mathematics,
CITIC and ITMATI
University of A Coruña
A Coruña
Spain

Juan Romo
Department of Statistics
Carlos III University of Madrid
Getafe
Spain

Wenceslao González Manteiga
Faculty of Mathematics
University of Santiago de Compostela
Santiago de Compostela
Spain

ISSN 2194-1009 ISSN 2194-1017 (electronic)
Springer Proceedings in Mathematics & Statistics
ISBN 978-3-319-41581-9 ISBN 978-3-319-41582-6 (eBook)
DOI 10.1007/978-3-319-41582-6

Library of Congress Control Number: 2016942534

Mathematics Subject Classification (2010): 62G05, 62G07, 62G08, 62G09, 62G10, 62G15, 62G20, 62G30, 62G32, 62G35, 62G99

Printed on acid-free paper

This Springer imprint is published by Springer Nature
The registered company is Springer International Publishing AG Switzerland

Preface

This book provides a selection of papers developed from talks presented at the Second Conference of the International Society for Nonparametric Statistics (ISNPS), held in Cádiz (Spain) during June 12–16, 2014. The papers cover a wide spectrum of subjects within nonparametric and semiparametric statistics, including theory, methodology, applications and computational aspects. Some of the topics in this volume include nonparametric curve estimation, regression smoothing, dependent and time series data, varying coefficient models, symmetry testing, robust estimation, additive models, statistical process control, reliability, generalized linear models and nonparametric filtering.

ISNPS was founded in 2010 "to foster the research and practice of nonparametric statistics, and to promote the dissemination of new developments in the field via conferences, books and journal publications." ISNPS has a distinguished Advisory Committee that includes R. Beran, P. Bickel, R. Carroll, D. Cook, P. Hall, R. Johnson, B. Lindsay, E. Parzen, P. Robinson, M. Rosenblatt, G. Roussas, T. SubbaRao, and G. Wahba; an Executive Committee comprising M. Akritas, A. Delaigle, S. Lahiri and D. Politis and a Council that includes P. Bertail, G. Claeskens, R. Cao, M. Hallin, H. Koul, J.-P. Kreiss, T. Lee, R. Liu, W. González Manteiga, G. Michailidis, V. Panaretos, S. Paparoditis, J. Racine, J. Romo and Q. Yao.

The second conference included over 300 talks (keynote, special invited, invited and contributed) with presenters coming from all over the world. After the success of the first and second conferences, the third conference has recently taken place in Avignon, France, during June 11–16, 2016, with more than 350 participants. More information on the ISNPS and the conferences can be found at http://www.isnpstat.org/
.

Ricardo Cao
Wenceslao González-Manteiga
Juan Romo
Co-Editors of the book and
Co-Chairs of the Second ISNPS Conference

Contents

Contributors

Anestis Antoniadis University Cape Town, Cape Town, South Africa; University Joseph Fourier, Grenoble, France

D. Bagkavos Accenture, Athens, Greece

Giacomo Benini Geneva School for Economics and Management, Université de Genéve, Geneva, Switzerland

Eva Boj Facultat d'Economia i Empresa, Universitat de Barcelona, Barcelona, Spain

Stefano Bonnini Department of Economics and Management, University of Ferrara, Ferrara, Italy

Xavier Brossat EDF R&D, Clamart, France

Sutirtha Chakraborty National Institute of Biomedical Genomics, Kalyani, India

Teresa Costa Facultat d'Economia i Empresa, Universitat de Barcelona, Barcelona, Spain

V. Dalla National and Kapodistrian University of Athens, Athens, Greece

Somnath Datta University of Florida, Gainesville, FL, USA

Alexander V. Dobrovidov V.A. Trapeznikov Institute of Control Sciences of Russian Academy of Sciences, Moscow, Russia

María Paz Espinosa Departamento de Fundamentos del Análisis Económico II, BRiDGE, BETS, University of the Basque Country, Bilbao, Spain

Eva Ferreira Departamento de Economía Aplicada III & BETS, University of the Basque Country, Bilbao, Spain

Yannig Goude EDF R&D, Clamart, France; University Paris-Sud, Orsay, France

Ruite Guo Department of Statistics, Florida State University, Tallahassee, USA

J. Hidalgo London School of Economics, London, UK

Šárka Hudecová Department of Probability and Mathematical Statistics, Charles University of Prague, Prague 8, Czech Republic

Marie Hušková Department of Probability and Mathematical Statistics, Charles University of Prague, Prague 8, Czech Republic

Gennady Koshkin National Research Tomsk State University, Tomsk, Russia

Ignacio López-de-Ullibarri Universidade da Coruña. Escola Universitaria Politécnica, Ferrol, Spain

Natalia Markovich V.A. Trapeznikov Institute of Control Sciences of Russian Academy of Sciences, Moscow, Russia

Simos Meintanis Department of Economics, National and Kapodistrian University of Athens, Athens, Greece; Unit for Business Mathematics and Informatics, North-West University, Potchefstroom, South Africa

Antonio Meneses Universidad Nacional de Chimborazo, Riobamba, Ecuador

Salvador Naya Universidade da Coruña. Escola Politécnica Superior, Ferrol, Spain

P.N. Patil Department of Mathematics and Statistics, Mississippi State University, Mississippi, USA

Vic Patrangenaru Department of Statistics, Florida State University, Tallahassee, USA

Jean-Michel Poggi University Paris-Sud, Orsay, France; University Paris Descartes, Paris, France

Maria L. Rizzo Department of Mathematics and Statistics, Bowling Green State University, Bowling Green, OH, USA

Valery Smagin National Research Tomsk State University, Tomsk, Russia

Stefan Sperlich Geneva School for Economics and Management, Université de Genéve, Geneva, Switzerland

Winfried Stute Mathematical Institute, University of Giessen, Giessen, Germany

Gábor J. Székely National Science Foundation, Arlington, VA, USA

Javier Tarrío-Saavedra Universidade da Coruña. Escola Politécnica Superior, Ferrol, Spain

Raoul Theler Geneva School for Economics and Management, Université de Genéve, Geneva, Switzerland

Vincent Thouvenot Thales Communication & Security, Gennevilliers, France; University Paris-Sud, Orsay, France

A.T.A. Wood School of Mathematical Sciences, The University of Nottingham, Nottingham, UK

K. David Yao Department of Mathematics, Florida State University, Tallahassee, USA

A Numerical Study of the Power Function of a New Symmetry Test

D. Bagkavos, P.N. Patil and A.T.A. Wood

Abstract A new nonparametric test for the null hypothesis of symmetry is proposed. A necessary and sufficient condition for symmetry, which is based on the fact that under symmetry the covariance between the probability density and cumulative distribution functions of the underlying population is zero, is used to define the test statistic. The main emphasis here is on the small sample power properties of the test. Through simulations with samples generated from a wide range of distributions, it is shown that the test has a reasonable power function which compares favorably against many other existing tests of symmetry. It is also shown that the defining feature of this test is "the higher the asymmetry higher is the power".

Keywords Asymmetry · Skewness · Nonparametric estimation · Correlation

1 Introduction

The notion of symmetry or skewness of a probability density function (p.d.f.) is frequently met in the literature and in applications of statistical methods either as an assumption or as the main objective of study. Essentially the literature so far has been focused on assessing symmetry and skewness through characteristic properties of symmetric distributions (e.g., [5, 16]) or more recently through asymmetry functions (e.g., [4, 6, 8, 15]). See [6, 10] for an overview of the various measures, hypothesis tests, and methodological approaches developed so far. One aspect of asymmetry which did not receive much attention in the literature is its quantification. In this

D. Bagkavos (✉)
Accenture, Rostoviou 39–41, 11526 Athens, Greece
e-mail: dimitrios.bagkavos@gmail.com

P.N. Patil
Department of Mathematics and Statistics, Mississippi State University, Mississippi, USA

A.T.A. Wood
School of Mathematical Sciences, The University of Nottingham, Nottingham, UK

R. Cao et al. (eds.), *Nonparametric Statistics*, Springer Proceedings in Mathematics & Statistics 175, DOI 10.1007/978-3-319-41582-6_1

direction, [18, 19], proposed a weak and strong asymmetry measure, respectively, having as a purpose to measure the degree of asymmetry of a p.d.f. on a scale from -1 to 1.

Taking this work a step further, a hypothesis test for the null hypothesis of symmetry based on the weak asymmetry measure of [19] is developed in [17]. A similar test based on the strong asymmetry measure of [18] is under development in [3] and here the focus is to study the power function of the same test. Specifically the objective here is to discuss the practical implementation of the test, study its power function for various distributions and compare its performance against the tests of symmetry that have been proposed before.

The evidence arising from the simulation study of the present work is that the test compares favorably against the existing tests. Except for the tests proposed in [17], most of the tests of symmetry are designed mainly to detect departures from symmetry and do not necessarily make use of the size of symmetry in their construction. A consequence of this, as discussed in [17], is that their power does not reflect the size of asymmetry. In contrast, besides having as good or better power than existing tests, the main characteristic of the test considered here is that "the higher the asymmetry higher is the power".

The rest of the paper is organized as follows. Section 2 discusses the development of the test and provides the test statistic. Section 3 contains details on the practical implementation of the test. Numerical evidence on the power of the test and its comparison with the powers of other tests is given in Sect. 4.

2 Motivation and Test Statistic

Let f and F denote the probability density and the cumulative distribution function, respectively, associated with a random variable X. We wish to test the null hypothesis of symmetry,

$$H_0 : f(\theta - x) = f(\theta + x)\ \forall\, x\ \in \mathbb{R} \text{ vs} \qquad H_1 : f(\theta - x) \neq f(\theta + x) \text{ for at least one } x \in \mathbb{R}. \tag{1}$$

To test the hypothesis in (1), a basis for constructing a test statistic is provided by the fact that for a symmetric random variable X, $\mathbb{C}\mathrm{ov}(f(X), F(X)) = 0$. In [19] it is noted that this is a necessary but not sufficient condition and in [18] this is then modified to the following necessary and sufficient condition. A density function f is symmetric if and only if

$$\int_{-\infty}^{\xi_p} f^2(x)\, dx = \int_{\xi_{1-p}}^{+\infty} f^2(x)\, dx$$

for all $p \in (1/2, 1)$ where ξ_p is such that $F(\xi_p) = p$. Which is equivalent to the necessary and sufficient condition that $f(x)$ is symmetric if and only if $\delta_p + \delta_p^* = 0$ for every $1/2 \leq p < 1$, where

$$\delta_p = p^{-3}\left[\int_{-\infty}^{\xi_p} f^2(x)F(x)dx - \frac{p}{2}\int_{-\infty}^{\xi_p} f^2(x)dx\right]$$

$$\delta_p^* = p^{-3}\left[-\int_{\xi_{1-p}}^{\infty} f^2(x)\big(1-F(x)\big)dx + \frac{p}{2}\int_{\xi_{1-p}}^{\infty} f^2(x)dx\right].$$

Thus one can take the measure of asymmetry to be maximum of $\delta_p + \delta_p^*$ over p. However, note that the definitions of δ_p and δ_p^* result from the fact that they both represent

$$\delta_p = \mathbb{C}\mathrm{ov}_{f_p}\left(f_p(X), F_p(X)\right), \quad \delta_p^* = \mathbb{C}\mathrm{ov}_{f_p^*}\left(f_p^*(X), F_p^*(X)\right), \tag{2}$$

where

$$f_p(x) = \begin{cases} \frac{f(x)}{p} & \text{if } x \leq \xi_p, \\ 0 & \text{otherwise} \end{cases}, \quad f_p^*(x) = \begin{cases} \frac{f(x)}{p} & \text{if } x \geq \xi_{1-p}, \\ 0 & \text{otherwise} \end{cases} \tag{3}$$

and the distribution functions corresponding to f_p and f_p^* are defined by,

$$F_p(x) = \begin{cases} \frac{F(x)}{p} & \text{if } x \leq \xi_p, \\ 1 & \text{if } x \geq \xi_p \end{cases}, \quad F_p^*(x) = \begin{cases} 0 & \text{if } x \leq \xi_{1-p}, \\ 1 - \frac{1-F(x)}{p} & \text{if } x \geq \xi_{1-p}. \end{cases}$$

Since considering the correlation rather than the covariance has an advantage of turning the resulting measure into a scale and location invariant, we define

$$\rho_p = \frac{2\sqrt{3}}{p}\frac{\left[\int_{-\infty}^{\xi_p} f^2(x)F(x)dx - \frac{p}{2}\int_{-\infty}^{\xi_p} f^2(x)dx\right]}{\left[p\int_{-\infty}^{\xi_p} f^3(x)dx - (\int_{-\infty}^{\xi_p} f^2(x)dx)^2\right]^{1/2}}, \tag{4}$$

$$\rho_p^* = \frac{2\sqrt{3}}{p}\frac{\left[-\int_{\xi_{1-p}}^{\infty} f^2(x)(1-F(x))dx + \frac{p}{2}\int_{\xi_{1-p}}^{\infty} f^2(x)dx\right]}{\left[p\int_{\xi_{1-p}}^{\infty} f^3(x)dx - (\int_{\xi_{1-p}}^{\infty} f^2(x)dx)^2\right]^{1/2}}. \tag{5}$$

Therefore, in [18] the measure of asymmetry is defined as

$$\eta(X) = -\frac{1}{2}\,\mathrm{sign}(\rho_1)\max_{\frac{1}{2}\leq p\leq 1}|\rho_p + \rho_p^*|, \tag{6}$$

which is zero if and only if $f(x)$ is symmetric. Further, the values of $\eta(X)$ range from -1 (for most negatively asymmetric densities) to +1 (most positively asymmetric densities). Therefore a sample analogue of $\eta(X)$ can be used to test the null hypothesis of symmetry as $\eta(X) = 0$ implies H_0 in (1). On the contrary, values of $\eta(X) \neq 0$, implies H_1.

Remark 1 Note that $-1 \times \rho_1$ is the asymmetry coefficient of [19] which we denote by $\eta_w(X)$. It corresponds to the necessary but not sufficient condition for symmetry that $\mathrm{Cov}(f(x), F(x)) = 0$. Also note that $|\eta_w(X)| \leq |\eta(X)|$.

Remark 2 It may be noted that η does satisfy the properties that one is likely to ask of a measure of symmetry, i.e.,

- For a symmetric random variable X, $\eta(X) = 0$.
- If $Y = aX + b$ where $a > 0$ and b is any real number, then $\eta(X) = \eta(Y)$.
- If $Y = -X$, $\eta(X) = -\eta(Y)$.

3 Practical Implementation

Let $X_1, X_2, \cdots, X_n$ be a random sample from a continuous density function $f(x)$. First note that to estimate η, for various values of nonnegative integers k and l, one needs to estimate

$$\int_a^b f^{k+1}(x)F^l(x)\,dx = E\big[f^k(X)F^l(X)I[a < X < b]\big], \tag{7}$$

where I is an indicator function and, $-a$ and/or b could be ∞. Therefore, an estimator of η can be obtained by plugging in the sample counterparts of f and F, in a simple unbiased estimator of the last quantity given by

$$\frac{1}{n}\sum_{i=1}^{n} f^k(X_i)F^l(X_i)I[a < X_i < b].$$

For this, a simple approach is to estimate the density $f(x)$ by the standard kernel density estimate

$$\hat{f}(x) = (nh)^{-1}\sum_{i=1}^{n} K\left(\frac{x - X_i}{h}\right), \tag{8}$$

where K is a second order kernel function and h denotes the bandwidth parameter. Popular bandwidth selection rules include the solve-the-equation and direct plug-in rules of [21] and Silverman's rule of thumb ([22], (3.31)) which is already implemented in `R` through the `bw.nrd0` routine and is used throughout this work. The distribution function $F(x)$ is estimated by the standard kernel distribution function estimate

$$\hat{F}(x) = \int_{-\infty}^{x} \hat{f}(u)\,du.$$

Thus, (7) is estimated by

$$\hat{\psi}_{kl}(a,b) = \frac{1}{n}\sum_{i=1}^{n}(\hat{f}(X_i))^k(\hat{F}(X_i))^l I_{[a,b]}(X_i).$$

Then the estimators of ρ_p and ρ_p^* based on $\hat{f}(x)$ and $\hat{F}(x)$ are

$$\hat{\rho}_p = \frac{2\sqrt{3}}{p}\frac{\hat{\psi}_{11}(-\infty,\hat{\xi}_p) - \frac{p}{2}\hat{\psi}_{10}(-\infty,\hat{\xi}_p)}{\left[p\hat{\psi}_{20}(-\infty,\hat{\xi}_p) - \hat{\psi}_{10}^2(-\infty,\hat{\xi}_p)\right]^{1/2}}, \text{ for } 1/2 \le p < 1,$$
$$\hat{\rho}_p^* = \frac{2\sqrt{3}}{p}\frac{-\hat{\psi}_{10}(\hat{\xi}_{1-p},+\infty) + \hat{\psi}_{11}(\hat{\xi}_{1-p},+\infty) + \frac{p}{2}\hat{\psi}_{10}(\hat{\xi}_{1-p},+\infty)}{\left[p\hat{\psi}_{20}(\hat{\xi}_{1-p},+\infty) - \hat{\psi}_{10}^2(\hat{\xi}_{1-p},+\infty)\right]^{1/2}},$$

for $1/2 \le p < 1$, and thus η is estimated by

$$\hat{\eta} = -\frac{1}{2}\,\text{sign}(\hat{\rho}_1)\max_{\frac{1}{2}\le p<1}|\hat{\rho}_p + \hat{\rho}_p^*|.$$

It may be helpful to note here that $\hat{\eta}$ could be shown to be consistent by arguments similar to that in [11]. Also, throughout this work $\hat{\eta}$ is implemented by simply ignoring the denominators in both $\hat{\rho}_p$ and $\hat{\rho}_p^*$ as the objective is only to test for symmetry and not to provide a scaled measure of asymmetry.

4 Numerical Evaluation of the Test's Power

In this section, finite sample distributional data is used to exhibit the performance of the proposed test's power properties for various sample sizes. Nine different classes of probability models are used for this purpose. These are the standard Normal, the Cauchy, the Lognormal, the Folded normal, the Exponential, mixtures of Normals, the skew Normal (defined in [2]), the Sinh–arcsinh family (defined in [14]) and the Fernadez and Steel (defined in [9]) families of distributions. The p.d.f. of the normal mixture family is given by

$$f_{NM}(x; s, \mu_1, \mu_2, \sigma_1^2, \sigma_2^2) = s\,N(\mu_1, \sigma_1^2) + (1-s)\,N(\mu_2, \sigma_2^2)$$

where $\mu_1 = 0, \sigma_1^2 = 1, \mu_2 = 2, \sigma_2^2 = 2$. Here, four different versions of this family are implemented, defined by $s = 0.945, 0.872, 0.773, 0.606$ respectively. The p.d.f. of the skew Normal family is given by

$$f_{SN}(x;\lambda) = 2\phi(x)\Phi(\lambda x) \quad -\infty \le x \le +\infty$$

where ϕ and Φ denote the standard normal p.d.f. and c.d.f., respectively. Obviously, $\lambda = 0$ reduces $f_{SN}(x;\lambda)$ to the symmetric standard normal distribution. When $\lambda > 0$, $f_{SN}(x;\lambda)$ is skewed to the right and $\lambda < 0$ corresponds to left skewness. Eight different versions are used here. These correspond to parameters

$$\lambda = 1.2135, 1.795, 2.429, 3.221, 4.310, 5.970, 8.890, 15.570.$$

The Sinh–arcsinh distributions are defined by the p.d.f.

$$f_{SAS}(x;\epsilon,\delta) = \frac{1}{\sqrt{2\pi}} \frac{\delta C_{\epsilon,\delta}(x)}{\sqrt{1+x^2}} \exp\left\{-\frac{1}{2} S^2_{\epsilon,\delta}(x)\right\}, \quad \varepsilon \in \mathbb{R},\ \delta > 0,$$

where

$$C_{\epsilon,\delta}(x) = \cosh\left[\epsilon + \delta \sinh^{-1}(x)\right],$$
$$S^2_{\epsilon,\delta}(x) = \sinh\left[\epsilon + \delta \sinh^{-1}(x)\right].$$

Here ε controls skewness while δ controls the weight of the tails. The eight versions of $f_{SAS}(x;\epsilon,\delta)$ are implemented with $\delta = 1$ and

$$\varepsilon = 0.1, 0.203, 0.311, 0.430, 0.565, 0.727, 0.939, 1.263.$$

The Fernandez and Steel family has p.d.f.

$$f_{FAS}(x;\gamma,\nu) = \frac{2}{\gamma + \frac{1}{\gamma}} \left\{ f_t\left(\frac{x}{\gamma};\nu\right) I_{\{x\geq 0\}} + f_t(\gamma x;\nu) I_{\{x<0\}} \right\}, \tag{9}$$

where the parameter $\gamma \in (0,+\infty)$ controls the skewness of the distribution. From [9], f_t can be any symmetric unimodal distribution so for $\gamma = 1$, f_{FAS} is symmetric. Here, in contrast to [17], $f_t(x;\nu)$ is the p.d.f. of the (symmetric, unimodal) t distribution with $\nu = 5$ degrees of freedom. In the present implementation, eight different versions of this family are realized with parameters

$$\gamma = 1.111, 1.238, 1.385, 1.564, 1.791, 2.098, 2.557, 3.388.$$

The critical region which determines acceptance or rejection of the null is based on approximating the distribution of the test statistic under the null by calculating its value on $k = 10{,}000$ i.i.d. samples from the standard normal distribution. Different regions are calculated for samples of size $n = 30, 50, 70$. The standardized version of the test statistic, $S_i = \hat{\eta}_i/\mathrm{sd}(\hat{\eta})$ with $\mathrm{sd}(\hat{\eta})$ being the sample standard deviation of $\hat{\eta}$ as this results from its 10,000 values, is used for determining its distribution. Then, definition 7 of [7], readily implemented in R via the `quantile()` function, is applied to deduce data driven estimates of $-q_{a/2}$ and $q_{a/2}$, so as to construct the critical region $D = (-\infty, -q_{a/2}) \cup (q_{a/2}, +\infty)$. This yields a critical region of the form $\hat{D} = (-\infty, \alpha) \cup (\beta, +\infty)$ with $\alpha < 0, \beta > 0$. Values of $S_i \in \hat{D}$ signify rejection of the null.

The size function of the test is approximated as follows. For the three different sample sizes, 10,000 i.i.d. samples are generated from the Cauchy, and the symmetric versions of the Sinh–arcsinh and Fernandez and Steel p.d.f.'s. Note here that the symmetric versions of the f_{NM} and f_{SN} p.d.f's reduce to the standard normal distribution for which $\hat{D}$ is already calculated and for this reason does not make any sense to consider those too. Then, $S_i, i = 1, \ldots, k$ is computed and the value of $\{\#S_i \in \hat{D}\}/10{,}000$ is used as an approximation of the probability $\mathbb{P}(\hat{\eta}/\mathrm{sd}(\hat{\eta}) \in \hat{D}|H_0)$, which defines the size of the test.

On the other hand, computation of $S_i, i = 1, \ldots, k$ and subsequently calculation of $\{\#S_i \in \hat{D}\}/10{,}000$ for all the other (nonsymmetric) distributions mentioned above leads to a numerical approximation of $\mathbb{P}(\hat{\eta}/\mathrm{sd}(\hat{\eta}) \in \hat{D}|H_1)$ i.e. the power function of the test. It has to be noted here that the present formulation highlights the fact that skewness and asymmetry are two different concepts under the alternative. At the same time it corroborates with the fact that skewness and asymmetry are the same concept and equal to zero under the null.

Implementation of $\hat{\eta}$ in practice is discussed in detail in Sect. 3. For comparison purposes, four symmetry tests are used to benchmark the performance of $\hat{\eta}/\mathrm{sd}(\hat{\eta})$. The tests are

$$S_1 = \sqrt{n}\frac{\bar{x} - \tilde{\theta}}{s},$$

where $\bar{x}$, $\tilde{\theta}$ and s are the sample mean, sample median and sample standard deviation respectively. This test was proposed by [5] and large values of S_1 signify departure from symmetry. The second test is given by $S_2 = \mathscr{R}(0)$ where

$$\mathscr{R}(a) = \frac{1}{\sqrt{n}} \sum_{i=1}^{n} G_a\left(\frac{R(|X_i - \tilde{\theta}|)}{2(n+1)}\right) \mathrm{sign}(X_i - \tilde{\theta}),$$

$$G_a(x) = \min\left(x, \frac{1}{2} - a\right)$$

and $R(X_i)$ is the rank of X_i in the sample. This test was proposed by [1] and as in the case of S_1, here too large values of the test statistic signify departure from symmetry. The third test is the 'triples' test of [20], given by

$$S_3 = \frac{1}{3}\binom{N}{3}^{-1} \sum_{i<j<k} \Big\{\mathrm{sign}(X_i + X_j - 2X_k) + \mathrm{sign}(X_i + X_k - 2X_j) + \mathrm{sign}(X_j + X_k - 2X_i)\Big\}$$

where a triple of observations (X_i, X_j, X_k) is defined as the right triple if the middle observation is closer to the smallest observation than it is to the largest observation and vice versa for the left triple. Again, large values of S_3 indicate departure from symmetry. The fourth test is the test of [12] with test statistic

$$S_4 = \frac{\frac{1}{n}\sum_{i=1}^{n}(X_i - \bar{X})^3}{\left(\frac{1}{n-1}\sum_{i=1}^{n}(X_i - \bar{X})^2\right)^{\frac{3}{2}}}.$$

As with the previous three tests construction of S_4 is based on detecting departure from asymmetry through skewness. Thus significantly large values of $|S_4|$ indicate departure from symmetry.

The empirical powers of $S_1 - S_4$ for the same sample sizes as used here ($n = 30, 50, 70$) can be found on [17]. It has to be noted that more tests are available for comparison with $\hat{\eta}$ in [17]. However, the focus here is put on $S_1 - S_4$; the reason is that these four tests are designed to detect departure from symmetry and hence comparison with them sheds light on the benefits yield by focusing on quantification of asymmetry as suggested by $\hat{\eta}$.

The results for $\hat{\eta}/\mathrm{sd}(\hat{\eta})$ are displayed in Table 1. The first outcome is that for the normal mixtures, the skew normal, the sinh-arcsinh and the Fernandez and Steel families, the test is very sensitive in capturing departure from symmetry. This insight is derived by the figures of the power function for the first parameters of each distribution where the test is much more effective in detecting the asymmetry of the p.d.f. compared to its competitors. Also, as expected the power of the test is rising as sample size and the amount of asymmetry is increasing. Another outcome is that the test compares favorably in terms of power to the other four tests, with S_3 being its closest competitor. More importantly, as mentioned in the Introduction, $S_1 - S_4$ are designed to detect the departure from symmetry and do not necessarily make use of the size of symmetry in their construction. A consequence of this is that their power does not reflect the size of asymmetry. A case in point are the Log-normal and Folded normal distributions where the simulation results indicate that the test detects asymmetry in Folded normal with less power than in the Log-normal case, even though the latter is less asymmetric than the former. One reason for this is the fact that the reflection method for boundary correction ([13]) works better for the Lognormal than for the Folded normal distribution.

Now, the test based on $\hat{\eta}$ not only has as good a power as other tests, but also its power to detect the asymmetry in Folded normal is higher than its power to detect asymmetry in Lognormal distribution. In general, from empirical powers in Table 1 higher the asymmetry higher is the power of the test based on $\hat{\eta}$.

Table 1 Empirical powers (in %) for $\hat{\eta}/\mathrm{sd}(\hat{\eta})$ for $a = 5\,\%$

η	Distribution	Power $\hat{\eta}$		
		$n = 30$	$n = 50$	$n = 70$
0	N(0, 1)	7	4.3	5.3
0	Cauchy	5.4	4.3	4.7
0.1	$f_{NM}(x; 0.945, 0, 2, 1, 2)$	10.7	12.8	15.2
0.2	$f_{NM}(x; 0.872, 0, 2, 1, 2)$	18.3	29.8	40.4
0.3	$f_{NM}(x; 0.773, 0, 2, 1, 2)$	28.1	52.7	67.5
0.4	$f_{NM}(x; 0.606, 0, 2, 1, 2)$	39.9	74.1	88.7
0.1	$f_{SN}(x; 1.2135)$	15.7	14.4	15.7
0.2	$f_{SN}(x; 1.795)$	24.8	26.6	36.3
0.3	$f_{SN}(x; 2.429)$	39.8	44.3	51.1
0.4	$f_{SN}(x; 3.221)$	54.9	65.2	81.8
0.5	$f_{SN}(x; 4.310)$	71.8	84.2	94.8
0.6	$f_{SN}(x; 5.970)$	88.5	95	99.1
0.7	$f_{SN}(x; 8.890)$	93.9	98.8	99.8
0.8	$f_{SN}(x; 15.570)$	97.4	99.7	100
0	$f_{SAS}(x; 0, 1)$	6.6	6.2	5.9
0.1	$f_{SAS}(x; 0.1, 1)$	24.7	39.6	48.8
0.2	$f_{SAS}(x; 0.203, 1)$	38.4	56.3	62.4
0.3	$f_{SAS}(x; 0.311, 1)$	51.7	72.3	74.5
0.4	$f_{SAS}(x; 0.430, 1)$	64.2	82.5	87.2
0.5	$f_{SAS}(x; 0.565, 1)$	78.6	96.3	98.6
0.6	$f_{SAS}(x; 0.727, 1)$	89.1	97.2	99.1
0.7	$f_{SAS}(x; 0.939, 1)$	90.6	98.2	100
0.8	$f_{SAS}(x; 1.263, 1)$	93.4	98.3	100
0	$f_{FAS}(x; 1, 5)$	6.4	5.9	5.8
0.1	$f_{FAS}(x; 1.111, 5)$	24.2	34.1	39.6
0.2	$f_{FAS}(x; 1.238, 5)$	31.6	49.4	53.9
0.3	$f_{FAS}(x; 1.385, 5)$	45.6	59.6	64.5
0.4	$f_{FAS}(x; 1.564, 5)$	69.1	82.6	90.4
0.5	$f_{FAS}(x; 1.791, 5)$	70.3	88.3	90.7
0.6	$f_{FAS}(x; 2.098, 5)$	80.6	89.4	94.7
0.7	$f_{FAS}(x; 2.557, 5)$	80.5	93.6	95.4
0.8	$f_{FAS}(x; 3.388, 5)$	91.2	98.2	100
0.91	LogNormal	87.6	94.3	96.2
0.95	Folded Normal	72.2	86.4	94.3
1	Exponential(1)	88.6	92.2	100

References

1. Antille, A., Kersting, G., Zucchini, W.: Testing symmetry. Journal of the Am. Stat. Assoc. **77**, 639–646 (1982)
2. Azzalini, A.: A class of distributions which includes the normal ones. Scand. J. Stat. **12**, 171–178 (1985)
3. Bagkavos, D., Patil, P.N., Wood, A.T.A.: Tests of symmetry and estimation of asymmetry based on a new coefficient of asymmetry. In preparation
4. Boshnakov, G.N.: Some measures for asymmetry of distributions. Stat. Prob. Lett. **77**, 1111–1116 (2007)
5. Cabilio, P., Massaro, J.: A simple test of symmetry about an unknown median. Can. J. Stat. **24**, 349–361 (1996)
6. Critchley, F., Jones, M.C.: Asymmetry and gradient asymmetry functions: density-based skewness and kurtosis. Scand. J. Stat. **35**, 415–437 (2008)
7. Hyndman, R., Fan, Y.: Sample quantiles in statistical packages. Am. Stat. **50**, 361–365 (1996)
8. Ekström, M., Jammalamadaka, S.R.: An asymptotically distribution-free test of symmetry. J. Stat. Plann. Infer. **137**, 799–810 (2007)
9. Fernandez, C., Steel, M.F.J.: On bayesian modeling of fat tails and skewness. J. Am. Stat. Assoc. **93**, 359–371 (1998)
10. Ghosh, K.: A New Nonparametric Test of Symmetry. Advances in Directional and Linear, Statistics, pp. 69–83 (2011)
11. Giné, E., Mason, D.: Uniform in bandwidth estimation of integral functionals of the density function. Scand. J. Stat. **35**, 739–761 (2008)
12. Gupta, M.K.: An asymptotically nonparametric test of symmetry. Ann. Math. Stat. **38**, 849–866 (1967)
13. Jones, M.C.: Simple boundary correction for kernel density estimation. Stat. Comput. **3**, pp. 135–146 (1993)
14. Jones, M.C., Pewsey, A.: Sinh-arcsinh distributions. Biometrika **96**(4), 761–780 (2009)
15. Maasoumi, E., Racine, J.S..: A robust entropy-based test of asymmetry for discrete and continuous processes. **28**, Econom. Rev. 246–261 (2008)
16. MacGillivray, H.L.: Skewness and asymmetry: measures and orderings. Ann. Stat. **14**, 994–1011 (1986)
17. Parlett, C., Patil, P.N.: Measuring asymmetry and testing symmetry. Ann. Inst. Stat. Math. *To appear*. doi:10.1007/s10463-015-0547-4
18. Patil, P.N., Bagkavos, D., Wood, A.T.A.: A measure of asymmetry based on a new necessary and sufficient condition for symmetry. Sankhya Ser. A **76**, 123–145 (2014)
19. Patil, P.N., Patil, P., Bagkavos, D.: A measure of symmetry. Stat. Papers **53**, 971–985 (2012)
20. Randles, R.H., Flinger, M.A., Policello, G.E., Wolfe, D.A.: An asymptotically distribution free test for symmetry versus asymmetry. J. Am. Stat. Assoc. **75**, 168–172 (1980)
21. Sheather, S.J., Jones, M.C.: A reliable data-based bandwidth selection method for kernel density estimation. J. Roy. Stat. Soc. Ser. B **53**, 683–690 (1991)
22. Silverman, B.W.: Density Estimation. Chapman and Hall, London (1986)

Nonparametric Test on Process Capability

Stefano Bonnini

Abstract The study of process capability is very important in designing a new product or service and in the definition of purchase agreements. In general we can define capability as the ability of the process to produce conforming products or deliver conforming services. In the classical approach to the analysis of process capability, the assumption of normality is essential for the use of the indices and the interpretation of their values make sense but also to make inference on them. The present paper focuses on the two-sample testing problem where the capabilities of two processes are compared. The proposed solution is based on a nonparametric test. Hence the solution may be applied even if normality or other distributional assumptions are not true or not plausible and in the presence of ordered categorical variables. The good power behaviour and the main properties of the power function of the test are studied through Monte Carlo simulations.

Keywords Process capability · Permutation test · Two-sample test

1 Introduction

To ensure a high quality of product or service, the production process or service delivery process should be stable and a continuous quality improvement should be pursued. Control charts are the basic instruments for a statistical process control (SPC). One of the main goals of these and other statistical techniques consists in studying and controlling the capability of the process. A crucial aspect which should be studied and controlled is the process variability.

Every process, even if well-designed, presents a natural variability due to unavoidable random factors. In the presence of specific factors that cause systematic variability, the process is out of control and its performances are unacceptable. In these situations the process variability is greater than the natural variability and high per-

S. Bonnini (✉)
Department of Economics and Management, University of Ferrara,
Via Voltapaletto 11, Ferrara, Italy
e-mail: stefano.bonnini@unife.it

R. Cao et al. (eds.), *Nonparametric Statistics*, Springer Proceedings
in Mathematics & Statistics 175, DOI 10.1007/978-3-319-41582-6_2

centage of outputs (products, services, etc) could be nonconforming, that is the process would produce high percentages of waste. In other words, when the process is in control, most of the values of the response variable under monitoring falls between the specification limits. When the process is out of control, the probability that the response variable takes values outside the specification limits is high. Hence the main purpose of a SPC is to minimize the process variability.

The study of process capability is very important in designing a new product or service and in the definition of purchase agreements. In general we can define capability as the ability of the process to produce conforming products/services. In other words the greater the probability of observing values of the response in the interval [*LSL*, *USL*], the greater the process capability, where *LSL* and *USL* are lower specification limit and upper specification limit respectively.

In the statistical literature several works have been dedicated to process capability indices. For a deep discussion see, among the others, [5, 6, 9–11, 14, 15].

By assuming normality for the response, a simple way of measuring the process capability is based on the index

$$C_p = (USL - LSL)/(6\sigma), \tag{1}$$

where σ is the standard deviation of the response. For a non centred process, that is when the central tendency of the distribution of the response is not centred in the specification interval, a more appropriate measure of process capability is provided by

$$C_{pk} = \min[(USL - \mu), (\mu - LSL)]/(3\sigma), \tag{2}$$

where μ is the process mean. C_p can be considered as potential capacity of the process, while C_{pk} can be considered as actual capacity. When the process is centred $C_p = C_{pk}$. If $LSL \leq \mu \leq LSL$ then $C_{pk} \geq 0$ and when $\mu = LSL$ or $\mu = USL$ we have $C_{pk} = 0$.

The assumption of normality is essential for the use of the indices and the interpretation of their values make sense. Some approaches, proposed in the presence of non normal data, are based on a suitable transformation of data. Alternative solutions consist in defining general families of distributions like those of Pearson and Johnson (see [14]).

When the capabilities of two or more processes are compared, we should consider that a given value of C_{pk} could correspond to one process with centred mean and high variability or to another process with less variability and non centred mean. As a consequence, high values of C_{pk} may correspond to a non centred process with low variability. To take into account the centering of the process we should jointly consider C_p and C_{pk}. An alternative is represented by the following index of capability

$$C_{pkm} = (USL - LSL)/(6\sqrt{\sigma^2 + (\mu - T)^2}), \tag{3}$$

where T is the target value for the response. It is worth noting that $C_{pkm} = C_p/\sqrt{1+\theta^2}$, where $\theta = (\mu - T)/\sigma$.

Under the assumption of normality, it is possible to compute confidence intervals for the capability indices by means of point estimates of μ and σ. Common and very useful testing problems consider the null hypothesis $H_0 : C = C_0$ against the alternative $H_1 : C > C_0$, where C is a given index of capability and C_0 is a specific reference value for C (see for example [9]). We wish to focus on the two-sample testing problem where the capabilities of two processes, C_1 and C_2 are compared. The goal consists in testing the null hypothesis $H_0 : C_1 = C_2$ against the alternative $H_1 : C_1 > C_2$. Typical situations are related to the comparison between sample data drawn from a given process under study and sample data from an in-control process or to the comparison between the capabilities of the processes associated to different industrial plants, operators, factories, offices, corporate headquarters, etc. Some interesting contributions about capability testing are provided by [7, 8, 12, 13].

The proposal of the present paper is based on a nonparametric solution. Hence the test may be applied even if normality or other distributional assumptions are not true or not plausible. The method is based on a permutation test and neither requires distributional assumptions nor needs asymptotic properties for the null distribution of the test statistic. Hence, it is a very robust procedure and can also be applied for small sample sizes and for ordered categorical data.

The basic idea is to transform the continuous response variable into a categorical variable through a suitable transformation of the support of the original response into a set of disjoint regions and to perform a test for comparing the heterogeneities of two categorical distributions. In Sect. 2 the procedure is described. Section 3 presents the results of a simulation study for proving the good power behaviour of the proposed test. Final conclusions are given in Sect. 4.

2 Permutation Test on Capability

Let X be a continuous random variable representing the response under study in the SPC. The probability that X takes values in the region $R \in \Re$ is

$$\pi_R = P[X \in R] = \int_R f(x)dx, \tag{4}$$

where $f(x)$ is the (unknown) density function of X. Let us define $R_T = [LSL, USL]$ the target region for X, $R_L = (-\infty, LSL)$ and $R_U = (USL, +\infty)$. A reasonable assumption, unless the process is severely out of control, is that most of the probability mass is concentrated in R_T, i.e., the probability that X falls in the target region is greater than the probability than X takes values in the lower tail or in the upper tail. Formally

$$\pi_{R_T} = \max(\pi_{R_L}, \pi_{R_T}, \pi_{R_U}), \tag{5}$$

with $\pi_{R_L} + \pi_{R_T} + \pi_{R_U} = 1$. The ideal situation, when the process is in control, is that the probability of producing waste is null, that is $\pi_{R_L} = \pi_{R_U} = 0$ and $\pi_{R_T} = 1$. The worst situation, when π_{R_T} takes its absolute minimum under the constrain defined in Eq. 5, consists in the uniform distribution, where $\pi_{R_L} = \pi_{R_T} = \pi_{R_U} = 1/3$. Hence a suitable index of capability could be the one's complement of a normalized measure of heterogeneity for categorical variables. A solution could be based on the use of the index of *Gini*

$$C^{(G)} = 1 - (3/2)[1 - (\pi_{R_L}^2 + \pi_{R_T}^2 + \pi_{R_U}^2)]. \tag{6}$$

The famous entropy of *Shannon* may be also considered for computing a normalized index of capability

$$C^{(S)} = 1 + (\pi_{R_L} \ln \pi_{R_L} + \pi_{R_T} \ln \pi_{R_T} + \pi_{R_U} \ln \pi_{R_U}) / \ln 3. \tag{7}$$

Other alternatives can be provided by the family of indices proposed by *Rényi*

$$C^{(\omega)} = 1 - (1 - \omega)^{-1} \ln(\pi_{R_L}^{\omega} + \pi_{R_T}^{\omega} + \pi_{R_U}^{\omega}) / \ln 3. \tag{8}$$

Each normalized index of heterogeneity takes value 1 in case of maximum heterogeneity (uniform distribution), value 0 in case of minimum heterogeneity (degenerate distribution) and greater values when moving from the degenerate towards the uniform distribution (see [4]). Hence the greater the value of the index of heterogeneity the lower the capability of the process because the capability is non decreasing function of the probability concentration. For this reason, if the probabilities were known, the comparison of two process capabilities could be done by comparing the cumulative ordered probabilities $\Pi_{1(s)} = \sum_{t=1}^{s} \pi_{1(t)}$ and $\Pi_{2(s)} = \sum_{t=1}^{s} \pi_{2(t)}$ with $\pi_{jR_T} = \pi_{j(1)} \geq \pi_{j(2)} \geq \pi_{j(3)}$, $j = 1, 2$, $s = 1, 2, 3$. Thus the hypotheses of the problem are

$$H_0 : [C_1 = C_2] \equiv [\Pi_{1(s)} = \Pi_{2(s)} \forall s], \tag{9}$$

and

$$H_1 : [C_1 > C_2] \equiv [\Pi_{1(s)} \geq \Pi_{2(s)} \forall s \text{ and } \exists s \text{ s.t. } \Pi_{1(s)} > \Pi_{2(s)}]. \tag{10}$$

Under the null hypothesis, when the cumulative ordered probabilities are equal, exchangeability holds. But $\pi_{j(t)}$, $j = 1, 2$, $t = 1, 2, 3$ are unknown parameters of the distribution and need to be estimated by using the observed ordered frequencies $\hat{\pi}_{j(t)} = n_{j(t)}/n_j$, where $n_{j(t)}$ is the tth ordered absolute frequency for the j-th sample and n_j is the size of the j-th sample. Hence the real ordering of the probabilities is estimated and the exchangeability under H_0 is approximated and not exact.

[1, 3] suggest that a test statistic for the similar problem of two-sample test on heterogeneity may be based on the difference of the sampling estimates of the indices of heterogeneity. By adapting this approach to our specific problem, we suggest to

use, as test statistic, the difference of the sampling estimates of the process capabilities under comparison: $T = \hat{C}_1 - \hat{C}_2$, where $\hat{C}_j$ is computed like C_j but by replacing $\pi_{j(t)}$ with $\hat{\pi}_{j(t)}$, with $j = 1, 2$ and $t = 1, 2, 3$. Hence we have $T_G = \hat{C}_1^{(G)} - \hat{C}_2^{(G)}$, $T_S = \hat{C}_1^{(S)} - \hat{C}_2^{(S)}$ and $T_{R_\omega} = \hat{C}_1^{(\omega)} - \hat{C}_2^{(\omega)}$.

An alternative solution could be based on the combination of more than one statistic, by considering the information provided by different indices. For example, according to the additive combining rule, we have

$$T_C = T_G + T_S + T_{R_3} + T_{R_\infty}, \tag{11}$$

where T_{R_3} and T_{R_∞} are the test statistics based on the indices of *Rényi* of order 3 and ∞ respectively. Whatever the statistics used for the problem, the null hypotheses must be rejected for large values of this statistic.

The first step of the testing procedure consists of the computation of the observed ordered table, that is $\{n_{j(t)}; j = 1, 2; t = 1, 2, 3\}$ and the observed value of the test statistic $T^{(0)}$. By performing B independent permutations of the dataset, then obtaining B permuted ordered tables $\{n^*_{j(t)}; j = 1, 2; t = 1, 2, 3\}$ and B corresponding permutation values of the test statistic $T^{*(1)}, \ldots, T^{*(B)}$, the p-value, according to the permutation distribution, can be computed as

$$p = \sum_{b=1}^{B} I(T^{*(b)} \geq T^{(0)})/B, \tag{12}$$

where $I(E) = 1$ iff the event E is true, and $I(E) = 0$ otherwise. An alternative resampling strategy may be based on a bootstrap approach but [2] proves that this solution is usually not as powerful as the permutation one.

3 Monte Carlo Simulation Study

To analyze the power behaviour of the proposed tests, a Monte Carlo simulation study was performed. Data for the j-th sample were randomly generated by the following variable:

$$X_j = 1 + int[3U^{\gamma_j}], \tag{13}$$

where U is a uniform random variable, and $\gamma_j \in (0, 1]$ is the heterogeneity parameter: the greater γ_j the higher the heterogeneity of X_j (thus the lower C_j), hence $C_1 > C_2$ iff $\gamma_1 < \gamma_2$. For each specific setting, defined in terms of γ_1, γ_2, n_1 and n_2 values, $CMC = 1000$ datasets where generated and, for each dataset, $B = 1000$ permutations were performed to estimate the p-values and compute the rejection rates of the tests. The estimated power (rejection rates) of the tests on capability based on the indices of *Gini*, *Shannon*, *Rényi* (order 3 and order ∞) and on the direct (additive) combination of the four mentioned tests were computed.

Table 1 Simulation results under $H_0 : C_1 = C_2$, $n_1 = n_2 = 50$, $\alpha = 0.05$, $B = 1000$, $CMC = 1000$

			Rejection rates				
γ_1	γ_2	$\gamma_1 - \gamma_2$	T_G	T_S	T_{R_3}	T_{R_∞}	T_C
0.2	0.2	0.0	0.055	0.054	0.055	0.055	0.055
0.4	0.4	0.0	0.053	0.052	0.053	0.054	0.054
0.6	0.6	0.0	0.040	0.048	0.040	0.025	0.040

Table 2 Simulation results under $H_1 : C_1 > C_2$, $n_1 = n_2 = 50$, $\alpha = 0.05$, $B = 1000$, $CMC = 1000$

			Rejection rates				
γ_1	γ_2	$\gamma_1 - \gamma_2$	T_G	T_S	T_{R_3}	T_{R_∞}	T_C
0.8	1.0	0.2	0.044	0.042	0.047	0.060	0.052
0.6	1.0	0.4	0.375	0.399	0.372	0.314	0.360
0.4	1.0	0.6	0.945	0.955	0.933	0.863	0.935

In Table 1, the rejection rates under the null hypothesis of equality in capability are reported with samples sizes equal to 50. Three different capability levels are considered. The powers of all the tests seem to increase with the capability: as a matter of fact capability is negatively related to heterogeneity, hence lower capability implies greater heterogeneity and greater heterogeneity means greater uncertainty. Table 1 shows that all the tests are well approximated, because the rejection rates are very similar to the nominal α level, even if, in the presence of high capabilities, the tests tend to be slightly anticonservative. The test based on the *Rényi* index of order ∞ is less stable than the others because of its very low power in the presence of low capabilities.

Table 2 shows the estimated power of the tests under H_1, when the capability of the second process is at the minimum level and for three different capability levels of the first process. As expected, the greater the difference in capability, the greater the power of the tests. When the difference in capability is low, the most powerful tests are those based on the direct combination and on the *Rényi* index of order ∞. Instead, when the difference in capability is high, the latter test is the less powerful, the power performance of the others is similar and the test based on the *Shannon* index is slightly preferable.

In Table 3 the behaviour of the rejection rates as function of the sample sizes, when the parameter difference is equal to 0.4, can be appreciated. The consistency of the tests is evident because larger sample sizes correspond to higher power. Again the power behaviours of the tests are very similar and, for small sample sizes, the test based on the *Rényi* index of order ∞ is the most powerful but for large sample sizes this test is the less powerful.

Table 4 focuses on the power comparison of the tests for different sample sizes when the difference between the capabilities is small. Even in this case, the test based on *Rényi* index of order ∞ is the best in the presence of small sample sizes

Table 3 Simulation results under $H_1 : C_1 > C_2,\ \gamma_1 = 0.6,\ \gamma_2 = 1.0,\ \alpha = 0.05,\ B = 1000,$ $CMC = 1000$

			Rejection rates				
n_1	n_2	$n_1 - n_2$	T_G	T_S	T_{R_3}	T_{R_∞}	T_C
20	20	0	0.109	0.109	0.108	0.114	0.108
60	60	0	0.440	0.455	0.424	0.365	0.426
100	100	0	0.759	0.769	0.742	0.640	0.740

Table 4 Simulation results under $H_1 : C_1 > C_2,\ \gamma_1 = 0.8,\ \gamma_2 = 1.0,\ \alpha = 0.05,\ B = 1000,$ $CMC = 1000$

			Rejection rates				
n_1	n_2	$n_1 - n_2$	T_G	T_S	T_{R_3}	T_{R_∞}	T_C
20	20	0	0.019	0.019	0.020	0.028	0.019
40	40	0	0.046	0.050	0.045	0.048	0.054
60	60	0	0.052	0.052	0.048	0.058	0.058
100	100	0	0.109	0.113	0.104	0.109	0.109

and this is not true in the presence of large sample sizes. In the intermediate case of sample sizes equal to 40, the most powerful test seems to be the one based on direct combination.

4 Conclusions

The two-sample nonparametric test on process capability is a robust solution and allows inferential comparative analysis of process capabilities even when distributional assumptions (e.g., normality) do not hold or cannot be tested. Under the null hypothesis of equality in heterogeneity, data exchangeability is not exact but the good approximation of the permutation test is proved by the Monte Carlo simulation study.

According to this proposal, the test statistic is based on the comparison of the two-sample heterogeneities, computed by using suitable indices of heterogeneity, like the *Gini* index, the *Shannon* entropy, the *Rényi* family of indices, or a suitable combination of test statistics based on different indices, for example on the sum of these different test statistics.

The Monte Carlo simulation study proves that the power of all the tests seems to increase with the capability: as a matter of fact capability is negatively related to heterogeneity, hence lower capability implies greater heterogeneity and consequently greater uncertainty.

All the considered tests are well approximated, because under the null hypothesis of equality in capability, the rejection rates are very similar to the nominal α level.

Under the alternative hypothesis, when the difference in capability is low, the most powerful tests are those based on the direct combination and on the *Rényi* index of order ∞. Instead, when the difference in capability is high, the latter test is the less powerful and the test based on the *Shannon* index is slightly preferable.

The tests are consistent because if sample sizes increase then power increases. For small sample sizes the test based on the *Rényi* index of order ∞ is the most powerful but for large sample sizes it is the less powerful. In the presence of small difference in the capabilities of the two compared processes, again the test based on the *Rényi* index of order ∞ is the best in the presence of small sample sizes but not in the presence of large sample sizes. In the case of intermediate sample sizes, the test based on the direct combination seems to be the most powerful. Hence, if we consider the instability of the *Rényi* index of order ∞, the test based on the direct combination is the best solution under the alternative hypothesis, when it is difficult to detect the difference in the capabilities of the two processes, i.e., near the null hypothesis.

References

1. Arboretti, G.R., Bonnini, S., Pesarin, F.: A permutation approach for testing heterogeneity in two-sample problems. Stat. Comput. **19**, 209–216 (2009)
2. Bonnini, S.: Testing for heterogeneity for categorical data: permutation solution vs. bootstrap method. Commun. Stat. A-Theor. **43**(4), 906–917 (2014)
3. Bonnini, S.: Combined tests for comparing mutabilities of two populations. In: Topics in Statistical Simulation. Book of Proceedings of the Seventh International Workshop on Simulation 2013, Rimini, 21–25 May 2013, pp. 67–78. Springer, New York (2014)
4. Bonnini, S., Corain, L., Marozzi, M., Salmaso, L.: Nonparametric Hypothesis Testing: Rank and Permutation Methods with Applications in R. Wiley, Chichester (2014)
5. Boyles, R.A.: The Taguchi capability index. J. Qual. Technol. **23**, 17–26 (1991)
6. Chan, L.K., Cheng, S.W., Spiring, F.A.: A new measure of process capability: Cpm. J. Qual. Technol. **20**, 162–175 (1988)
7. Chen, J.P., Tong, L.I.: Bootstrap confidene interval of the difference between two process capability indices. Int. J. Adv. Manuf. Tech. **21**, 249–256 (2003)
8. Choi, Y.M., Polansky, A.M., Mason, R.L.: Transforming non-normal data to normality in statistical process control. J. Qual. Technol. **30**(2), 133–141 (1998)
9. Kane, V.E.: Process capability indices. J. Qual. Technol. **18**, 41–52 (1986)
10. Kotz, S., Johnson, N.L.: Process Capability Indices. Chapman & Hall, London (1993)
11. Pearn, W.L., Kotz, S., Johnson, N.L.: Distributional and inferential properties of process capability indices. J. Qual. Technol. **24**, 216–231 (1992)
12. Pearn, W.L., Lin, P.C.: Testing process performance based on capability index C_{pk} with critical values. Comput. Ind. Eng. **47**, 351–369 (2004)
13. Polansky, A.M.: Supplier selection based on bootstrap confidence regions of process capability indices. Int. J. Rel. Qual. Saf. Eng. **10**, 1 (2003). doi:10.1142/S0218539303000968
14. Rodriguez, R.N.: Recent developments in process capability analysis. J. Qual. Technol. **24**, 176–187 (1992)
15. Vannman, K.: A unified approach to capability indices. Stat. Sin. **5**, 805–820 (1995)

Testing for Breaks in Regression Models with Dependent Data

J. Hidalgo and V. Dalla

Abstract The paper examines a test for smoothness/breaks in a nonparametric regression model with dependent data. The test is based on the supremum of the difference between the one-sided kernel regression estimates. When the errors of the model exhibit strong dependence, we have that the normalization constants to obtain the asymptotic Gumbel distribution are data dependent and the critical values are difficult to obtain, if possible. This motivates, together with the fact that the rate of convergence to the Gumbel distribution is only logarithmic, the use of a bootstrap analogue of the test. We describe a valid bootstrap algorithm and show its asymptotic validity. It is interesting to remark that neither subsampling nor the sieve bootstrap will lead to asymptotic valid inferences in our scenario. Finally, we indicate how to perform a test for k breaks against the alternative of $k + k_0$ breaks for some k_0.

Keywords Nonparametric regression · Breaks/smoothness · Strong dependence · Extreme-values distribution · Frequency domain bootstrap algorithms

1 Introduction

The literature on breaks/continuity on parametric regression models is both extensive and exhaustive in both econometric and statistical literature, see [23] for a survey. Because as in many other situations an incorrect specification of the model can lead to misleading conclusions, see for instance [14], it is of interest to develop tests which do not rely on any functional specification of the regression model. Although some work has been done in the nonparametric setup, the literature appears to focus mostly on the estimation of the break point, see for instance [22], Chu and Wu (1992) and [8], rather than on the testing of its existence. With this view, the purpose of this paper is to fill this gap by looking at testing for the hypothesis of continuity against the

J. Hidalgo (✉)
London School of Economics, Houghton Street, London WC2A 2AE, UK
e-mail: F.J.Hidalgo@lse.ac.uk

V. Dalla
National and Kapodistrian University of Athens, Sofokleous 1, 10559 Athens, Greece

R. Cao et al. (eds.), *Nonparametric Statistics*, Springer Proceedings
in Mathematics & Statistics 175, DOI 10.1007/978-3-319-41582-6_3

alternative of the existence of (at least) one discontinuity point in a nonparametric regression model, although we shall indicate how to perform a test for k breaks against the alternative of $k + k_0$ breaks for some k_0.

More specifically, we consider the regression model

$$y_t = r(x_t) + u_t; \quad t = 1, \ldots, n, \tag{1.1}$$

where we assume that the homoscedastic errors $\{u_t\}_{t \in \mathbb{Z}}$ follow a covariance stationary linear process exhibiting possibly strong dependence, to be more precise in Condition $C1$ below. We shall assume that x_t is deterministic, say a time trend. A classical example of interest in time series is a polynomial trend, that is $x_t = \left(t, t^2, \ldots, t^p\right)$, and/or when regressors are of the type "$\cos t\lambda_0$" and/or "$\sin t\lambda_0$", where $\lambda_0 \neq 0$. The latter type of regressors can be convenient when the practitioner suspects that the data may exhibit some cyclical behavior. Hence, one possible hypothesis of interest is to know if such a deterministic trend and/or cyclical behavior has breaks. Our results are a natural extension to those obtained in [1] when the errors $\{u_t\}_{t \in \mathbb{Z}}$ are a sequence of independent and identically (iid) distributed random variables. Of course, we can allow for stochastic covariates x, however, this is beyond the scope of this paper as the technical aspects are quite different than those with deterministic regressors.

Our main goal is to test the null hypothesis $r(x) =: E(y \mid x)$ is continuous being the alternative hypothesis that there exists a point in $\mathcal{X}$ such that $r(x)$ is not continuous, and where herewith $\mathcal{X}$ denotes the domain of the variable x. We are also very much interested into the possible consequence of assuming that the errors u_t exhibit strong dependence, as opposed to weak dependence, and in particular, the consequence on the asymptotic distribution of the test.

In this paper, the methodology that we shall follow is based on a direct comparison between two "alternative" estimates of $r(x)$. More specifically, based on a sample $\{y_t, x_t\}_{t=1}^n$, the test is based on global measures of discrepancy between nonparametric estimates of $E(y \mid x)$ when we take only observations at the right and left of the point $x \in \mathcal{X}$. For that purpose, we have chosen the supremum norm, e.g., a Kolmogorov–Smirnov type of test. Alternatively, we could have employed the $L_2 - norm$, see among others [2].

One of our main findings of the paper is that the constant ζ_n used to normalize the statistic (see Theorem 1 below) depends on the so-called strong dependent parameter of the error term. However, due to the slow rate of convergence to the Gumbel distribution and that the implementation of the test can be quite difficult for a given data set, we propose and describe a bootstrap algorithm. So in our setup bootstrap algorithms are not only necessary because they provide more reliable inferences, but due to our previous comment regarding its implementation. The need to use resampling/subsampling algorithm leads to a rather surprising result. In our context, subsampling is not a valid method to estimate the critical values of the test. The reason being, as Theorem 1 below illustrates, see also the comments after Theorem 2 in Sect. 4, the implementation of the test requires the estimation of some normalization constants which subsampling is not able to compute consistently. Because the well-known possible problems of the moving block bootstrap with strong dependence data, and that the sieve bootstrap is neither consistent when we allow for strong dependence, we will propose an algorithm in the frequency domain which overcomes the problem.

The paper is organized as follows. In the next section, we describe the model and test. Also, we present the regularity conditions and the one-sided kernel estimators of the regression function. Section 3 presents the main results of the paper. Due to the nonstandard results obtained in Sects. 3 and 4 describes and examines a bootstrap algorithm, showing the validity in our context. The bootstrap is performed in the frequency domain and it extends results to the case when the errors are not necessarily weakly dependent. A Monte Carlo experiment is presented in Sect. 5. Section 6 gives the proofs of the results which rely on a series of lemmas in Sect. 7.

2 The Model and Test. Regularity Conditions

As we mentioned in the introduction, our main concern is to test the null hypothesis that $r(x)$ is continuous being the alternative hypothesis that there exists a point in $\mathcal{X}$ such that the function $r(x)$ is not continuous. So, noting that continuity of $r(x)$ means that $\forall x \in \mathcal{X}$, $r_+(x) = r_-(x)$, where $r_\pm(x) = \lim_{z\to x\pm} r(z)$, we can set our null hypothesis H_0 as

$$H_0 : r_+(x) = r_-(x), \quad \forall x \in \mathcal{X}, \tag{2.1}$$

being the alternative hypothesis the negation of the null.

The null hypothesis in (2.1) and the nonparametric nature of $r(x)$ suggests that we could base the test for the null hypothesis H_0 in (2.1) on the difference between the kernel regression estimates of $r_+(x)$ and $r_-(x)$. To that end, we shall employ one-sided kernels as proposed by [26] since in our context they appear necessary since the implementation of the test requires the estimation of $r_+(\cdot)$ and $r_-(\cdot)$, that is estimates of $r(z)$ at $z+$ and $z-$, respectively. Denoting by $K_+(x)$ and $K_-(x)$ one-sided kernels, that is, kernel functions taking values for $x > 0$ and $x < 0$, respectively, we estimate $r_+(x)$ and $r_-(x)$ at points $x_q = q/n$, $q \in \mathcal{Q}_n$, where $\mathcal{Q}_n = \{q : \check{n} < q \le n - \check{n}\}$, by

$$\widehat{r}_{a,+}(q) := \widehat{r}_{a,+}(x_q) = \frac{1}{\check{n}} \sum_{t=q}^{n} y_t K_{+,t-q}, \quad \widehat{r}_{a,-}(q) := \widehat{r}_{a,-}(x_q) = \frac{1}{\check{n}} \sum_{t=1}^{q} y_t K_{-,t-q}, \tag{2.2}$$

where henceforth we abbreviate $K_\pm\left(\frac{t}{\check{n}}\right)$ by $K_{\pm,t}$, $\check{n} = [na]$ and $a = a(n)$ is a bandwidth parameter such that $a \to 0$ as n increases to infinity, and where for notational simplicity we shall take $x_t = t/n$ henceforth. Thus, the test for H_0 in (2.1) becomes

$$\mathcal{T}_d = \sup_{q \in \mathcal{Q}_n} \left| \widehat{r}_{a,+}(q) - \widehat{r}_{a,-}(q) \right|. \tag{2.3}$$

Remark 1 It is worth mentioning that to take the supremum on [0, 1] or at point j/n, for integer j, is the same as $\widehat{r}_{a,+}(x_q) = \widehat{r}_{a,+}(x)$ for all $x \in (x_{q-1}, x_q]$.

Next, let us introduce the following regularity conditions:

C1: $\{u_t\}_{t\in\mathbb{Z}}$ is a covariance stationary linear process defined as

$$u_t = \sum_{j=0}^{\infty} \vartheta_j \varepsilon_{t-j}; \ \sum_{j=0}^{\infty} \vartheta_j^2 < \infty, \ \text{with } \vartheta_0 = 1,$$

where $\{\varepsilon_t\}_{t\in\mathbb{Z}}$ is an iid sequence with $E(\varepsilon_t) = 0$, $E\left(\varepsilon_t^2\right) = \sigma_\varepsilon^2$, $E\left(|\varepsilon_t|^\ell\right) = \mu_\ell < \infty$ for some $\ell > 4$. Also, the spectral density function of $\{u_t\}_{t\in\mathbb{Z}}$, denoted $f(\lambda)$, can be factorized as

$$f(\lambda) = \frac{\sigma_\varepsilon^2}{2\pi} g(\lambda) h(\lambda), \tag{2.4}$$

where $g(\lambda) = \left|1 - e^{ij\lambda}\right|^{-2d}$, $h(\lambda) = |B(\lambda)|^2$, $B(\lambda) = \sum_{j=0}^{\infty} b_j e^{-ij\lambda}$; and $\sum_{k=0}^{\infty} k^2 |b_k| < \infty$.

The case $d = 0$ refers to weak dependence, whereas the case $0 < d < \frac{1}{2}$ refers to strong dependence. One model satisfying (2.4) is the $FARIMA(p, d, q)$ process $(1-L)^d \Phi_p(L) u_t = \Theta_q(L) \varepsilon_t$, where $(1-L)^{-d} = \sum_{k=0}^{\infty} \overline{\vartheta}_k L^k$ with $\overline{\vartheta}_k = \Gamma(k+d) / (\Gamma(d)\Gamma(k+1))$, where $\Gamma(\cdot)$ denotes the gamma function such that $\Gamma(c) = \infty$ for $c = 0$ and $\Gamma(0)/\Gamma(0) = 1$, and $\Phi_p(L)$ and $\Theta_q(L)$ are the autoregressive and moving average polynomials with no common roots and outside the unit circle. The latter implies that $\Phi_p^{-1}(L)\Theta_q(L) = \sum_{j=0}^{\infty} b_j L^j$ with $b_j = O\left(j^{-c}\right)$ for any $c > 0$. The condition $\sum_{k=0}^{\infty} k^2 |b_k| < \infty$ implies that $h(\lambda)$ is twice continuously differentiable for all $\lambda \in [0, \pi]$. We finish pointing out that the sole motivation to assume homoscedastic errors is only for notational simplicity as well as to shorten the arguments of the already technical proofs and eases some of the arguments for the proof of the validity of the bootstrap described in Sect. 4 below.

C2: For all $x \in [0, 1]$, $r(x)$ satisfies

$$\lim_{y\to x} \left| \frac{r(y) - r(x) - R(x)}{|x-y|^\tau} \right| = o(1),$$

where $0 < \tau \leq 2$ and $R(x)$ is a polynomial of degree $[\tau - 1]$ with $[z]$ denoting the integer part of z.

Condition $C2$ is only slightly stronger than functions $r(x)$ which are Lipschitz continuous of order τ if $0 < \tau \leq 1$, or $r(x)$ is differentiable with derivative satisfying a Lipschitz condition of degree $\tau - 1$, if $1 < \tau \leq 2$. For instance, when $\tau = 2$, $C2$ means that $r(x)$ is twice continuously differentiable.

C3: $K_+ : [0, 1] \to \mathbb{R}$ and $K_- : [-1, 0] \to \mathbb{R}$, where $K_+(x) = K_-(-x)$, $\int_0^1 K_+(x)\,dx = 1$ and $\int_0^1 x K_+(x)\,dx = 0$.

Kernels $K_+(x)$, and therefore $K_-(x)$, satisfying $C3$ can be obtained from any function $\upsilon(x)$ with domain in $[0,1]$ as $K_+(x) = \upsilon(x)(c_1 + c_2 x)$, where c_1 and c_2 are the solutions to $\int_0^1 K_+(x)\,dx = 1$ and $\int_0^1 x K_+(x)\,dx = 0$. As an example let $\upsilon(x) = x(x+1)$, then $K_+(x) = 12x(1-x)(3-5x)$, see [8].

Our next condition deals with the bandwidth parameter a.

C4: As $n \to \infty$, (i) $(na)^{-1} \to 0$ and (ii) $(na)^{\frac{1}{2}-d} a^\tau \le D < \infty$, with τ as in $C2$.

Part (i) is standard in kernel regression estimation, whereas part (ii) needs more explanation. The latter differs from the analogue assumed by [29]. Contrary to the latter work, we do not need to assume that $\check{n}^{\frac{1}{2}-d} a^\tau \to 0$ as $n \to \infty$. This allows us to choose the optimal bandwidth parameter a, in the sense of being the value a which minimizes the MSE of the nonparametric regression estimator. More precisely, suppose that $d = 0$ and $\tau = 2$. Then, it is known that the optimal choice of a satisfies $a = Dn^{-1/5}$ for some finite positive constant D, which corresponds to the choice of the bandwidth parameter by, say, cross-validation. Also, note that for a given degree of smoothness on $r(x)$, that is τ in $C2$, the bandwidth parameter converges to zero slower as d increases. That is, given a particular bandwidth it requires less smoothness in $r(x)$.

We finish indicating how we can extend our testing procedure to the case where we know that there exist k breaks and we want to test the existence of k_0 additional ones. That is, our null hypothesis is that

$$r(x) = \begin{cases} r_1(x) & x < x^1 \\ r_2(x) & x^1 \le x < x^2 \\ & \cdots \\ r_{k+1}(x) & x^k \le x, \end{cases}$$

where the functions $r_i(x)$ are continuous being the alternative hypothesis that there exist k_0 points in $\mathcal{X}$ for which $r_i(x)$ are not continuous, for some $i = 1, \ldots, k+1$. We now describe or envisage how we can modify our test in (2.3). To that end, let

$$\widehat{r}_{a,q} =: \left|\widehat{r}_{a,+}(x_q) - \widehat{r}_{a,-}(x_q)\right|, \quad q \in \widetilde{Q}_n,$$

where

$$\widetilde{Q}_n = \left\{q : q \in Q_n \backslash \cup_{p=1}^k \widetilde{Q}_n^{(p)}\right\}$$

with $\widetilde{Q}_n^{(p)} = \left\{q : x^p - \check{n} < q \le x^p + \check{n}\right\}$, $p = 1, \ldots, k$. That is $\widetilde{Q}_n$ is the set of points $q \in Q_n$ which do not belong to the set $\cup_{p=1}^k \widetilde{Q}_n^{(p)}$. Next, denote $\widehat{r}_{(q)}$ the qth-order statistic of $\left\{\widehat{r}_{a,q}\right\}_{q=1}^{\mathring{Q}_n}$, so that $\widehat{r}_{(1)} = \min_{q \in \widetilde{Q}_n} \widehat{r}_{a,q}$ and $\widehat{r}_{\left(\mathring{Q}_n\right)} = \max_{q \in \widetilde{Q}_n} \widehat{r}_{a,q}$, where $\mathring{Q}_n = \#\left\{\widetilde{Q}_n\right\}$. Then, if k_0 is a known a priory positive integer, the test can be based on

$$\mathcal{T}_d^{k_0} =: \widehat{r}_{\left(\mathring{Q}_n - (k_0 - 1)\right)}.$$

To examine the asymptotic behavior of the test is beyond the scope of this paper and it will be discussed in a different manuscript.

3 Results

Before we examine the properties of $\mathcal{T}_d$ in (2.3), we shall first examine the covariance of $\widetilde{r}_a(q)$ at two points $q_1 \le q_2 \in \mathcal{Q}_n$, where in what follows $\widetilde{r}_a(q) =: \widehat{r}_{a,+}(q) - \widehat{r}_{a,-}(q)$. Also define $b(q_1, q_2) =: (q_2 - q_1)/\check{n}$ and $\vartheta(d) = 2\Gamma(1-2d)\cos\left(\pi\left(\frac{1}{2} - d\right)\right)$.

Proposition 1 *Assuming C1–C4, under H_0, for any $\check{n} < q_1 \le q_2 \le n - \check{n}$, as $n \to \infty$,*

$$\check{n}^{1-2d} Cov\left(\widetilde{r}_a(q_1), \widetilde{r}_a(q_2)\right) \to \rho(b;d) =: \rho_+(b;d) + \rho_-(b;d) - \rho_\pm(b;d) - \rho_\mp(b;d),$$

where $b := \lim_{n\to\infty} b\left(q_1, q_2\right)$ is finite and

(a) if $0 < d < \frac{1}{2}$,

$$\rho_+(b;d) = h(0)\vartheta(d) \int_0^1 \int_b^{1+b} |v-w|^{2d-1} K_+(v) K_+(w-b)\, dvdw,$$
$$\rho_-(b;d) = h(0)\vartheta(d) \int_{-1}^0 \int_{b-1}^{b} |v-w|^{2d-1} K_-(v) K_-(w-b)\, dvdw,$$
$$\rho_\pm(b;d) = h(0)\vartheta(d) \int_0^1 \int_{b-1}^{b} |v-w|^{2d-1} K_+(v) K_-(w-b)\, dvdw,$$
$$\rho_\mp(b;d) = h(0)\vartheta(d) \int_{-1}^0 \int_b^{1+b} |v-w|^{2d-1} K_-(v) K_+(w-b)\, dvdw.$$

(b) if $d = 0$,

$$\rho_+(b;d) = \rho_-(b;d) = 4^{-1}\sigma_u^2 \int_0^1 K_+(v) K_+(v-b)\, dv \quad \rho_\pm(b;d) = \rho_\mp(b;d) = 0.$$

Proof The proof is omitted since it proceeds as that in [29]. □

Proposition 1 indicates that the covariance structure is independent of the points at which $r_\pm(x)$ is estimated and only depends on the distance among the points where we estimate $r_\pm(x)$.

The next proposition deals with the correlation structure of $\widetilde{r}_a(q)$ as $b(q_1, q_2) \to 0$ and when $b(q_1, q_2) \to \infty$ as $n \to \infty$. In what follows, D will denote a positive finite constant.

Proposition 2 *Under C1–C4, for some $\alpha \in (0, 2]$, as $n \to \infty$,*

$$(a)\ \frac{\rho(b(q_1, q_2); d)}{\rho(b(q_1, q_1); d)} - 1 = -D|b(q_1, q_2)|^{\alpha} + o(|b(q_1, q_2)|^{\alpha}) \ \ as\ \ b(q_1, q_2) \to 0,$$

$$(b)\ \rho(b(q_1, q_2); d)\log(b(q_1, q_2)) = o(1) \ \ as \ \ b(q_1, q_2) \to \infty.$$

Proof The proof of this proposition or any other result is confined to Sect. 6 below. □

Proposition 3 *Assuming C1–C4, for any finite collection q_j, $j = 1, \ldots, p$, such that $q_j \in Q_n$ and for any z such that $|q_{j_1} - q_{j_2}| \geq nz > 0$, as $n \to \infty$,*

$$\check{n}^{\frac{1}{2}-d}\rho^{-\frac{1}{2}}(0; d)\left(\widetilde{r}_a(q_j)\right)_{j=1}^{p} \xrightarrow{d} \mathcal{N}(0, diag(1, \ldots, 1)).$$

First of all, we observe that the lack of asymptotic bias when the bandwidth parameter a is chosen optimally. This is in clear contrast to standard kernel regression estimation results, for which a bias term appears in the asymptotic distribution, when a is chosen to minimize the MSE, e.g., when a is chosen as in $C4$. Moreover, the latter result together with Proposition 1 implies that $\widetilde{r}_a(q)$ has asymptotically stationary increments, which are key to obtain the asymptotic distribution of $\mathcal{T}_d$.

Before we present our main result, we shall give a proposition which may be of independent interest.

Proposition 4 *Let $\overline{u}_t = \sum_{j=0}^{\infty}\vartheta_j\overline{\varepsilon}_{t-j}$ and $\{\overline{\varepsilon}_t\}_{t\in\mathbb{Z}}$ is a zero mean iid sequence of standard normal random variables. Then under $C1$ and $C3$, we have that*

$$\sup_{1\leq s\leq n}\left|\sum_{t=1}^{s}K_{\pm,t}(0)\,u_t - \sum_{t=1}^{s}K_{\pm,t}(0)\,\overline{u}_t\right| = o_p\left(n^{d+1/4}\right). \tag{3.1}$$

We now give the main result of this section. Let $\upsilon_n = (-2\log a)^{1/2}$.

Theorem 1 *Assuming C1–C4, under H_0,*

$$Prob\left\{\upsilon_n\left(\check{n}^{\frac{1}{2}-d}\rho^{-\frac{1}{2}}(0; d)\,\mathcal{T}_d - \zeta_n\right) \leq x\right\} \underset{n\uparrow\infty}{\to} \exp\left(-2e^{-x}\right), \ \ for\ x > 0,$$

where (a) If $0 < d < 1/2$, then

$$\zeta_n = \upsilon_n + \upsilon_n^{-1}\left\{\left(\frac{1}{2} - \frac{1}{\alpha}\right)\log\log a^{-1} + \log\left((2\pi)^{-\frac{1}{2}}\,2^{\frac{2-\alpha}{2\alpha}}E^{\frac{1}{\alpha}}\,\mathcal{J}_\alpha\right)\right\}$$

for some $0 < E < \infty$, *where* α *is as given in Proposition 2,*

$$0 < \mathcal{J}_\alpha \equiv \lim_{a \to 0} \int_0^\infty e^s \Pr\left\{ \sup_{0 \le t \le [a]^{-1}} \mathcal{Y}(t) > s \right\} ds < \infty$$

and $\mathcal{Y}(t)$ *is a stationary mean zero Gaussian process with covariance structure*

$$Cov\left(\mathcal{Y}(t_1), \mathcal{Y}(t_2)\right) = |t_1|^\alpha + |t_2|^\alpha - |t_2 - t_1|^\alpha .$$

(*b*) *If* $d = 0$, *then* $\zeta_n = \upsilon_n + \upsilon_n^{-1} \log\left((2\pi)^{-1} \left(\int_0^1 (\partial K_+(x)/\partial x)^2 dx \right)^{1/2} \right)$.

The results of our Theorem 1 are a natural extension to those obtained in [1] when the errors $\{u_t\}_{t\in\mathbb{Z}}$ are a sequence iid distributed random variables. Their result is included in part (*b*) of our previous theorem together with Proposition 1 part (*b*), where we consider $d = 0$. In fact their results are exactly the same as ours when $d = 0$, as the scaling constant $\rho(0; 0)$ is the same regardless the u_t is an iid sequence or not, i.e., it depends on the variance of the errors u_t.

3.1 Power of the Test

A desirable and important characteristic of any test is its consistency, that is under the alternative hypothesis the probability of rejection converges to 1 as $n \to \infty$. In addition to examine the limiting behavior under local alternatives enables to make comparisons between different consistent tests. We begin with the latter. To that end, we consider the following sequence of local alternatives

$$H_a: \ \exists x^0 \in [0, 1] \ \text{such that}\ r_+\left(x^0\right) = r_-\left(x^0\right) + r_n\left(x^0\right), \tag{3.2}$$

where $r_n\left(x^0\right) = \check{n}^{d-1/2} |2 \log a|^{-1/2} r$ with $r \neq 0$ and $r(x)$ satisfies $C2$ for $x \neq x^0$. Then, we have the following:

Corollary 1 *Assuming C1–C4, under* H_a *in (3.2)*

$$Prob\left\{ \upsilon_n \left(\check{n}^{\frac{1}{2}-d} \rho^{-\frac{1}{2}}(0; d)\, \mathcal{T}_d - \zeta_n \right) \le x \right\} \underset{n\uparrow\infty}{\to} \exp\left(-2e^{-\left(x - \frac{|r|\varrho(K_+)}{\rho^{1/2}(0;d)}\right)} \right), \ x > 0,$$

where ζ_n *was given in Theorem 1 and* $\varrho(K_+) = \max_{\ell=1,\dots,\hbar} \int_{\ell/\check{n}}^1 K_+(v)\, dv$.

Note that $\varrho(K_+)$ is not necessarily equal to 1 as would be the case if $K_+(\cdot)$ were nonnegative. This is because the condition $\int_0^1 x K_+(x)\, dx = 0$ implies that $K_+(\cdot)$ takes negative values in some subset of [0, 1].

From Corollary 1, one would expect that for fixed alternatives

$$H_1: \ \exists x^0 \in [0,1] \text{ such that } r_+\left(x^0\right) = r_-\left(x^0\right) + r; \quad |r| > 0,$$

and $r(x)$ satisfies $C2$ for $x \neq x^0$, we should have

$$\lim_{n\to\infty} \text{Prob}\left\{ \upsilon_n \left(\check{n}^{\frac{1}{2}-d} \rho^{-\frac{1}{2}}(0;d)\,\mathcal{T}_d - \zeta_n \right) \leq x \right\} = 0$$

that is, the test is consistent. This is confirmed in the next corollary.

Corollary 2 *Assuming C1–C4, $\mathcal{T}_d$ is consistent.*

Although Theorem 1 gives asymptotic justification for our test $\mathcal{T}_d$ under H_0, we observe that the normalization constant ζ_n depends not only on d but more importantly on $\mathcal{J}_\alpha$. The latter quantity is very difficult to compute except for the special cases $\alpha = 1$ or 2, see [24], where $\mathcal{J}_2 = \upsilon_n + \upsilon_n^{-1} \log\left(\pi^{-1}(E/2)^{1/2}\right)$ and $\mathcal{J}_1 = \upsilon_n + \upsilon_n^{-1} \log\left\{(E/\pi)^{1/2} + 2^{-1} \log\log a^{-1}\right\}$, where E is a constant which depends on K_+ although easy to obtain. More specifically, in our context, although d can be estimated, we face one potential difficulty when implementing the test. As we observe from (the proof of) Proposition 2, α depends on K_+ and d, so that to obtain $\mathcal{J}_\alpha$ does not seem an easy task. Under these circumstances, a bootstrap algorithm appears to be a sensible way to proceed.

4 The Bootstrap Approach

The comments made at the end of Sect. 3 and in the introduction suggest that to perform the test we need the help of bootstrap algorithms. In a context of time series, several approaches have been described in the literature. However, as we indicated in the introduction and after Corollary 2, the subsampling is not an appropriate method, neither the sieve bootstrap of [6] as the latter is not consistent for the sample mean of the error term with strong dependent data. Recall that in our context the statistical properties of the sample mean plays an important role into the asymptotic distribution of the test.

Due to this, in this section we describe and examine a bootstrap algorithm in the frequency domain similar to that proposed by [17], although they did not provide its justification and our conditions are significantly weaker than theirs. Two differences of our bootstrap procedure with moving block bootstrap (MBB) described in [20], say, are that (a) it is not a subset of the original data, and (b) the bootstrap data, say $\left\{u_t^*\right\}_{t=1}^n$, is covariance stationary as we have that $Cov^*\left(u_t^*, u_s^*\right)$ is a function of $|t-s|$. Herewith, by $Cov^*(z_1, z_2)$ or, say $E^*(z)$, we mean the covariance or expectation conditional on the data.

We now describe our main ingredients of the bootstrap and its justification. Suppose that in $C1$, $d = 0$, that is $u_t = \sum_{k=0}^{\infty} b_k \varepsilon_{t-k}$. Then, using the identity

$$u_t = \frac{1}{n^{1/2}} \sum_{j=1}^{n} e^{it\lambda_j} w_u\left(\lambda_j\right), \tag{4.1}$$

which can be considered as a "discrete" Cràmer representation of $\{u_t\}_{t=1}^n$, and Bartlett's approximation of $w_u\left(\lambda_j\right)$, see [5] Theorem 10.3.2, we obtain that

$$u_t \approx \frac{1}{n^{1/2}} \sum_{j=1}^{n} e^{it\lambda_j} B\left(-\lambda_j\right) w_\varepsilon\left(\lambda_j\right),$$

where "$\approx$" should be read as "approximately". Because $C1$ allows for strong dependence, the previous arguments suggests the approximation

$$w_u\left(\lambda_j\right) \approx \left(1 - e^{-i\lambda_j}\right)^{-d} B\left(-\lambda_j\right) w_\varepsilon\left(\lambda_j\right). \tag{4.2}$$

However, the lack of smoothness of $\left(1 - e^{-i\lambda_j}\right)^{-d}$ around $\lambda_j = 0$ and results given in [27] Theorem 1 at frequencies λ_j for fixed j indicate that for those frequencies the approximation in (4.2) seems to be invalid. Observe that these frequencies are precisely the more relevant ones when examining the asymptotic behavior of $\widehat{r}_{a,\pm}(q)$ in (2.2). So we consider

$$u_t \approx \widetilde{u}_t =: \frac{1}{n^{1/2}} \sum_{j=1}^{n} e^{it\lambda_j} \widetilde{g}^{1/2}\left(-\lambda_j; d\right) B\left(-\lambda_j\right) w_\varepsilon\left(\lambda_j\right), \tag{4.3}$$

where

$$\widetilde{g}^{1/2}\left(-\lambda_j; d\right) = \left| \sum_{\ell=-n+1}^{n-1} \gamma_\ell(d) e^{-i\ell\lambda_j} \right|^{1/2}, \tag{4.4}$$

with $\gamma_\ell(d) = \frac{(-1)^\ell \Gamma(1-2d)}{\Gamma(\ell-d+1)\Gamma(1-\ell-d)}$. It is easy to show that the right side of (4.3) preserves (asymptotically) the covariance structure of $\{u_t\}_{t\in\mathbb{Z}}$.

We now describe the bootstrap in the following *6 STEPS*.

STEP 1: Let $\widehat{t} = \arg\max_{t\in Q_n} \left|\widehat{r}_{a,+}(t) - \widehat{r}_{a,-}(t)\right|$, and obtain the centered residuals $\widehat{u}_t = \widetilde{u}_t - n^{-1} \sum_{t=1}^{n} \widetilde{u}_t$, $t = 1, \ldots, n$, where $\widetilde{u}_t = y_t - \widehat{r}_a(t)$ with

$$\widehat{r}_a(t) = \begin{cases} \widehat{r}_{a,+}(t), & t \le \check{n} \\ \widehat{r}_{a,-}(t), & \check{n} < t < \widehat{t} \\ \widehat{r}_{a,+}(t), & \widehat{t} \le t \le n - \check{n} \\ \widehat{r}_{a,-}(t), & t > n - \check{n}, \end{cases} \tag{4.5}$$

and $\widehat{r}_{a,+}(t)$ and $\widehat{r}_{a,-}(t)$ given in (2.2).

It is worth indicating that we could have computed the residuals using an estimate of the regression model under the null hypothesis of continuity, i.e.,

$$\widehat{r}_a(t) = \begin{cases} \widehat{r}_{a,+}(t), & t \le \check{n} \\ \frac{1}{2}\left(\widehat{r}_{a,+}(t) + \widehat{r}_{a,-}(t)\right), & \check{n} < t \le n - \check{n} \\ \widehat{r}_{a,-}(t), & t > n - \check{n}. \end{cases} \tag{4.6}$$

However, as it is well known, it is always preferable to obtain the residuals under the alternative hypothesis, as in (4.5), than under the null hypothesis.

STEP 2: We estimate d by [28] GSE,

$$\widehat{d} = \arg \min_{d \in [0, \Delta]} \widetilde{R}(d), \tag{4.7}$$

where $0 < \Delta < 1/2$, and

$$\widetilde{R}(d) = \log\left(\frac{1}{m}\sum_{j=1}^{m} \lambda_j^{2d} I_{\widehat{u}\widehat{u}}(\lambda_j)\right) - 2d\sum_{j=1}^{m} \log \lambda_j$$

for integer $m \in [1, \tilde{n})$, with $\tilde{n} = [n/2]$, and where $I_{\widehat{u}\widehat{u}}(\lambda) = |w_{\widehat{u}}(\lambda)|^2/(2\pi)$ is the periodogram of $\{\widehat{u}_t\}_{t=1}^n$, with $m^{-1} + mn^{-1} \to 0$.

We define our estimator of $2\pi h(\lambda) = |B(\lambda)|^2$ by

$$\widehat{h}(\lambda) = \frac{1}{2m+1}\sum_{j=-m}^{m}\left|1 - e^{-i(\lambda+\lambda_j)}\right|^{2\widehat{d}} I_{\widehat{u}\widehat{u}}(\lambda + \lambda_j).$$

Our third step describes how to obtain $w_\varepsilon^*(\lambda_j)$, $j = 1, \ldots, \tilde{n}$.

STEP 3: Let $\{\varepsilon_t^*\}_{t=1}^n$ be a random sample from standard normal and obtain its *discrete Fourier transform*,

$$\eta_j^* := w_\varepsilon^*(\lambda_j) = \frac{1}{n^{1/2}}\sum_{t=1}^{n} \varepsilon_t^* e^{-it\lambda_j}, \quad j = 1, \ldots, \tilde{n},$$

with $\eta_{n-j}^* = \overline{\eta}_j^*$, $j = 1, \ldots, \tilde{n}$, and $\overline{z}$ denoting the conjugate of z.

STEP 4: Compute

$$u_t^* = \frac{1}{n^{1/2}}\sum_{j=1}^{n} e^{it\lambda_j}\widetilde{g}^{1/2}(\lambda_j; \widehat{d})\,\widehat{A}(\lambda_j)\,\eta_j^*, \quad t = 1, \ldots, n,$$

where $\widetilde{g}^{1/2}(\lambda_j; \widehat{d})$ is given in (1.1),

$$\widehat{A}(\lambda_j) = \exp\left\{\frac{1}{2}\widehat{c}_0 + \sum_{r=1}^{M}\widehat{c}_r e^{-ir\lambda_j}\right\}, \quad j = 1, \ldots, \tilde{n} \tag{4.8}$$

with $\widehat{A}\left(\lambda_{n-j}\right) = \overline{\widehat{A}\left(\lambda_j\right)}$, and for $r = 0, \ldots, M = [n/4m]$, $\widehat{c}_r = \tilde{n}^{-1} \sum_{\ell=1}^{\tilde{n}} \log\left(\widehat{h}\left(\lambda_\ell\right)\right) \cos\left(r\lambda_\ell\right)$.

Remark 2 (**i**) It is worth mentioning that the way to obtain the bootstrap observations u_t^* in *STEP 4*, together with the definition of $\widehat{h}\left(\lambda\right)$ in (2.4), has some similarities with the autoregressive-aided bootstrap in [19].

(**ii**) There are doubts that *STEP 3* can be modified to allow $\left\{\varepsilon_t^*\right\}_{t=1}^n$ to be a random sample from the empirical distribution of $\{\widehat{u}_t\}_{t=1}^n$, following arguments in Huskova et al. (2008). However the latter will lengthen the arguments and the proof considerably and it is then beyond the scope of this paper.

The modulus square of $\widehat{A}\left(\lambda\right)$ in (4.8) is an estimator of $h\left(\lambda\right)$ in (2.4) and it comes from the so-called canonical spectral decomposition of $h\left(\lambda\right)$, see for instance [4, p.78–79].

STEP 5: Compute $\widehat{r}_a\left(t\right)$ as in (4.6) and then

$$y_t^* = \widehat{r}_a\left(t\right) + u_t^*; \quad t = 1, \ldots, n.$$

STEP 5 employs the same bandwidth as that in *STEP 1*, so that the standard requirement of an additional bandwidth e, such that $a = o\left(e\right)$, in computing the bootstrap analogue of (1.1), see for instance [13], is not needed. The reason comes from the observation that the bias of the nonparametric estimator of $r_+\left(\cdot\right) - r_-\left(\cdot\right)$ is $o\left(a^2\right)$ instead of the usual $O\left(a^2\right)$. Our final step is:

STEP 6: Compute $\widehat{r}_{a,+}^*\left(q\right)$ and $\widehat{r}_{a,-}^*\left(q\right)$, $q \in \mathcal{Q}_n$, as in (2.2) but with y_t replaced by y_t^* and the same bandwidth parameter a employed in *STEP 1*. Then we compute the bootstrap version of $\mathcal{T}_d$ as

$$\mathcal{T}_d^* = \sup_{q \in \mathcal{Q}_n} \left|\widehat{r}_{a,+}^*\left(q\right) - \widehat{r}_{a,-}^*\left(q\right)\right| .$$

The next proposition examines the behavior of $\widehat{d}$ given in (4.7).

Proposition 5 *Under* $C1$ *and* $C3$, $\left|\widehat{d} - d\right| = O_p\left(\left(ma\right)^{-1} + a^4 n \left(m/n\right)^{2d}\right)$.

Proof The proof is omitted as it follows step by step that of [29] Theorem 3, after noting that in our case we do not have his terms $I_{\xi\xi}$ and $I_{\chi\chi}$. □

Let us introduce the following condition on the smoothing parameter m and the bandwidth parameter a.

C5: As $n \to \infty$, (i) $D^{-1}n^{-1/3} < a < Dn^{-1/4}$ and (ii) $D^{-1}n^{3/5} < m < Dn^{3/4}$.

The next proposition discusses the bias of the nonparametric estimator $\widetilde{r}_a^*\left(q\right) = \widehat{r}_{a,+}^*\left(q\right) - \widehat{r}_{a,-}^*\left(q\right)$.

Proposition 6 *Assuming C1–C3, with* $\tau = 2$ *there, and C5, under* $H_0 \cup H_1$, *as* $n \to \infty$, $E^*\widetilde{r}_a^*\left(q\right) = o_p\left(\check{n}^{d-1/2}\right)$.

Proposition 7 *Denote* $E(u_\ell u_0) = \delta_{|\ell|}$. *Assuming C1–C3 with* $\tau = 2$ *there, and C5, we have that for* $q \in Q_n$,

$$\frac{1}{\check{n}^{2d}} \sum_{t=q+1}^{\check{n}+q} \left| E^*\left(u_t^* u_{t+\ell}^*\right) - \delta_{|\ell|} \right| = o_p(1). \tag{4.9}$$

Proof The proof proceeds as that of [15] Proposition 4.2 and thus it is omitted. □

Theorem 2 *Under the same conditions of Proposition 7 and* $H_0 \cup H_1$, *as* $n \to \infty$

$$\Pr\left\{ \upsilon_n \left(\check{n}^{\frac{1}{2}-\widehat{d}} \rho^{-\frac{1}{2}}\left(0;\widehat{d}\right) \mathcal{T}_d^* - \zeta_n \right) \le x \middle| \mathcal{Y} \right\} \to \exp\left(-2e^{-x}\right), \quad \textit{for } x > 0,$$

where υ_n *and* ζ_n *were defined in Theorem 1.*

We now comment on Theorem 2 and the important reason why subsampling is not a valid procedure as we now argue. To obtain a critical value, say $x(\beta)$, for which $\exp\left(-2e^{-x(\beta)}\right) = 1 - \beta$, is the same as to find the value, say $z_n(\beta)$, which obeys the equality

$$\lim_{n\to\infty} \Pr\{\mathcal{T}_d \le z_n(\beta)\} = 1 - \beta.$$

However the value $z_n(\beta)$ depends on both ζ_n and υ_n and thus, indirectly, on the choice of the bandwidth parameter a. In fact, since the constants ζ_n and υ_n are not possible to be computed, in practice we would only hope to obtain the critical values via $z_n(\beta)$. So, when employing the bootstrap sample, we are bound to obtain $z_n^*(\beta)$, for which

$$\lim_{n\to\infty} \Pr\left\{\mathcal{T}_d^* \le z_n^*(\beta) \middle| \mathcal{Y}\right\} = 1 - \beta.$$

As with the original data, the value $z_n^*(\beta)$ depends on both ζ_n and υ_n and thus on the choice of a. The latter has thus to be kept in mind when computing $z_n^*(\beta)$. But recall that one requirement for the bootstrap to be (asymptotically) valid is that $z_n^*(\beta)$ needs to satisfy $\left|z_n^*(\beta)/z_n(\beta) - 1\right| \xrightarrow{P} 0$. It is obvious that, for the latter expression to hold true, we need the constants ζ_n and υ_n to be the same for both $\mathcal{T}_d$ and $\mathcal{T}_d^*$, or that their ratio converges to one in probability. This is obviously possible only if the bandwidth parameter a is the same when estimating the regression function with both the original $\{y_t\}_{t=1}^n$ and bootstrap $\left\{y_t^*\right\}_{t=1}^n$ data.

5 Monte-Carlo

We perform a Monte Carlo experiment to assess the performance of the $\mathcal{T}_d$ statistic (2.3) and the validity of the bootstrap algorithm in Sect. 4. We take sample size $n = 256$ and generate 999 simulations from the model $y_t = 1 + \delta_0 I(t > n/2) + (1 + \delta_1 I(t > n/2))\, t/n + u_t$, where u_t is generated as Gaussian $FARIMA(0, d, 0)$

Table 1 Size and power of $\mathcal{T}_d$ statistic

	$d = 0$		$d = 0.3$	
$a =$	$0.5n^{-1/4}$	$0.75n^{-1/4}$	$0.5n^{-1/4}$	$0.75n^{-1/4}$
$\delta_0 = 0, \delta_1 = 0$	5.11	3.70	7.01	6.81
$\delta_0 = 0, \delta_1 = 1$	30.03	40.24	17.02	21.32
$\delta_0 = 1, \delta_1 = 0$	88.09	97.10	69.77	78.88

process with $d = 0, 0.3$ and standard deviation of 0.25. To examine the size of our test we set $\delta_0 = \delta_1 = 0$ and for the power we consider $\delta_0 = 1, \delta_1 = 0$ and $\delta_0 = 0, \delta_1 = 1$. The kernel is set to $K_+ (x) = (1 - x)(6 - 2x)$ and the bandwidth to $a = 0.5n^{-1/4}, 0.75n^{-1/4}$. In the estimation of d by [28] GSE,we choose $m = n/8$.

Given that bootstrap methods are computationally demanding in Monte Carlo studies, we employ the warp–speed method of [10] in order to calculate critical values for the $\mathcal{T}_d$ statistic. With this method, rather than computing critical values based on 999 bootstrap replications for each Monte Carlo sample, we generate only one bootstrap replication for each Monte Carlo sample and compute the bootstrap test statistic $\mathcal{T}_d^*$ for that sample. We perform 999 replications and collect the 999 bootstrap statistics $\mathcal{T}_d^*$. Then, the bootstrap critical value at α significance level is the $(1 - \alpha)$ percentile of these bootstrap statistics.

Table 1 reports the rejection frequencies in percentages at significance level $\alpha = 5\,\%$. The statistic $\mathcal{T}_d$ has satisfactory size and power. The power is stronger when there is a break in the constant rather than a break in the slope. The presence of long memory worsens the performance of the statistic.

6 Proofs of the Main Results

Proof of Proposition 2. We shall begin with the case $d > 0$. Abbreviating $b(q_1, q_2)$ by b, Proposition 1 implies that

$$\rho(b; d) = \rho_+ (b; d) + \rho_- (b; d) - \rho_\pm (b; d) - \rho_\mp (b; d) + o(1) . \tag{6.1}$$

Noting that for $\ell > 0$ and $d > 0$, $\ell^{2d-1} = \frac{2}{\pi}\Gamma(2d)\cos(d\pi)\int_0^\infty \lambda^{-2d}\cos(\ell\lambda)\,d\lambda$, proceeding as in [15], we have that the first term on the right of (6.1) is

$$\rho_+ (b; d) = \int_{-\infty}^{\infty} |\lambda|^{-2d} \cos(|\lambda|\, b) \left| \int_0^1 K_+ (v)\, e^{i\lambda v} dv \right|^2 d\lambda \tag{6.2}$$

for finite b. Likewise the last three terms on the right of (6.1) are, respectively,

$$\int_{-\infty}^{\infty} |\lambda|^{-2d} \cos(|\lambda| b) \left| \int_0^1 K_+(v) e^{-iv\lambda} dv \right|^2 d\lambda$$
$$\frac{1}{2} \int_{-\infty}^{\infty} |\lambda|^{-2d} \int_0^1 \int_0^1 \left(\left(e^{-i(v+w)\lambda} e^{-ib\lambda} + e^{i(v+w)\lambda} e^{ib\lambda} \right) K_+(v) K_+(w) dv dw \right) d\lambda$$
$$\frac{1}{2} \int_{-\infty}^{\infty} |\lambda|^{-2d} \int_0^1 \int_0^1 \left(\left(e^{i(v+w)\lambda} e^{-ib\lambda} + e^{-i(v+w)\lambda} e^{ib\lambda} \right) K_+(v) K_+(w) dv dw \right) d\lambda,$$

after noticing that by $C3$, $K_+(v) = K_-(-v)$. Hence, gathering (6.2) and the last displayed expressions, we conclude that

$$\rho(b; d) = \int_{-\infty}^{\infty} |\lambda|^{-2d} \cos(|\lambda| b) \left| \int_0^1 K_+(v) \left(e^{i\lambda v} - e^{-i\lambda v} \right) dv \right|^2 d\lambda.$$

The proof now proceeds as that of [15] Proposition 3.2.

Next, when $\mathbf{d} = \mathbf{0}$, the proof is omitted as it follows by standard arguments. See for instance [2] Theorem B.1. ■

Proof of Proposition 3. Proceeding as in [29] Theorem 1 and Lemma 1, it suffices to show that

$$E\left(\widetilde{r}_a(q)\right) = \begin{cases} o\left(a^\tau\right) & \text{if } 0 < \tau \le 1 \\ o\left(a^\tau + n^{-1}\right) & \text{if } 1 < \tau \le 2. \end{cases} \tag{6.3}$$

Observe that by uniform integrability of u_t^2 and that Propositions 1 and 2 imply that the covariance of $\widetilde{r}_a(q) = \widehat{r}_{a,+}(q) - \widehat{r}_{a,-}(q)$ at two points q_1 and q_2 converges to zero when $|q_1 - q_2| \ge nz > 0$, for any $z > 0$, we conclude that the covariance of the asymptotic distribution of the estimators is zero by Theorem A of [30, p.14].

On the other hand, under H_0 and standard kernel manipulations, we obtain that

$$\begin{aligned} E\left(\widehat{r}_{a,+}(q) - r(q/n)\right) &= \int_0^1 K_+(v) \left(r(av + q/n) - r(q/n) \right) dv \\ &= \int_0^1 K_+(v) R(av + q/n) dv + \begin{cases} o\left(a^\tau\right) & \text{if } 0 < \tau \le 1 \\ o\left(a^\tau + n^{-1}\right) & \text{if } 1 < \tau \le 2 \end{cases} \end{aligned}$$

by $C2$. Similarly,

$$E\left(\widehat{r}_{a,-}(q) - r(q/n)\right) = \int_{-1}^0 K_-(v) R(av + q/n) dv + \begin{matrix} o\left(a^\tau\right) & \text{if } 0 < \tau \le 1 \\ o\left(a^\tau + n^{-1}\right) & \text{if } 1 < \tau \le 2. \end{matrix}$$

From here and after an obvious change of variables, (6.3) holds true because $Q(x)$ is twice continuously differentiable and $\int_0^1 x K_+(x) dx = 0$ by $C3$. ■

Proof of Proposition 4. We shall consider the case that $K_{+,t}(0) = 1$, $\overline{u}_{t,n} = \sum_{j=0}^{\infty} \vartheta_j \overline{\varepsilon}_{t-j}$ the general case follows after observing that by Abel summation by parts

$$\left|\sum_{t=1}^{s} K_{+,t}(0)(u_t - \overline{u}_t)\right| \leq \left|\sum_{t=1}^{s} \left(K_{+,t}(0) - K_{+,t+1}(0)\right) \sum_{l=1}^{t} (u_l - \overline{u}_l)\right| + \left|K_{+,s}(0) \sum_{l=1}^{s} (u_l - \overline{u}_l)\right|$$

and then that $\sum_{t=1}^{s} \left|K_{+,t}(0) - K_{+,t+1}(0)\right| < D$ by $C3$. First we observe that we can write u_t as follows

$$u_t = \sum_{j=0}^{t-1} \vartheta_j \varepsilon_{t-j} + \sum_{j=0}^{\infty} \vartheta_{j+t} \varepsilon_{-j} := u_{1,t} + u_{2,t}.$$

So, it suffices to show (3.1) when u_t is replaced by $u_{1,t}$ and $u_{2,t}$. That is,

$$\sup_{1 \leq s \leq n} \left|\sum_{t=1}^{s} u_{j,t} - \sum_{t=1}^{s} \overline{u}_{j,t}\right| = o_p\left(n^{d+1/4}\right), \quad j = 1, 2 \tag{6.4}$$

and $\overline{u}_{1,t} = \sum_{j=0}^{t-1} \vartheta_j \overline{\varepsilon}_{t-j}$ and $\overline{u}_{2,t} = \sum_{j=0}^{\infty} \vartheta_{j+t} \overline{\varepsilon}_{-j}$.

We shall prove (6.4) for $j = 2$ first. After standard algebra and inequalities, for some $\chi \in (1, 2)$, we have that the left side is bounded by

$$\sup_{1 \leq s \leq n} \left|\sum_{t=1}^{s} \sum_{j=s^{\chi}+1}^{\infty} \vartheta_{j+t}\left(\varepsilon_{-j} - \overline{\varepsilon}_{-j}\right)\right| + \sup_{1 \leq s \leq n} \left|\sum_{t=1}^{s} \sum_{j=1}^{s^{2-\chi}} \vartheta_{j+t}\left(\varepsilon_{-j} - \overline{\varepsilon}_{-j}\right)\right| \tag{6.5}$$

$$+ \sup_{1 \leq s \leq n} \left|\sum_{t=1}^{s} \sum_{j=s^{2-\chi}+1}^{s^{\chi}} \vartheta_{j+t}\left(\varepsilon_{-j} - \overline{\varepsilon}_{-j}\right)\right|.$$

The expression inside the absolute value of first term of (6.5) is

$$\sum_{j=s^{\chi}+1}^{\infty} \left(\sum_{t=1}^{s} \vartheta_{j+t}\right)\left(\varepsilon_{-j} - \overline{\varepsilon}_{-j}\right) = \sum_{j=s^{\chi}+1}^{\infty} \left\{\sum_{t=1}^{s} \left(\vartheta_{j+t} - \vartheta_{j+t+1}\right)\right\} \sum_{p=s^{\chi}+1}^{j} \left(\varepsilon_{-p} - \overline{\varepsilon}_{-p}\right).$$

But by well-known results due to Komlós, Major and Tusnady, $\sup_{s^{\chi}+1 \leq p \leq j} \left|\sum_{p=s^{\chi}+1}^{j} \left(\varepsilon_{-p} - \overline{\varepsilon}_{-p}\right)\right| = o_p\left(j^{1/4}\right)$, so that the right side of the last displayed equality becomes

$$o_p(1) \sum_{j=s^{1+\chi}+1}^{\infty} j^{1/4} \sum_{t=1}^{s} (j+t)^{d-2} = o_p\left(s^{d+1/4}\right),$$

because $\left|\vartheta_j - \vartheta_{j+1}\right| \leq D j^{d-2}$ by $C1$, and hence the first term of (6.5) is $o_p\left(n^{d+1/4}\right)$.

Next, proceeding similarly, we have that by Abel summation by parts, the expression inside the absolute value of the second term of (6.5) is

$$\sum_{j=1}^{s^{2-\chi}}\left\{\sum_{t=1}^{s}\left(\vartheta_{j+t}-\vartheta_{j+t+1}\right)\right\}\sum_{p=1}^{j}\left(\varepsilon_{-p}-\overline{\varepsilon}_{-p}\right)+\sum_{t=1}^{s}\vartheta_{s^{2-\chi}+t}\sum_{p=1}^{s^{2-\chi}}\left(\varepsilon_{-p}-\overline{\varepsilon}_{-p}\right)=o_p\left(s^{d+1/4}\right),$$

because $\chi > 1$ implies that

$$\sum_{j=1}^{s^{2-\chi}} j^{1/4}\sum_{t=1}^{s}(j+t)^{d-2}\leq K\sum_{j=1}^{s^{2-\chi}} j^{-1+d+1/q}\sum_{t=1}^{s}(j+t)^{-1}=o\left(s^{d+1/4}\right)$$

and $\sum_{t=1}^{s}\vartheta_{s^{2-\chi}+t}=O\left(s^{d}\right)$. So, it remains to examine the third term of (6.5). Proceeding as before we have, by Abel summation by parts, that it is

$$o_p(1)\sup_{1\leq s\leq n}\sum_{j=s^{2-\chi}+1}^{s^{\chi}} j^{1/4}\sum_{t=1}^{s}(j+t)^{d-2}=o_p(1)\sum_{t=1}^{n}\left(s^{2-\chi}+t\right)^{d-1+1/4}=o_p\left(n^{d+1/4}\right)$$

by standard manipulations. This completes the proof of (6.4) for $j=2$.

We now show (6.4) for $j=1$. First, $\sum_{t=1}^{s}u_{1,t}-\sum_{t=1}^{s}\overline{u}_{1,t}$ is

$$\sum_{t=1}^{s}\left(\sum_{j=0}^{t-1}\vartheta_j\left(\varepsilon_{t-j}-\overline{\varepsilon}_{t-j}\right)\right)=\sum_{\ell=1}^{s}\left(\sum_{j=0}^{s-\ell}\vartheta_j\right)\left(\varepsilon_{\ell}-\overline{\varepsilon}_{\ell}\right)=\sum_{\ell=1}^{s}\left(\sum_{j=0}^{s-\ell}\vartheta_j\right)\left\{\left(S_{\ell}-S_{\ell-1}\right)-\left(\overline{S}_{\ell}-\overline{S}_{\ell-1}\right)\right\},$$

where $S_{\ell}=\sum_{p=1}^{\ell}\varepsilon_p$ and $\overline{S}_{\ell}=\sum_{p=1}^{\ell}\overline{\varepsilon}_p$. From here the proof follows as in Lemma 5 of [21] since $\Lambda_s=\sum_{j=0}^{s}\vartheta_j$ satisfies their Assumption 1. ∎

Proof of Theorem 1. Because the asymptotic independence of the distributions of $\max_q$ and $\min_q$ and the asymptotic distributions of $\sup_i X_i$ and $\inf_i -X_i$ are the same, it suffices to show that, for $x>0$,

$$\Pr\left\{\upsilon_n\left(\sup_{\check{n}<q\leq n-\check{n}}\check{n}^{\frac{1}{2}-d}\rho^{-\frac{1}{2}}(0;d)\left(\widetilde{r}_a(q)\right)-\zeta_n\right)\leq x\right\}\to\exp\left(2e^{-x}\right). \tag{6.6}$$

To that end, we will show that $\check{n}^{\frac{1}{2}-d}\widetilde{r}_a(q)$ converges to a Gaussian process $\mathcal{G}(u)$ in $\mathbb{D}[0,\infty)$, whose correlation structure satisfies conditions (v) and (vi) of [2] Theorem A1, for some $\alpha>0$. See also [24] equations (1.2) and (2.1). From here and Proposition 4, the limiting distribution in (6.6) holds by [2] Theorem 1, for some $\alpha>0$.

First by standard arguments, Propositions 3 and 4 implies that the finite-dimensional distributions converge to those of a Gaussian process $\mathcal{G}(u)$, whereas, by Proposition 2, the correlation structure of $\mathcal{G}(u)$ satisfies the conditions in [2] or [24]. So, to complete the proof it suffices to show the tightness condition for the process $\check{n}^{\frac{1}{2}-d}\left(\widetilde{r}_a(q)\right)$. To that end, we shall denote

$$X_{\pm,n}(\widetilde{q})=\frac{1}{\check{n}^{\frac{1}{2}+d}}\sum_{t=1}^{n}u_tK_{\pm}\left(\frac{t}{\check{n}}-\widetilde{q}\right),\quad \widetilde{q}=\frac{1}{\check{n}},\frac{2}{\check{n}},\ldots,[a]^{-1}.$$

So, we have that $X_{+,n}(\widetilde{q})$, say, is a process in $\mathbb{D}\left[0,[a]^{-1}\right]$ equipped with Skorohod's metric, where we extend $\mathbb{D}\left[0,[a]^{-1}\right]$ to $\mathbb{D}[0,\infty)$ by writing $X_{+,n}(\infty)=X_{+,n}\left([a]^{-1}\right)$. Then Pollard (1981, Ch. V) implies that we need only to show tightness in $\mathbb{D}[0,D]$ for any finite $D>0$. To that end, let

$$\check{n}^{\frac{1}{2}-d}\left\{\left(\widehat{r}_{a,+}(q)-E\widehat{r}_{a,+}(q)\right)-\left(\widehat{r}_{a,-}(q)-E\widehat{r}_{a,-}(q)\right)\right\}:=X_{+,n}(\widetilde{q})+X_{-,n}(\widetilde{q}).$$

Next Proposition 2 implies that the process $X_{+,n}(\widetilde{q})$ has independent and stationary increments, that is for $\widetilde{q}\in[c_1,d_1]$ and $\widetilde{q}\in[c_2,d_2]$ and $[c_1,d_1]\cap[c_2,d_2]=\emptyset$, $X_{+,n}(\widetilde{q})$ are (asymptotically) independent with the same finite dimensional distributions.

Because $\mathcal{G}(\bullet)$ has continuous paths, by [3] Theorem 15.6, it suffices to show the Kolmogorov's moment condition

$$E\left(\left|X_{+,n}(\widetilde{q}_2)-X_{+,n}(\widetilde{q})\right|^{\beta}\left|X_{+,n}(\widetilde{q})-X_{+,n}(\widetilde{q}_1)\right|^{\beta}\right)\leq D\left|\widetilde{q}_2-\widetilde{q}\right|^{\frac{1+\delta}{2}}\left|\widetilde{q}-\widetilde{q}_1\right|^{\frac{1+\delta}{2}}$$

for some $\delta>0$, $\beta>0$ and where $0\leq\widetilde{q}_1<\widetilde{q}<\widetilde{q}_2\leq D$. Observe that we can consider only the situation for which $\check{n}^{-1}<\widetilde{q}_2-\widetilde{q}_1$, since otherwise the left side is trivially zero. Because for any $0\leq a<b<c\leq D$, $|c-b|\,|b-a|\leq|c-a|^2$ by Cauchy–Schwarz inequality, the last displayed inequality holds true if

$$E\left|X_{+,n}(\widetilde{q}_2)-X_{+,n}(\widetilde{q}_1)\right|^{2\beta}\leq D\left|\widetilde{q}_2-\widetilde{q}_1\right|^{1+\delta}. \tag{6.7}$$

It suffices to consider $|\widetilde{q}_2-\widetilde{q}_1|<1$, the case $|\widetilde{q}_2-\widetilde{q}_1|\geq1$ is trivial since the left side of (6.7) is bounded provided that $\beta\leq1$.

By definition, $X_{+,n}(\widetilde{q}_2)-X_{+,n}(\widetilde{q}_1)$ is

$$\frac{1}{\check{n}^{\frac{1}{2}+d}}\left\{\sum_{t=\check{n}-(q_2-q_1)+1}^{\check{n}}u_{t+q_2}K_{+,t}+\sum_{t=1}^{\check{n}-(q_2-q_1)}u_{t+q_2}\left(K_{+,t}-K_{+,t+q_1-q_2}\right)-\sum_{t=1}^{q_2-q_1}u_{t+q_1}K_{+,t}\right\}. \tag{6.8}$$

Choose $\beta=1$ in (6.7). Because $\check{n}\widetilde{q}=q$, $C3$ implies that $K_{+,t}=D\left(t/\check{n}\right)(1+o(1))$, and the second moment of the third term of (6.8) is bounded by

$$\frac{D}{\check{n}^{1+2d}}\sum_{t,s=1}^{q_2-q_1}|t-s|^{2d-1}\frac{t}{\check{n}}\frac{s}{\check{n}}\leq D\left|\widetilde{q}_2-\widetilde{q}_1\right|^{3+2d},$$

so that we have that the last term of (6.8) satisfies the inequality (6.7). Similarly, because $K_{+,t}=D\left(1-t/\check{n}\right)(1+o(1))$ as $t\to\check{n}$ by $C3$, we obtain that the second term of (6.8) is bounded by $D\left|\widetilde{q}_2-\widetilde{q}_1\right|^2(1-(1-(\widetilde{q}_2-\widetilde{q}_1)))^{2d}\leq D\left|\widetilde{q}_2-\widetilde{q}_1\right|^{2+2d}$ because $0<\widetilde{q}_2-\widetilde{q}_1<1$. Finally, by continuous differentiability of $K_+(u)$ for $u\in(0,1)$, we obtain that the second moment of the middle term in (6.8) is bounded by $D(\widetilde{q}_2-\widetilde{q}_1)^2\frac{1}{\check{n}^{1+2d}}\sum_{t,s=1}^{\check{n}-(q_2-q_1)}|t-s|^{2d-1}\leq D\left|\widetilde{q}_2-\widetilde{q}_1\right|^2$ because

$0 \le d < 1/2$. So, (6.7) holds true choosing $\beta = 1$ and $\delta = 1$ and hence $X_{+,n}(\widetilde{q})$ is tight. By identical arguments, $X_{-,n}(\widetilde{q})$ is also tight, which implies that the process $\breve{n}^{\frac{1}{2}-d}\left(\widehat{r}_{a,+}(q) - \widehat{r}_{a,-}(q)\right)$ is tight. This concludes the proof of the theorem. ■

Proof of Corollary 1. From the proof of Theorem 1, we only need to show that

$$\sup_{\breve{n} < q < n - \breve{n}} \breve{n}^{\frac{1}{2}-d} \upsilon_n \left| E\left(\widetilde{r}_a(q)\right)\right| \to r\varrho\left(K_+\right).$$

But this is the case because by standard kernel manipulations and that $C3$ implies that $K_+(x) = K_-(-x)$, we obtain that under H_a given in (3.2),

$$\breve{n}^{\frac{1}{2}-d} E\left(\widetilde{r}_a(q_0)\right) = \frac{r_n\left(x^0\right)}{\breve{n}^{\frac{1}{2}-d}} \frac{1}{\breve{n}} \sum_{t=|q-q_0|}^{\breve{n}} K_{+,t} = \frac{r}{\upsilon_n} \int_{|q-q_0|/\breve{n}}^{1} K_+(x)\, dx\,(1 + o(1)),$$

by [4, p.15] as $\int_0^1 |\partial K_+(u)/\partial u|\, du < \infty$ and where q_0/n is the closest point to x^0. The conclusion is standard because $\sup_{q \in \mathcal{Q}_n} \int_{|q-q_0|/\breve{n}}^{1} K_+(x)\, dx \to \varrho(K_+)$. But under H_a, $\upsilon_n \breve{n}^{\frac{1}{2}-d} r_n\left(x^0\right) = r$, so following the arguments preceding (6.6), it suffices to show that

$$\Pr\left\{\upsilon_n\left(\sup_{q\in\mathcal{Q}_n} \breve{n}^{\frac{1}{2}-d} \rho^{-\frac{1}{2}}(0;d)\, \widetilde{r}_a(q) - \frac{r\varrho\left(K_+\right)}{\upsilon_n \rho^{1/2}(0;d)} - \zeta_n\right) \le x\right\} \to \exp\left(-2e^{-x}\right),$$

which is the case as we now argue. Proceeding as with Theorem 1, the last expression holds true because: (a) the finite dimensional distributions of $\breve{n}^{\frac{1}{2}-d}\widetilde{r}_a(q) - r\varrho(K_+)/\upsilon_n$ converge to those of a Gaussian process with correlation structure $Corr(b)$; (b) the process $\breve{n}^{\frac{1}{2}-d}\widetilde{r}_a(q)$ is tight proceeding as in Theorem 1. ■

Proof of Corollary 2. Since for any sequence of random variables, $X_1, \ldots, X_n$, $\Pr\left\{\max_{i \le n} X_i > x\right\} \ge \Pr\left\{\max_{i=k,\ldots,n-\ell} X_i > x\right\}$, it suffices to show that there exists $q \in \mathcal{Q}_n$, such that

$$\Pr\left\{\upsilon_n\left(\breve{n}^{\frac{1}{2}-d}\rho^{-\frac{1}{2}}(0;d)\left(\widetilde{r}_a(q)\right) - \zeta_n\right) > x\right\} \to 1$$

for all $x > 0$. Choose $q = q_0$, with q_0 as in Corollary 1. Proceeding as in the proof of Proposition 3, we have that $\breve{n}^{\frac{1}{2}-d}\left|\widetilde{r}_a(q_0) - r\right| = O_p(1)$, and hence $\left|\widetilde{r}_a(q_0)\right| = O_p\left(\breve{n}^{d-\frac{1}{2}}\right) + |r|(1 + o(1))$. So, we obtain that

$$\upsilon_n\left(\breve{n}^{\frac{1}{2}-d}\left|\widetilde{r}_a(q_0)\right| - \zeta_n\right) \to \infty$$

because $C4$ and that $d < 1/2$ imply that $\breve{n}^{\frac{1}{2}-d}\zeta_n^{-2} = D\breve{n}^{\frac{1}{2}-d}\log^{-1} n \to \infty$. The conclusion now follows by standard arguments. ■

Proof of Proposition 6. By definition and because $E^{*}u_{t}^{*}=0$,

$$E^{*}\widetilde{r}_{a}^{*}(q)=\frac{1}{\check{n}}\sum_{t=1}^{n}\widehat{r}_{a}(t)\left(K_{+,t-q}-K_{-,t-q}\right).$$

So, we need to show that the right side is $o_{p}\left(\check{n}^{d-1/2}\right)$. Because $K_{+}\left(t/\check{n}\right)=K_{-}\left(-t/\check{n}\right)$ by $C3$, it suffices to show that

$$(a)\ \frac{1}{\check{n}}\sum_{t=1}^{n}\left(E\left(\widehat{r}_{a}(t)\right)-r\left(\frac{q}{n}\right)\right)\left(K_{+,t-q}-K_{-,t-q}\right)=o\left(\check{n}^{d-1/2}\right)$$

$$(b)\ \frac{1}{\check{n}}\sum_{t=1}^{n}\left(\widehat{r}_{a}(t)-E\left(\widehat{r}_{a}(t)\right)\right)\left(K_{+,t-q}-K_{-,t-q}\right)=o_{p}\left(\check{n}^{d-1/2}\right).$$

We begin with (a). By Proposition 3, we have that $E\left(\widehat{r}_{a}(t)\right)-r(t/n)=o\left(a^{2}\right)=o\left(\check{n}^{d-1/2}\right)$ because by $C5$, $(na)^{1/2-d}a^{2}\leq D$. On the other hand, because $d<1/2$, and $\tau=2$, the proof of Proposition 3 implies that

$$\check{n}^{-1}\sum_{t=1}^{n}\left(r(t/n)-r(q/n)\right)\left(K_{+,t-q}-K_{-,t-q}\right)=o\left(a^{2}\right)=o\left(\check{n}^{d-1/2}\right)$$

by $C5$ and that $d<1/2$. Next we show part (b). By definition, it equals

$$\frac{1}{\check{n}}\sum_{t=1}^{n}\left\{\frac{1}{\check{n}}\sum_{s=1}^{n}u_{s}K_{t-s}\right\}\left\{K_{+,t-q}-K_{-,t-q}\right\}.$$

where $K_{t-s}=\frac{1}{2}\left(K_{+,t-s}+K_{-,t-s}\right)$. Let us examine the contribution due to $K_{+,t-q}$, that from $K_{-,t-q}$ being similarly handled. The second moment is

$$\frac{\sigma_{u}^{2}}{\check{n}^{4}}\sum_{s_{1},s_{2}=1}^{n}\gamma_{u}\left(|s_{1}-s_{2}|\right)\sum_{t_{1},t_{2}=1}^{n}K_{t_{1}-s_{1}}K_{t_{2}-s_{2}}K_{+,t_{1}-q}K_{+,t_{2}-q}=o\left(\check{n}^{2d-1}\right).$$

From here we conclude that part (b) and the proof of the proposition. ∎

Proof of Theorem 2. As we argue with (6.6), we only need to show that

$$\Pr\left\{\upsilon_{n}\left(\sup_{\check{n}<q<n-\check{n}}\check{n}^{\frac{1}{2}-\widehat{d}}\rho^{-\frac{1}{2}}\left(0;\widehat{d}\right)\widetilde{r}_{a}^{*}(q)-\zeta_{n}\right)\leq x\,|\mathcal{Y}\right\}\rightarrow\exp\left(2e^{-x}\right),\qquad(6.9)$$

for $x>0$. To that end, we will show that $\check{n}^{\frac{1}{2}-\widehat{d}}\rho^{-\frac{1}{2}}\left(0;\widehat{d}\right)\widetilde{r}_{a}^{*}(q)$ converges, in bootstrap sense, to the Gaussian process $\mathcal{G}(q)$ in $\mathbb{D}[0,\infty)$, whose correlation structure is that given in Proposition 1. Proceeding as with the proof of Theorem 1, it suffices to show the tightness condition. To that end, denote

$$X^*_{+,n}(\widetilde{q}) = \frac{1}{\check{n}^{\frac{1}{2}+\widehat{d}}} \sum_{t=1}^{n} u^*_t K_+\left(\frac{t}{\check{n}} - \widetilde{q}\right), \quad \widetilde{q} = \frac{1}{\check{n}}, \frac{2}{\check{n}}, \ldots, [a]^{-1}.$$

Arguing as in the proof of Theorem 1, it suffices to show the moment condition

$$E^* \left|X^*_{+,n}(\widetilde{q}_2) - X^*_{+,n}(\widetilde{q}_1)\right|^{2\beta} \leq DH_n(\widetilde{q}_2, \widetilde{q}_1) |\widetilde{q}_2 - \widetilde{q}_1|^{1+\delta} \tag{6.10}$$

with $\check{n}^{-1} < \widetilde{q}_2 - \widetilde{q}_1$ and $H_n(\widetilde{q}_2, \widetilde{q}_1) = O_p(1)$. It suffices to consider $|\widetilde{q}_2 - \widetilde{q}_1| < 1$, the case $|\widetilde{q}_2 - \widetilde{q}_1| \geq 1$ is trivial since the left side of (6.10) is bounded in probability.

By definition, $X^*_{+,n}(\widetilde{q}_2) - X^*_{+,n}(\widetilde{q}_1)$ is

$$\frac{1}{\check{n}^{\frac{1}{2}+\widehat{d}}} \left\{ \sum_{t=\check{n}-(q_2-q_1)+1}^{\check{n}} u^*_{t+q_2} K_{+,t} + \sum_{t=1}^{\check{n}-(q_2-q_1)} u^*_{t+q_2} \left(K_{+,t} - K_{+,t+q_1-q_2}\right) - \sum_{t=1}^{q_2-q_1} u^*_{t+q_1} K_{+,t} \right\}. \tag{6.11}$$

Choosing $\beta = 1$ and because $\check{n}\widetilde{q} = q$, the contribution into the left of (6.10) due to the third term of (6.11) is

$$\left| \frac{1}{\check{n}^{1+2\widehat{d}}} \sum_{t,s=1}^{q_2-q_1} E^*\left(u^*_{t+q_1} u^*_{s+q_1}\right) K_{+,t} K_{+,s} \right| \leq DH_n(\widetilde{q}_2, \widetilde{q}_1) |\widetilde{q}_2 - \widetilde{q}_1|^{3+2d},$$

because by $C3$, $\left|K_{+,t}\right| \leq Dt/\check{n}$ and Proposition 6 implies that

$$\sum_{t,s=1}^{q} \left|E^*\left(u^*_{t+q_1} u^*_{s+q_1}\right)\right| = D \sum_{t,s=1}^{q} |E(u_t u_s)| \left(1 + o_p(1)\right) = Dq^{1+2d}\left(1 + o_p(1)\right).$$

Observe that $H_n(\widetilde{q}_2, \widetilde{q}_1) = D\check{n}^{2(d-\widehat{d})}$, which is $O_p(1)$ because by Proposition 5 and $C5$, $\left|\widehat{d} - d\right| = o_p\left(\log^{-1} n\right)$. So, the last term of (6.11) satisfies the inequality (6.10). Similarly, we obtain that

$$\begin{aligned} E^* \left| \frac{1}{\check{n}^{\frac{1}{2}+\widehat{d}}} \sum_{t=\check{n}-(q_2-q_1)+1}^{\check{n}} u^*_{t+q_2} K_{+,t} \right|^2 &\leq D\check{n}^{2(d-\widehat{d})} |\widetilde{q}_2 - \widetilde{q}_1|^2 \left(1 - (1 - (\widetilde{q}_2 - \widetilde{q}_1))\right)^{2d} \\ &\leq DH_n(\widetilde{q}_2, \widetilde{q}_1) |\widetilde{q}_2 - \widetilde{q}_1|^{2+2d}, \end{aligned}$$

because $0 < \widetilde{q}_2 - \widetilde{q}_1 < 1$ and choosing $H_n(\widetilde{q}_2, \widetilde{q}_1) = D\check{n}^{2(d-\widehat{d})}$. Finally, the continuous differentiability of $K_+(u)$ for $u \in (0, 1)$ implies that the bootstrap second moment of the middle term in (6.11) is bounded by

$$D\left(\widetilde{q}_2-\widetilde{q}_1\right)^2 \frac{1}{\check{n}^{1+2\widehat{d}}} \sum_{t,s=1}^{\check{n}-(q_2-q_1)} |t-s|^{2d-1} \leq DH_n\left(\widetilde{q}_2,\widetilde{q}_1\right)\left|\widetilde{q}_2-\widetilde{q}_1\right|^2,$$

because $d<1/2$. So, (6.10) holds true choosing $\delta=1$ and hence $X_{+,n}(\widetilde{q})$ is tight. On the other hand, proceeding similarly as with $X^*_{+,n}(\widetilde{q})$,

$$X^*_{-,n}(\widetilde{q}) = \frac{1}{\check{n}^{\frac{1}{2}+d}} \sum_{t=1}^{n} u^*_t K_-\left(\widetilde{q}-\frac{t}{\check{n}}\right), \quad \widetilde{q}=1/\check{n}, 2/\check{n}, \ldots, [a]^{-1},$$

is also tight. So, $\check{n}^{\frac{1}{2}-\widehat{d}}\left(\widehat{r}^*_+(q)-\widehat{r}^*_-(q)\right)$ is tight, which concludes the proof of the theorem because by Lemma 6, we have that the correlation structure converges in probability to that given in Proposition 2. ∎

7 Auxiliary Lemmas

In what follows $\varphi\left(\lambda_j\right)$ will be abbreviated as φ_j for a generic $\varphi(\lambda)$ function. Let us introduce the following notation. Let $\check{h}_\ell(d)=\frac{1}{2m+1}\sum_{j=-m}^{m}\psi_{\ell+j}^{2d} I_{uu,\ell+j}$, where $\psi_j=\left|2\sin\left(\lambda_j/2\right)\right|$. With this notation, we have that Taylor's expansion up to the βth term implies that

$$\check{h}_\ell\left(\widehat{d}\right)-\check{h}_\ell(d) = \sum_{p=1}^{\beta-1}\left\{\frac{1}{2m+1}\sum_{j=-m}^{m}\psi_{\ell+j}^{2d}\phi_{\ell+j}(p)\, I_{uu,\ell+j}\right\} + D\left|d-\widehat{d}\right|^{\beta}\frac{\log^{\beta} n}{2m+1}\sum_{j=-m}^{m}\psi_{\ell+j}^{2\widetilde{d}} I_{uu,\ell+j} \tag{7.1}$$

$$\widehat{h}_\ell-\check{h}_\ell\left(\widehat{d}\right) = \sum_{p=0}^{\beta-1}\left\{\frac{1}{2m+1}\sum_{j=-m}^{m}\psi_{\ell+j}^{2d}\phi_{\ell+j}(p)\left(I_{\widehat{u}\widehat{u},\ell+j}-I_{uu,\ell+j}\right)\right\} + D\left|d-\widehat{d}\right|^{\beta}\frac{\log^{\beta} n}{2m+1}\sum_{j=-m}^{m}\psi_{\ell+j}^{2\widetilde{d}}\left(I_{\widehat{u}\widehat{u},\ell+j}-I_{uu,\ell+j}\right), \tag{7.2}$$

where $\widetilde{d}$ is an intermediate point between d and $\widehat{d}$, and $\phi_j(p)=\frac{2^p\left(d-\widehat{d}\right)^p}{p!}\log^p\psi_j$. Denote $q_\ell(p)=(2m+1)^{-1}\sum_{j=-m}^{m}\phi_{\ell+j}(p)\frac{h_{\ell+j}}{h_\ell}\left(I_{\varepsilon\varepsilon,\ell+j}-\frac{\sigma_\varepsilon^2}{2\pi}\right)$.

Lemma 1 *Assuming C1′, C2 and C5, uniformly in $r \leq M$,*

$$\frac{r}{\tilde{n}}\sum_{\ell=1}^{\tilde{n}}\frac{\check{h}_\ell\left(\widehat{d}\right)-\check{h}_\ell(d)}{h_\ell}\cos(r\lambda_\ell)=\sum_{p=1}^{\beta-1}\frac{r}{\tilde{n}}\sum_{\ell=1}^{\tilde{n}}q_\ell(p)\cos(r\lambda_\ell)+O_p\left(\left|\widehat{d}-d\right|\log r\right). \tag{7.3}$$

Proof The proof proceeds as that of [15] Lemma 7.1, and so it is omitted. ■

Lemma 2 *Let υ_n be a sequence of random variables such that $E|\upsilon_n| = O\left(M^{-1}\right)$. Assuming C1′, C2, C3 and C5, for all $r \leq M$ and uniformly in $\ell = 1,\ldots,\tilde{n}$,*

$$\textbf{(a)}\ \frac{1}{\tilde{n}}\sum_{\ell=1}^{\tilde{n}}\left(\widehat{h}_\ell-\check{h}_\ell\left(\widehat{d}\right)\right)\cos(r\lambda_\ell)=O_p\left(\left|\widehat{d}-d\right|\right)\upsilon_n,\quad \textbf{(b)}\ \sup_{\ell=1,\ldots,\tilde{n}}\left|\widehat{h}_\ell-\check{h}_\ell\left(\widehat{d}\right)\right|=O_p\left(\left|\widehat{d}-d\right|\right).$$

Proof We begin with (**a**). Writing $\varphi_j(p)=\psi_j^{2d}\log^p\psi_j$, the contribution of the first term on the right of (7.2) into the left of (7.3) has as typical term

$$\frac{2^p\left(\widehat{d}-d\right)^p}{p!}\frac{1}{\tilde{n}}\sum_{\ell=1}^{\tilde{n}}\left(\frac{1}{2m+1}\sum_{j=-m}^{m}\varphi_{\ell+j}(p)\left(I_{\widehat{u}\widehat{u},\ell+j}-I_{uu,\ell+j}\right)\right)\cos(r\lambda_\ell). \tag{7.4}$$

Now, by definition of $\widehat{u}_t$, we have that $I_{\widehat{u}\widehat{u},j}-I_{uu,j}$ is

$$\frac{1}{2\pi n}\sum_{t,s=1}^{n}\left\{u_t\left(r(s)-\widehat{r}_a(s)\right)+u_s\left(r(t)-\widehat{r}_a(t)\right)+\left(r(t)-\widehat{r}_a(t)\right)\left(r(s)-\widehat{r}_a(s)\right)\right\}e^{i(t-s)\lambda_j}.$$

On the other hand, because $r(t)-\widehat{r}_a(t)=\xi_t-\theta_t$, where $\xi_t=\check{n}^{-1}\sum_{q=1}^{n}(r(t)-r(q))\widetilde{K}_{t-q}$; $\theta_t=\check{n}^{-1}\sum_{q=1}^{n}u_q\widetilde{K}_{t-q}$, with an obvious notation for $\widetilde{K}_t$, we have that (7.4) is governed by $2^p\left(\widehat{d}-d\right)^p/p!$ times

$$\frac{1}{\tilde{n}}\sum_{\ell=1}^{\tilde{n}}\left(\frac{1}{2m+1}\sum_{j=-m}^{m}\varphi_{\ell+j}\frac{1}{2\pi n}\sum_{t,s=1}^{n}\xi_t\xi_s e^{i(t-s)\lambda_{\ell+j}}\right)\cos(r\lambda_\ell) \tag{7.5}$$

$$+\frac{1}{\tilde{n}}\sum_{\ell=1}^{\tilde{n}}\left(\frac{1}{2m+1}\sum_{j=-m}^{m}\varphi_{\ell+j}\frac{1}{2\pi n}\sum_{t,s=1}^{n}\theta_t\theta_s e^{i(t-s)\lambda_{\ell+j}}\right)\cos(r\lambda_\ell) \tag{7.6}$$

$$+\frac{1}{\tilde{n}}\sum_{\ell=1}^{\tilde{n}}\left(\frac{1}{2m+1}\sum_{j=-m}^{m}\varphi_{\ell+j}\frac{1}{2\pi n}\sum_{t,s=1}^{n}u_t\left(\xi_s-\theta_s\right)e^{i(t-s)\lambda_{\ell+j}}\right)\cos(r\lambda_\ell). \tag{7.7}$$

Because the expression inside the parenthesis in (7.5) is positive, $|\cos\lambda| \le 1$ and $n = 4mM$, the absolute value of (7.5) is bounded by

$$\frac{D}{n^2}\sum_{t,s=1}^{n}\xi_t\xi_s\sum_{j=1}^{\tilde{n}}\varphi_j e^{i(t-s)\lambda_j} = \frac{D}{n}\sum_{t,s=1}^{n}\frac{1}{|t-s|_+^{1+2d}}|\xi_t\xi_s| + \frac{D}{n^2}\sum_{t,s=1}^{n}|\xi_t\xi_s|$$

which is $O\left(a^4\right)$ because the integrability of $(\partial/\partial u)\,\varphi(u)$ and Brillinger ([4, p.15]) implies that $\left|n^{-1}\sum_{j=1}^{n}\varphi_j - \int_0^1\varphi(u)\,du\right| = O\left(n^{-1}\right)$, and using $\int_0^1\varphi(u)\,e^{itu}du = O\left(t_+^{-1-2d}\right)$ and that $|\xi_t| = O\left(a^2\right)$ by Proposition 3. Next we handle (7.6), which following step by step the proof in [29, pp. 2077–2078], the first absolute moment of (7.6) is bounded by

$$\frac{D}{nm}\sum_{\ell=1}^{\tilde{n}}\sum_{j=-m}^{m}\left\{\min\left(1,\frac{1}{(\ell+j)^2a^2}\right)+\frac{\log n}{\ell+j}\right\} = O\left(a^{-2}m^{-1}n^{-1}+\frac{\log^2 n}{n}\right).$$

So (7.5) + (7.6) $= O\left(M^{-1}\right)O_p\left(\left|\widehat{d}-d\right|\right)$ because Proposition 1 implies that $\left|\widehat{d}-d\right| = O_p\left(a^{-1}m^{-1}\right)$ and then $C5$ part (ii). On the other hand, (7.7) $= O\left(M^{-1}\right)O_p\left(\left|\widehat{d}-d\right|\right)$ by an obvious use of the Cauchy–Schwarz inequality and the previous arguments. Next, as in Lemma 7.1, the contribution of the second term on the right of (7.2) into the left of (7.3) is $O_p\left(n^{-1}\right)$ by choosing β large enough. This concludes the proof of part (**a**).

The proof of part (**b**) is obvious by part (**a**) and using the usual chaining rule after observing that $\sup_{\ell=p,\ldots,q}\left|\widehat{h}_\ell - \check{h}_\ell\left(\widehat{d}\right)\right| \le \sum_{\ell=p}^{q}\left|\widehat{h}_\ell - \check{h}_\ell\left(\widehat{d}\right)\right|$. ■

Let $\widetilde{h}_\ell = (2m+1)^{-1}\sum_{j=-m}^{m}h_{\ell+j}$ and define

$$c_{r,n} = \frac{1}{\tilde{n}}\sum_{\ell=1}^{\tilde{n}}\log\left(h_\ell\right)\cos\left(r\lambda_\ell\right);\quad \widetilde{c}_{r,n} = \frac{1}{\tilde{n}}\sum_{\ell=1}^{\tilde{n}}\log\left(\widetilde{h}_\ell\right)\cos\left(r\lambda_\ell\right).$$

Lemma 3 *Let v_n be as in Lemma 2. Assuming C1−C3 and C5, uniformly in $r \le M$*

$$(\mathbf{a})\ \widehat{c}_r - \widetilde{c}_{r,n} = \frac{1}{\tilde{n}}\sum_{\ell=1}^{\tilde{n}}\frac{\check{h}_\ell(d)-\widetilde{h}_\ell}{h_\ell}\cos\left(r\lambda_\ell\right) + O_p\left(\frac{1}{m}\right) + O_p\left(\left|\widehat{d}-d\right|\right)v_n. \quad (7.8)$$

$$(\mathbf{b})\ \widetilde{c}_{r,n} - c_{r,n} = O\left(M^{-2}\right);\quad (\mathbf{c})\ c_{r,n} - c_r = O\left(n^{-1}\right).$$

Proof The proof proceeds as that of [15] Lemma 7.3, and so it is omitted. ■

Lemma 4 *Assuming C1−C3 and C5, $E\left|\sum_{\ell=1}^{\tilde{n}}\frac{\check{h}_\ell(d)-\widetilde{h}_\ell}{h_\ell}\cos\left(r\lambda_\ell\right)\right|^2 = O\left(1\right)$.*

Proof The proof is a standard extension of [16] Theorem 1, so it is omitted. ∎

Let us define

$$\widetilde{A}_{\ell,n} = \exp\left\{\sum_{r=1}^{M-1} \widetilde{c}_{r,n} e^{-ir\lambda_\ell}\right\}; \quad A_{\ell,n} = \exp\left\{\sum_{r=1}^{M-1} c_{r,n} e^{-ir\lambda_\ell}\right\}; \quad A_\ell^* = \exp\left\{\sum_{r=1}^{M-1} c_r e^{-ir\lambda_\ell}\right\}.$$

Lemma 5 *Let v_n be such that $E|v_n| = O\left(M/n^{1/2}\right)$. Assuming C1–C3 and C5, uniformly in ℓ,*

$$(a)\ \widehat{A}_\ell - \widetilde{A}_{\ell,n} = O_p\left(\left|\widehat{d} - d\right|\right) + v_n, \quad (b)\ \widetilde{A}_{\ell,n} - A_{\ell,n} = O\left(M^{-2}\right); \quad A_{\ell,n} - A_\ell^* = O\left(m^{-1}\right).$$

Proof The proof follows as that of Lemma 7.5 of [15] and thus it is omitted. ∎

Lemma 6 *For any $q_1 \le q_2 \in \mathcal{Q}_n$, as $n \to \infty$, assuming C1–C3 and C5,*

$$\check{n}^{1-2d} Cov^*\left(\widehat{r}^*_{a,+}(q_1) - \widehat{r}^*_{a,-}(q_1), \widehat{r}^*_{a,+}(q_2) - \widehat{r}^*_{a,-}(q_2)\right) \xrightarrow{P} \rho(b; d),$$

where the right side is as defined in Proposition 1.

Proof By definition, $\check{n}^{1-2d} Cov^*\left(\widehat{r}^*_{a,+}(q_1) - \widehat{r}^*_{a,-}(q_1), \widehat{r}^*_{a,+}(q_2) - \widehat{r}^*_{a,-}(q_2)\right)$ is

$$\begin{aligned} &\check{n}^{1-2d} Cov^*\left(\widehat{r}^*_{a,+}(q_1), \widehat{r}^*_{a,+}(q_2)\right) + \check{n}^{1-2d} Cov^*\left(\widehat{r}^*_{a,-}(q_1), \widehat{r}^*_{a,-}(q_2)\right) \\ &- \check{n}^{1-2d} Cov^*\left(\widehat{r}^*_{a,+}(q_1), \widehat{r}^*_{a,-}(q_2)\right) - \check{n}^{1-2d} Cov^*\left(\widehat{r}^*_{a,-}(q_1), \widehat{r}^*_{a,+}(q_2)\right). \end{aligned} \tag{7.9}$$

As was done in the proof of Proposition 1, we will only examine the first term of (7.9), the other three terms follow similarly. This term is

$$\frac{1}{\check{n}^{1+2d}}\left\{\sum_{t=1}^{\check{n}} \sum_{s=q_2-q_1+1}^{\check{n}+q_2-q_1} \left\{E^*\left(u^*_{t+q_1} u^*_{s+q_1}\right) - \delta_{t-s}\right\} K_{+,t} K_{+,s+q_1-q_2} + \sum_{t=1}^{\check{n}} \sum_{s=q_2-q_1+1}^{\check{n}+q_2-q_1} \delta_{t-s} K_{+,t} K_{+,s+q_1-q_2}\right\}.$$

However, it suffices to show that the first term on the right of the last displayed expression converges to zero in probability because by Proposition 1, the second term converges to $\rho_+(b; d)$.

Because $\left|K_{+,t}\right| \le D$ by $C3$, we have that the first term of the last displayed expression is bounded in absolute value by

$$\frac{D}{\check{n}^{1+2d}} \sum_{t=1}^{\check{n}} \sum_{s=q_2-q_1+1}^{\check{n}+q_2-q_1} \left|E^*\left(u^*_{t+q_1} u^*_{s+q_1}\right) - \delta_{t-s}\right| = \frac{D}{\check{n}^{2d}} \sum_{t=q_2-q_1+1}^{\check{n}+q_2-q_1} \left|E^*\left(u^*_{t+1} u^*_1\right) - \delta_t\right|$$

is $o_p(1)$ by standard arguments and then by Proposition 7. ∎

Lemma 7 *Assuming C1–C3 and C5, we have that for all $\phi > 0$,*

$$\sum_{j=1}^{n} E^* \left(n^{-1} \check{n}^{-1-2\widehat{d}} \left| \varsigma_{\pm,q} \left(\lambda_j \right) \eta_j^* \right|^2 \mathcal{I} \left(n^{-1} \check{n}^{-1-2\widehat{d}} \left| \varsigma_{\pm,q} \left(\lambda_j \right) \eta_j^* \right|^2 > \phi \right) \right) \xrightarrow{P} 0,$$

where $\varsigma_{\pm q} \left(\lambda_j \right) = \widehat{k}_{\pm,q} \left(\lambda_j \right) \widetilde{g}^{1/2} \left(\lambda_j ; \widehat{d} \right) \widehat{B} \left(\lambda_j \right)$ *with* $\widehat{k}_{\pm,q} \left(\lambda_j \right) = \sum_{t=1}^{n} K_{\pm,t-q} e^{it\lambda_j}$.

Proof The proof is identical to that of [15] Lemma 7.9 and thus it is omitted. ∎

Acknowledgments We like to thank Marie Huskova for their comments on a previous version of the paper. Of course, any remaining errors are our sole responsibility.

References

1. Antoch, J., Gregoire, G., Huskova, M.: Tests for continuity of regression function. J. Stat. Plan. Inference **137**, 753–777 (2007)
2. Bilckel, P.J., Rosenblatt, M.: On some global measures of the deviations of density function estimates. Ann. Stat. **1**, 1071–1095 (1973)
3. Billingsley, P.: Convergence of Probability Measures. Wiley and Sons, New York (1968)
4. Brillinger, D.R.: Time Series, Data Analysis and Theory. Holden-Day, San Francisco (1981)
5. Brockwell, P.J., Davis, R.A.: Time Series: Theory and Methods. Springer, New York (1991)
6. Bühlmann, P.: Sieve-bootstrap for time series. Bernoulli **3**, 123–148 (1997)
7. Chu, J.S., Wu, C.K.: Kernel-type estimators of jump points and values of a regression function. Ann. Stat. **21**, 1545–1566 (1993)
8. Delgado, M.A., Hidalgo, J.: Nonparametric inference on structural breaks. J. Econom. **96**, 113–144 (2000)
9. Fuller, W.A.: Introduction to Statistical Time Series. Wiley and Sons, New York (1996)
10. Giacomini, R., Politis, D.N., White, H.: A warp-speed method for conducting Monte cargo experiments involving bootstrap estimators. Econom. Theory **29**, 567–589 (2013)
11. Giné, E., Zinn, J.: Necessary conditions for the bootstrap of the mean. Ann. Stat. **17**, 684–691 (1989)
12. Hall, P.: On convergence rates of suprema. Probab. Theory Relat. Fields **89**, 447–455 (1991)
13. Härdle, W., Marron, J.S.: Bootstrapping simultaneous error bars for nonparametric regression. Ann. Stat. **19**, 778–796 (1991)
14. Hidalgo, J.: A nonparametric conditional moment test for structural stability. Econom. Theory **11**, 671–698 (1995)
15. Hidalgo, J.: Specification testing for regression models with dependent data, STICERD EM/2007/518. J. Econom. **143**, 143–165 (2007). Published version
16. Hidalgo, J., Yajima, Y.: Prediction in the frequency domain under long-range processes with application to the signal extraction problem. Econom. Theory **18**, 584–624 (2002)
17. Hurvich, C.M., Zeger, S.L.: Frequency domain bootstrap methods for time series. Technical report **87-115**, Graduate School of Business Administration, New York University (1987)
18. Hušková, M., Kirch, C., Prášková, Z., Steinebach, J.: On the detection of changes in autoregressive time series, II. resampling procedures. J. Stat. Plan. Inference **138**, 1697–1721 (2008)
19. Kreiss, J.-P., Paparoditis, E.: Autoregressive-aided periodogram bootstrap for time series. Ann. Stat. **31**, 1923–1955 (2003)
20. Künsch, H.A.: The jackknife and the bootstrap for general stationary observations. Ann. Stat. **17**, 1217–1241 (1989)

21. Marinucci, D., Robinson, P.M.: Weak convergence of multivariate fractional processes. Stoch. Process. Appl. **80**, 103–120 (2003)
22. Müller, H.-J.: Change-points in nonparametric regression analysis. Ann. Stat. **20**, 737–761 (1992)
23. Perron, P.: Dealing with structural breaks. In: Mills T.C., Patterson, K. (eds.) Palgrave Handbook of Econometrics, Chapter 8, vol. 1 (2006)
24. Pickands, J.: Asymptotic properties of the maximum in a stationary gaussian process. Trans. Am. Math. Soc. **145**, 75–86 (1969)
25. Pollard, D.: Convergence Stochastic Processes. Springer, New York (1980)
26. Rice, J.: Boundary modification for kernel regression. Commun. Stat. Theory and Methods **13**, 893–900 (1984)
27. Robinson, P.M.: Log-Periodogram regression for time series with long range dependence. Ann. Stat. **23**, 1048–1072 (1995a)
28. Robinson, P.M.: Gaussian semiparametric estimation of long-range dependence. Ann. Stat. **23**, 1630–1661 (1995b)
29. Robinson, P.M.: Large-sample inference for nonparametric regression with dependent errors. Ann. Stat. **25**, 2054–2083 (1997)
30. Serfling, R.J.: Approximation Theorems of Mathematical Statistics. Wiley and Sons, New York (1980)
31. Woodroofe, M.: On the maximum deviation of the sample density. Ann. Math. Stat. **38**, 475–481 (1967)

Change Detection in INARCH Time Series of Counts

Šárka Hudecová, Marie Hušková and Simos Meintanis

Abstract In the present paper we develop an online procedure for detecting changes in the parameters of integer ARCH models of order one. The test statistic utilizes the notion of the empirical probability generating function. The asymptotic behavior of the test under the null hypothesis is derived.

Keywords Sequential monitoring · Time series of counts · Empirical probability generating function

1 Introduction

The detection of structural changes (or breaks) is an important problem in time series in that a structural change indicates that the underlying system can no longer be described by the current model and calls for remodeling of certain aspects of this model. There exist two basic types of procedures, often depending on the way that data become available: The so-called off-line (or retrospective) procedures whereby we have a certain set of data at hand and wish to know if there is a structural break in these data, and the online (or sequential) procedures which are performed with the data becoming available during the course of the statistical analysis.

Š. Hudecová (✉) · M. Hušková
Department of Probability and Mathematical Statistics, Charles University of Prague,
Sokolovská 83, 186 75 Praha 8, Czech Republic
e-mail: hudecova@karlin.mff.cuni.cz

M. Hušková
e-mail: huskova@karlin.mff.cuni.cz

S. Meintanis
Department of Economics, National and Kapodistrian University of Athens,
1 Sophocleous Str., and Aristidou Str., 105 59 Athens, Greece
e-mail: simosmei@econ.uoa.gr

S. Meintanis
Unit for Business Mathematics and Informatics, North-West University,
Potchefstroom, South Africa

R. Cao et al. (eds.), *Nonparametric Statistics*, Springer Proceedings
in Mathematics & Statistics 175, DOI 10.1007/978-3-319-41582-6_4

The objective here is to construct detector statistics for possible structural breaks in a given time series. While this problem has attracted much attention in time series with continuous observations (see for instance the volume by [4] and the review article by [18]), the same problem with time series of counts is relatively new in the literature.

Here we shall focus on online procedures for integer autoregressive conditionally heteroscedastic models (INARCH) models. Specifically we study sequential procedures for detection of changes in the parameters of Poisson autoregressions of order one, PAR(1). In the context of PAR models there exist relatively few works in the literature on this subject. The most relevant references are [14, 22, 24, 25, 36]; see also [11, 35]. In these works, the methods employed are based on score statistics or CUSUM statistics. In what follows we deviate from these procedures by considering detectors which utilize the probability generating function (PGF) and its empirical counterpart as the main tool. The remainder of this paper runs as follows: In Sect. 2 we introduce the model and the hypotheses of interest. In Sect. 3 the test statistic is introduced, while in Sect. 4 we study the asymptotic behavior of the proposed procedure under the null hypothesis. In Sect. 5 certain extensions of the new method are suggested. Sections 6 and 7 contain a short simulation study and a real data example. The proofs are provided in Sect. 8.

2 Model and Hypotheses

The INARCH was first put forward by [12]. It has been further investigated by [10, 28, 34].

Specifically observations $\{Y_t\}_t$ are said to follow the INARCH model of order one if

$$F_t = F(\lambda_t); \ \lambda_t = f(\theta; Y_{t-1}), \tag{1}$$

where F_t denotes the conditional distribution of Y_t given $Y_{t-1}, Y_{t-2} \ldots, F(\lambda)$ denotes the Poisson distribution with mean λ, θ is a vector of unknown parameters, and $f(\cdot;\cdot)$ is a fixed function satisfying certain mild conditions ensuring stationarity and ergodicity of the process. Specifically if $f(x, y)$ is Lipschitz-continuous in y for all x with Lipschitz constant strictly smaller than 1, then there exists a stationary ergodic solution which is β-mixing with exponential rate; confer [7, 13, 28], and the monograph by [6].

The model in Eq. (1) is also referred to as Poisson autoregression of order one, PAR(1) for short. An advantage of this model is that it mimics the dynamics of an ARCH model in the discrete world in that it includes a feedback mechanism yielding parsimony and thus has found applications in diverse areas of finance and biometry; see for example [5, 9, 11, 16]. Furthermore, estimation may be carried out by the standard methods such as LS and MLE, and the corresponding asymptotic properties for model (1) have been well studied. Also (1) admits natural extensions and generalizations to higher order, multivariate and non-Poisson models, although the corresponding theory does not readily generalize and requires further research.

We specify f in Eq. (1) to be a linear function of Y_{t-1}. In particular let $\theta = (\alpha, \beta)'$ and

$$f(\theta; Y_{t-1}) = \alpha + \beta Y_{t-1}, \tag{2}$$

with $\alpha > 0$ and $0 < \beta < 1$. In the context of structural breaks we shall consider a slightly more general PAR(1) model under which the parameter θ might change with time. Then we are interested in testing the null hypothesis that

$$\mathscr{H}_0 : F_t = F(\lambda_t), \quad \lambda_t = \alpha_0 + \beta_0 Y_{t-1}, \quad \forall t,$$

against the alternative

$$\begin{aligned} \mathscr{H}_1 : F_t &= F(\lambda_t), \quad \lambda_t = \alpha_0 + \beta_0 Y_{t-1}, \ t \le T + t_0; \\ F_t &= F(\lambda_t), \quad \lambda_t = \alpha^0 + \beta^0 Y_{t-1}, \ t > T + t_0, \end{aligned}$$

where $\alpha_0, \beta_0, \alpha^0, \beta^0$, as well as the change point t_0 are unknown, and where, as it is typical in the sequential setup, T is a known integer such that the *training data* $Y_1, ..., Y_T$, involve no change.

The proposed sequential test procedure for detecting changes in PAR(1) processes will be based on properties of the probability generating function (PGF) of the observed variables. In this connection recall that the PGF of a discrete random variable Y is defined as

$$g_Y(u) = \mathrm{E}(u^Y), \quad u \in [0, 1],$$

and that under very mild conditions this PGF uniquely determines the underlying distribution function of Y. The empirical counterpart of the PGF is defined by

$$\widehat{g}_{Y,n}(u) = \frac{1}{n} \sum_{t=1}^{n} u^{Y_t}, \quad u \in [0, 1].$$

Our motivation for considering the empirical PGF as our tool rests on prior experience about the performance of resulting methods. For i.i.d. observations the empirical PGF has been used for both estimation and testing; see for example [3, 8, 15, 27, 29, 30]. This idea was carried further to the context of time series of counts by [19, 26, 31, 32]. Moreover, in change point detection analogous methods utilizing the related notion of the empirical characteristic function were employed by [17, 21, 33]. In all the aforementioned works, methods based on the empirical PGF, apart from being convenient from the computational point of view, are reported to compare favorably and often outperform more standard methods such as Cramér-von Mises and CUSUM methods. Note also that the empirical PGF can be further used for the construction of detector statistics in count time series of a more general nature. Here, however, we focus on procedures for detecting changes in the parameters of PAR(1) processes.

3 Test Statistic

To motivate our procedure, we first notice that the marginal PGF $G(u)$ of the stationary distribution of $\{Y_t\}_{t\in\mathcal{N}}$ under the PAR(1) model specified by Eqs. (1)–(2) is given by

$$\begin{aligned} G(u) &= \mathrm{E}\left[u^{Y_t}\right] = \mathrm{E}\left[\mathrm{E}(u^{Y_t}|Y_{t-1})\right] \\ &= \mathrm{E}\left[e^{(\alpha+\beta Y_{t-1})(u-1)}\right]. \end{aligned}$$

In order to introduce our test statistic suppose that $\widehat{\theta}_T = (\widehat{\alpha}_T, \widehat{\beta}_T)'$ is an estimator of θ, based on the training data $\{Y_t,\ t = 1, ..., T\}$. Then the previous equation motivates us to introduce the test statistic:

$$\Delta_{T,t} = \int_0^1 \delta^2_{T,T+t}(u; \widehat{\theta}_T)du, \quad t = 1, \ldots, \tag{3}$$

with

$$\begin{aligned} \delta_{T,T+t}(u;\theta) = &\frac{1}{\sqrt{T}} \sum_{j=T+1}^{T+t} \left(u^{Y_j} - \exp\{(\alpha + \beta Y_{j-1})(u-1)\}\right) \\ &- \frac{t}{T}\frac{1}{\sqrt{T}} \sum_{j=2}^{T} \left(u^{Y_j} - \exp\{(\alpha + \beta Y_{j-1})(u-1)\}\right) \quad t = 1, \ldots . \end{aligned}$$

The advantage of $\Delta_{T,t}$ is that its limit distribution under the null hypothesis is the same as if the estimator $\widehat{\theta}_T$ is replaced by the true value of the parameter and consequently, we get functionals of partial sums of martingale differences for which the limit distribution can be easily obtained.

We note that the method presented in this section is fairly general in that it can be readily adapted to other related settings, such as higher order and non-Poisson autoregression models; see Sect. 5.

As already mentioned we consider online procedures whereby the test is applied sequentially on a dynamic data set which is steadily updated over time with the arrival of new observations. In this context, the null hypothesis is rejected when the value of a suitable detector statistic exceeds an appropriately chosen constant *for the first time*. Otherwise we continue monitoring. These statistics are commonly defined by a corresponding stopping rule. In order to define this stopping rule, and based on asymptotic considerations, we need to introduce a weight function in order to control the large-sample probability of type-I error. In particular we employ the detector statistics

$$Q_{T,t} = \frac{1}{q_\gamma^2\left(\frac{t}{T}\right)}\Delta_{T,t}, \quad t = 1, \ldots, \tag{4}$$

where $\Delta_{T,t}$ is defined by Eq. (3) and

$$q_\gamma(s) = (1+s)\left(\frac{s}{s+1}\right)^\gamma, \quad \gamma \in [0, 1/2). \tag{5}$$

The parameter γ figuring in (5) gives some flexibility to the resulting procedure. Specifically, if early changes are expected then the value of γ should be close to 1/2, while values closer to zero are appropriate for detecting changes occurring at later stages.

It is clear that since the training data $\{Y_1, ..., Y_T\}$ are assumed to involve no change, the monitoring period begins with time $t = T+1$. Typically this monitoring continues till time $T(m+1)$, where m denotes a fixed integer, and if $m < \infty$ we call the corresponding procedure close end. Otherwise (i.e., if $m = \infty$), we have an open end procedure. The corresponding stopping rule is specified as

$$\tau_T(\gamma, m) = \inf\{1 \le t \le mT : Q_{T,t} \ge c\},$$
$$\tau_T(\gamma, m) = \infty \quad if\ Q_{T,t} < c\ for\ all\ 1 \le t \le Tm,$$

for some fixed integer $m > 0$, where c is a constant that guarantees that the test has size equal to α, asymptotically. The corresponding delay in detecting the change can also be estimated, and following this we can estimate the location of the change point. In this regard one should proceed as in the regression setup, e.g., [1], but this is a topic of another paper.

The main problem is to find an approximation for critical value c and to investigate consistency of the test procedures. Particularly, we require that under $\mathscr{H}_0$ for a prechosen α

$$\lim_{T\to\infty} \mathrm{P}_{\mathscr{H}_0}(\tau_T < \infty) = \alpha \tag{6}$$

while under alternatives we want

$$\lim_{T\to\infty} \mathrm{P}_{\mathscr{H}_1}(\tau_T < \infty) = 1. \tag{7}$$

4 Asymptotics

Here we study the limit behavior of the test procedure. Particularly, we have to study the limit behavior of

$$M_T(\gamma) := \max_{1\le t\le mT} Q_{T,t},$$

where $Q_{T,t}$ is defined in (4), under $\mathscr{H}_0$. The limit is always for $T \to \infty$ and m fixed.

To this end assume that the estimator $\widehat{\theta}_T = (\widehat{\alpha}_T, \widehat{\beta}_T)'$ of the parameter $\theta = (\alpha, \beta)'$ based on training data $Y_1, \ldots, Y_T$, satisfies:

$$T\|\widehat{\theta}_T - \theta_0\|^2 = O_P(1), \tag{8}$$

where $\theta_0 = (\alpha_0, \beta_0)'$ denotes the true value of θ.

The main result under the null hypothesis reads as follows:

Theorem 4.1 *Let $\{Y_t\}$ follow the model (1)–(2) for $\alpha > 0, \beta \in (0, 1)$ and let $q_\gamma(\cdot)$, $\gamma \in [0, 1/2)$ be defined in Eq. (5). Assume further that (8) holds for the estimators of the parameters α, β. Then, under the null hypothesis $\mathscr{H}_0$ the limit distribution of $M_T(\gamma)$ as $T \to \infty$, is the same as that of*

$$\sup_{0<s<m/(m+1)} \frac{1}{s^{2\gamma}} \int_0^1 V^2(s, u)du$$

where $\{V(s, u);\ s \in (0, m/(m + 1)),\ u \in (0, 1)\}$ is a Gaussian process with zero mean and covariance structure described as

$$\text{cov}(V(s_1, u_1), V(s_2, u_2)) = \min(s_1, s_2)\sigma(u_1, u_2)$$

with

$$\sigma(u_1, u_2) = \text{E}\big(u_1^{Y_2} - \text{E}(u_1^{Y_2}|Y_1)\big)\big(u_2^{Y_2} - \text{E}(u_2^{Y_2}|Y_1)\big).$$

Proof It is postponed to Sect. 8. Q.E.D.

The explicit form of the limit null distribution is highly complicated and depends on unknown quantities. Nevertheless, one could try to approximate this distribution by replacing the unknown parameters and covariance structure by the respective estimators based on historical data and simulate the resulting process. Another possibility is to use parametric bootstrap by estimating (α, β) from the historical data and then generate bootstrap observations along Eqs. (1)–(2) with (α, β) replaced by their estimators. This possibility also leads to an asymptotically correct approximation of the limit null distribution of the test statistic. Also, in case that T is not large enough the estimators obtained from the training sample can have larger variance, and therefore it is advisable to modify the procedure by utilizing its sequential character along the line of [20, 23].

Next, we shortly discuss the limit behavior of our test statistic under fixed alternatives. We consider the following type of alternatives:

$\mathscr{H}_1^*$: there exists $t_0 = \lfloor T\nu_0 \rfloor$ for some $0 \le \nu_0 < m$ such that Y_t, $t \le T + t_0$, follows model (1)–(2) with $\theta = \theta_0$, and $\{Y_{T+t_0+t}\} =^d \{Y_t^0\}$, where $\{Y_t^0\}$ follows model (1)–(2) with $\theta = \theta^0$, $\theta_0 \neq \theta^0$.

The test is consistent for a large group of fixed alternatives of this type, or of a more general type. Specifically it may be shown that (7) holds provided that the quantity

$$\int_0^1 \left(\mathrm{E}\left(u^{Y_2^0} - \exp\{(\alpha_0 + \beta_0 Y_1^0)(u-1)\}\right)\right)^2 w(u)du, \tag{9}$$

is positive. Hence the test is consistent against alternatives with changes only in the parameters, as well as against alternatives that involve a structural change in the distribution of the observations. However, we will not pursue this issue any further here.

5 Extensions

We briefly consider extensions of the method presented in the previous section to related or more general settings. First consider the case of a nonlinear PAR(1). Following analogous reasoning as in Sect. 3, it becomes clear that our method readily extends to nonlinearity by considering the test statistic in Eq. (3) with

$$\begin{aligned}\delta_{T,T+t}(u;\theta) &= \frac{1}{\sqrt{T}} \sum_{j=T+1}^{T+t} \left(u^{Y_j} - \exp\{f(\theta; Y_{j-1})(u-1)\}\right) \\ &- \frac{t}{T}\frac{1}{\sqrt{T}} \sum_{j=2}^{T} \left(u^{Y_j} - \exp\{f(\theta; Y_{j-1})(u-1)\}\right) \quad t = 1, \ldots .\end{aligned}$$

We also consider the case of Negative Binomial autoregression as the most popular alternative to Poisson autoregression. Recall that the PGF of the negative binomial distribution is given by

$$g(u) = \frac{1}{[1+\rho(1-u)]^\lambda}, \quad \rho, \lambda > 0, \tag{10}$$

and that under this parameterization the mean is equal to $\lambda\rho$ and variance is $\lambda\rho(1+\rho)$. Considering the model in Eqs. (1)–(2) with the Poisson distribution replaced by the Negative Binomial distribution, we have

$$\begin{aligned}G(u) = \mathrm{E}\left[u^{Y_t}\right] &= \mathrm{E}\left[\mathrm{E}(u^{Y_t}|Y_{t-1})\right] = \mathrm{E}\left[\frac{1}{(1+\rho(1-u))^{\lambda_t}}\right] \\ &= \mathrm{E}\left[\frac{1}{(1+\rho(1-u))^{\alpha+\beta Y_{t-1}}}\right] = \left[\frac{1}{1+\rho(1-u)}\right]^\alpha \mathrm{E}\left[\frac{1}{(1+\rho(1-u))^{\beta Y_{t-1}}}\right].\end{aligned} \tag{11}$$

Consequently for the case of Negative Binomial autoregression we consider the test statistic in Eq. (3) with

$$\delta_{T,T+t}(u;\theta) = \frac{1}{\sqrt{T}} \sum_{j=T+1}^{T+t} \left(u^{Y_j} - \left[\frac{1}{1+\rho(1-u)}\right]^{\alpha} \frac{1}{(1+\rho(1-u))^{\beta Y_{t-1}}}\right)$$
$$- \frac{t}{T}\frac{1}{\sqrt{T}} \sum_{j=2}^{T} \left(u^{Y_j} - \left[\frac{1}{1+\rho(1-u)}\right]^{\alpha} \frac{1}{(1+\rho(1-u))^{\beta Y_{t-1}}}\right) \quad t = 1, \ldots .$$

Clearly and despite the fact that computations become somewhat cumbersome, the asymptotics for this model should follow analogous steps. Also, the case of PAR(2) and that of higher order models may be treated in an analogous manner. Finally, note that a Negative Binomial autoregression model was suggested by [37] with the parameter ρ being modeled using the autoregression. Here we suggest to model the parameter λ which, however, for fixed ρ has an equivalent statistical interpretation.

Finally, we point out that the test procedure suggested in Sect. 3 may be readily adapted to yield a test in the case of off-line break detection. Specifically the test procedure would then be based on

$$M_T = \max_{1<t\leq T} \int_0^1 \Delta_t^2(u)\, w(u)du,$$

with

$$\Delta_t(u) = \frac{1}{\sqrt{T}} \left| \sum_{j=2}^{t} \left(u^{Y_j} - \exp\{(\widehat{\alpha}_T + \widehat{\beta}_T Y_{j-1})(u-1)\}\right)\right|, \ 1 < t \leq T,$$

where $\widehat{\alpha}_T$ and $\widehat{\beta}_T$ are estimators based on all available observations.

6 Short Simulation Study

A short simulation study is presented as an illustration of the behavior of the proposed method. The significance of the test statistic $M_T(\gamma)$ is evaluated using the parametric bootstrap, mentioned at the end of Sect. 4. The behavior of the test is studied for a linear PAR(1) model with the choice $\alpha_0 = 6$, $\beta_0 = 0.4$, the training sample size $T = 50$, the monitoring period is set as mT for $m = 5, 8$, and $\gamma = 0$. Under the alternative, we take $t_0 = 25, 50$ and $\alpha^0 = 9$, $\beta^0 = 0.5$. The p-value of each test is computed from $B = 499$ bootstrap samples and the percentage of rejection is estimated from 500 repetitions. The estimator $\widehat{\theta}_T$ is the conditional least squares estimator.

The obtained results are presented in Table 1. The prescribed significance level α is slightly exceeded under the null hypothesis. The obtained power of the test is reasonable (over 90 % for $\alpha = 0.05$) under the considered alternative. As expected, the power is larger for the earlier change point appearance and the longer monitoring

Table 1 Estimated percentage of rejections of the bootstrap test under $\mathcal{H}_0$ and $\mathcal{H}_1$

	$\mathcal{H}_0$			$\mathcal{H}_1$: $t_0 = 25$			$\mathcal{H}_1$: $t_0 = 50$		
	α			α			α		
m	0.01	0.05	0.1	0.01	0.05	0.1	0.01	0.05	0.1
5	0.012	0.058	0.112	0.808	0.962	0.984	0.746	0.916	0.964
8	0.024	0.072	0.104	0.882	0.980	0.996	0.832	0.956	0.980

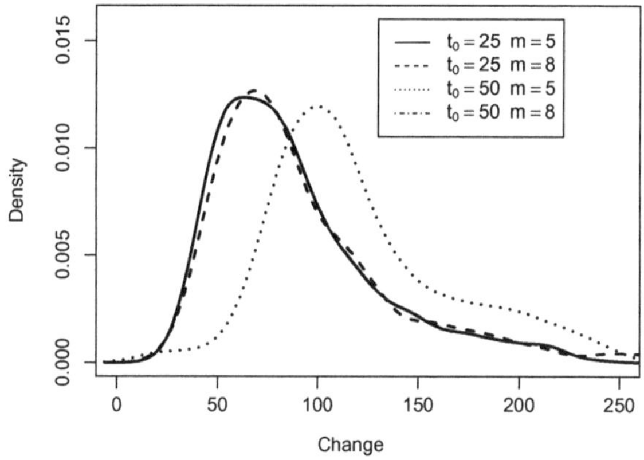

Fig. 1 Estimated density of τ_T for $\alpha = 0.05$ under the alternative

period. The estimated density of the stopping time τ_T for $\alpha = 0.05$ is plotted in Fig. 1. The medians of the estimated change points are 79, 81, 110, and 123, respectively (theoretical counterparts are 25, 25, 50, 50). This illustrates the delay in the detection.

7 Real Data Example

For an illustration of a practical usage of the presented methodology, the testing procedure is applied to a series of the number of transactions per minute for stock Ericsson B during July 3rd 2002. The series consists of 460 observations, see Fig. 2, and has been previously analyzed in [24]. The authors used a simple linear PAR(1) model for the series and tested for a presence of a change using an off-line procedure based on least squares scores. Together with the binary segmentation they detected three change points at observations 98, 148, and 305.

We analyze the series using the test statistic $M_T(\gamma)$ and the parametric bootstrap, with $\gamma = 0$, $B = 999$ bootstrap samples, and $\alpha = 0.05$ significance level. We consider the first $T = 50$ observations as the training sample, and the rest of the series is monitored sequentially. The null hypothesis $\mathcal{H}_0$ is rejected with p-value < 0.001 and the change point is estimated as $T + t_0 = 137$.

In order to compare our results with [24], we apply the whole procedure again on the rest of the series. Since the second change point detected by [24] comes relatively

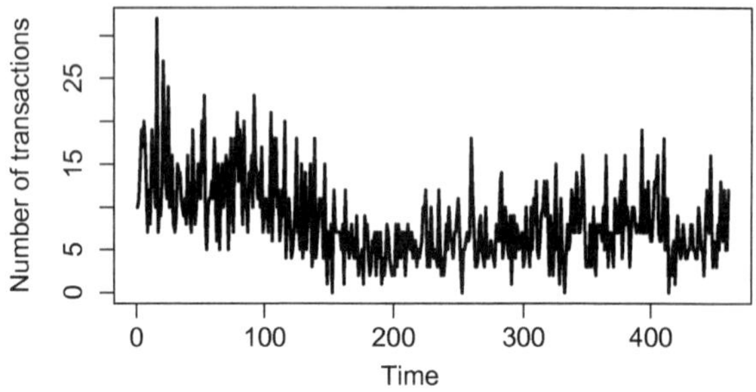

Fig. 2 Number of transactions per minute for the Ericsson B stock during July 3, 2002

shortly after our first estimated change point, we skip this segment and analyze the series starting from the 151st observation. Here the training sample size $T = 120$ seems to be reasonable. The null hypothesis $\mathscr{H}_0$ is again rejected with p-value 0.008 and the change point is estimated at the 349th observation of the original series.

Finally, we look at the last segment of the data, starting from the 350th observation. We take a training sample $T = 50$ and monitor till the end. The null hypothesis of no change is rejected with p-value 0.005 and the change point is estimated at the 420th observation of the original series. This indicates that the behavior of the market might change also at the end of the trading day.

Clearly our results differ from those in [24]. Specifically our estimated change points are delayed compared to the off-line procedure of [24] and the delay is approximately 40 observations (compare 98 in [24] with our 137, and 305 with 349). Moreover, our test detected an extra change point at the 420th observation of the series not captured by [24].

8 Proofs

Proof of Theorem 4.1 Due to a certain similarity to the proof of Theorem 4.1 in [19] we present only main steps of the proof of our Theorem.

By the Taylor expansion of $\delta_{T,T+t}(u, \widehat{\theta}_T)$ at θ_0 and by convergence properties of stationary sequences, we realize that under $\mathscr{H}_0$ the limit behavior of the statistic $\max_{1\le t\le mT} \Delta_{T,T+t}/q_\gamma^2(t/T)$ does not change if the estimator $\widehat{\theta}_T$ is replaced by the true value θ_0.

Since $\sum_{j=\ell+1}^{\ell+t} \Big(u^{Y_j} - \exp\{(\alpha + \beta Y_{j-1})(u-1)\}\Big)$, $t = 1, \ldots, \ell = 0, 1 \ldots$, are partial sums of bounded martingale differences, we can apply theorems on their limit behavior. The proof can be finished combining the arguments in the last part of the proof of Theorem 4.1 in [19] and the proof of Theorem 1 in [2].

Acknowledgments The research of Simos Meintanis was partially supported by grant Nr. 11699 of the Special Account for Research Grants (ELKE) of the National and Kapodistrian University of Athens. The research of Marie Hušková was partially supported by grant GAČR 15-09663S and AP research network grant Nr.P7/06 of the Belgian government (Belgian Science Policy). The research of Šárka Hudecová was supported by grant GAČR 15-04774Y.

References

1. Aue, A., Horváth, L.: Delay time in sequential detection of change. Statist. Probab. Lett. **67**, 221–231 (2004)
2. Aue, A., Horváth, L., Hušková, M., Kokoszka, P.: Change-point monitoring in linear models with conditionally heteroskedastic errors. Econom. J. **9**, 373–403 (2006)
3. Baringhaus, L., Henze, N.: A goodness of fit test for the Poisson distribution based on the empirical generating function. Statist. Probab. Lett. **13**, 269–274 (1992)
4. Csörgö, M., Horváth, L.: Limit Theorems in Change-point Analysis. Wiley, New York (1997)
5. Davis, R.A., Dunsmuir, W., Streett, S.B.: Observation-driven models for Poisson counts. Biometrika **90**, 777–790 (2003)
6. Dedecker, J., Doukhan, P., Lang, G., León, J.R., Louhichi, S., Prieur, C.: Weak dependence: with examples and applications. Springer, Berlin (2007)
7. Doukhan, P., Louhichi, S.: A new weak dependence condition and applications to moment inequalities. Stochastic Process. Appl. **84**, 313–342 (1999)
8. Epps, T.W.: A test of fit for lattice distributions. Comm. Statist. Theory Methods **24**, 1455–1479 (1995)
9. Ferland, R., Latour, A., Oraichi, D.: Integer-valued GARCH process. J. Time Ser. Anal. **27**, 923–942 (2006)
10. Fokianos, K.: Count time series models. In: Subba Rao, T., Subba Rao, S., Rao, C.R. (eds.) Handbook of Statistics: Time Series—Methods and Applications, vol. 30, pp. 315–348. Elsevier B.V., Amsterdam (2012)
11. Fokianos, K., Fried, R.: Interventions in INGARCH processes. J. Time Ser. Anal. **31**, 210–225 (2010)
12. Fokianos, K., Rahbek, A., Tjostheim, D.: Poisson autoregression. J. Amer. Statist. Assoc. **104**, 1430–1439 (2009)
13. Franke, J.: Weak dependence of functional INGARCH processes. Working paper, University of Kaiserslautern (2010)
14. Franke, J., Kirch, C., Tadjuidje Kamgaing, J.: Changepoints in time series of counts. J. Time Ser. Anal. **33**(5), 757–770 (2012)
15. Gürtler, N., Henze, N.: Recent and classical goodness-of-fit tests for the Poisson distribution. J. Statist. Plann. Infer. **90**, 207–225 (2000)
16. Heinen, A.: Modelling time series count data: an autoregressive conditional Poisson model. In: CORE discussion Paper 2003062 (2003)
17. Hlávka, Z., Hušková, M., Kirch, C., Meintanis, S.G.: Bootstrap procedures for online monitoring of changes in autoregressive models. Commun. Statist.—Simul. Comput. (2015). doi:10.1080/03610918.2014.904346
18. Horváth, L., Rice, G.: Extensions of some classical methods in change point analysis. Test **23**, 219–255 (2014)
19. Hudecová, Š., Hušková, M., Meintanis, S.: Tests for time series of counts based on the probability generating function. Statistics **49**(2), 316–337 (2015)
20. Hušková, M., Kirch, C.: Bootstrapping sequential change-point tests for linear regression. Metrika **75**, 673–708 (2011)
21. Hušková, M., Meintanis, S.G.: Change-point analysis based on empirical characteristic function of ranks. Sequent. Anal. **25**(421), 436 (2006)
22. Kang, J., Lee, S.: Parameter change test for Poisson autoregressive models. Scand. J. Statist. **41**(1136), 1152 (2014)
23. Kirch, C.: Bootstrapping sequential change-point tests. Seq. Anal. **27**(330), 349 (2008)

24. Kirch, C., Kamgaing, J.T.: Detection of change points in discrete valued time series. In: Davis, R.A., Holan, S.A., Lund, R.B., Ravishanker, N. (eds.) Handbook of Discrete Valued Time Series, Chapman and Hall (2015a)
25. Kirch, C., Tadjuidje Kamgaing, J.: On the use of estimating functions in monitoring time series for change points. J. Statist. Plann. Infer. **161**(25), 49 (2015b)
26. Meintanis, S.G., Karlis, D.: Validation tests for the innovation distribution in INAR time series models. Comput. Statist. **29**(5), 1221–1241 (2014)
27. Meintanis, S.G., Nikitin, YaYu.: A class of count models and a new consistent test for the Poisson distribution. J. Statist. Plann. Infer. **138**(3722), 3732 (2008)
28. Neumann, M.H.: Absolute regularity and ergodicity of Poisson count processes. Bernoulli **17**(1268), 1284 (2011)
29. Novoa-Muñoz, F., Jiménez Gamero, M.D.: Testing for the bivariate Poisson distribution. Metrika **77**(771), 793 (2014)
30. Rémillard, B., Theodorescu, R.: Inference based on the empirical probability generating function for mixtures of Poisson distributions. Statist. Decis. **18**, 349–366 (2000)
31. Schweer, S.: A goodness-of-fit test for integer-valued autoregressive processes. J. Time Ser. Anal. (2015). doi:10.1111/jtsa.12138.267.284
32. Schweer, S., Weiss, C.H.: Compound Poisson INAR(1) processes: stochastic properties and testing overdispersion. Comput. Statist. Dat. Anal. **77**(267), 284 (2014)
33. Steland, A., Rafajlowicz, E.: Decoupling change-point detection based on characteristic functions: methodology, asymptotics, subsampling and application. J. Statist. Plann. Infer. **145**(49), 73 (2014)
34. Tjostheim, D.: Some recent theory for autoregressive count time series. Test **21**(413), 438 (2012)
35. Weiss, C.H.: Modelling time series of counts with overdispersion. Stat. Methods Appl. **18**, 507–519 (2009)
36. Weiss, C.H., Testik, M.C.: Detection of abrupt changes in count data time series: cumulative sum derivations for INARCH(1) models. J. Qualit. Technol. **44**(249), 264 (2012)
37. Zhu, F.: A negative binomial integer-valued GARCH model. J. Time Ser. Anal. **32**(1), 54–67 (2011)

Varying Coefficient Models Revisited: An Econometric View

Giacomo Benini, Stefan Sperlich and Raoul Theler

Abstract Disaggregated data are characterized by a high degree of diversity. Nonparametric models are often flexible enough to capture it but they are hardly interpretable. A semiparametric specification that models heterogeneity directly creates the preconditions to identify causal links. Certainly, the presence of endogenous variables can destroy the ability of the model to distinguish correlation from causality. Triangular varying coefficient models that consider the returns as nonrandom functions, and at the same time exogeneize the problematic regressors are able to add to the flexibility of a semiparametric specification the causal interpretability. Moreover, they make the necessary assumptions much more credible than they typically are in the standard linear models.

Keywords Heterogeneity · Varying Coefficient · Endogeneity

1 The Causality Problem in the Presence of Heterogeneous Returns

Disentangling causality from correlation is one of the fundamental problems of data analysis.[1]Every time the experimental methodology—typical in some hard sciences—is not applicable, it becomes almost impossible to separate causality from observed correlations using non-simulated data. The only available alternative is to find a set of non-testable assumptions that allow to express the causal links as parameters or as functions, and to subsequently find consistent estimators for the conditional moments or distributions that describe the parameters (or functions) of interest. In

[1]We thank an anonymous referee and the participants of the ISNPS 2014 meeting in Cadiz for helpful comments and discussion.

G. Benini (✉) · S. Sperlich · R. Theler
Geneva School for Economics and Management, Université de Genéve,
Bd du Pont d'Arve 40, CH-1211 Geneva, Switzerland
e-mail: Giacomo.Benini@unige.ch

R. Cao et al. (eds.), *Nonparametric Statistics*, Springer Proceedings
in Mathematics & Statistics 175, DOI 10.1007/978-3-319-41582-6_5

particular, consider a response Y to be regressed on an explanatory variable W. The assumption that transforms a simple (cor)relation into a causal effect of W on Y, is often called 'exogeneity'.

Definition 1 A variable W is weakly exogenous for the parameter of interest ψ, if and only if there exists a re-parametrization λ for the joint density with parameter $\lambda = (\lambda_1, \lambda_2)$ such that

1. $f(y, w|\lambda_1, \lambda_2) = f(y|w; \lambda_1) f(w|\lambda_2)$.
2. ψ depends on λ_1 only.
3. (λ_1, λ_2) are variation free, i.e.: $(\lambda_1, \lambda_2) \in (\Lambda_1 \times \Lambda_2)$ for two given sets Λ_1, Λ_2.

The factorization presented in Definition 1 implies that the conditional density of Y given W is fully characterized by λ_1, while λ_2 is a so-called nuisance parameter [1]. In other words, if the causal impact of W on Y is the objective of interest, then the characterization of the distribution of W is unimportant. This convenient factorization allows to focus exclusively on the relationship between Y and W ignoring all the other associations.

In econometrics, an outcome equation that describes the relationship between Y and W often has a less restrictive moment specification than the one proposed by this definition. Usually, a factorization in the form of

$$E[YW|\lambda_1, \lambda_2] = E[E(Y|W; \lambda_1)W|\lambda_1, \lambda_2], \tag{1}$$

is sufficient to detect the causal impact of W on Y. The problem is that, even for simple economic situations, it is often hard to justify an assumption like (1).

Consider for example the case where an economist wants to study the demand function of soft drinks using the individual consumption of Coca-Cola (X), the individual consumption of Pepsi-Cola (Q), and their respective prices (p_1, p_2) (with $p_1 > p_2$). A typical dataset looks like the one in Fig. 1.

From the observation of the data, an econometrician would conjecture that, while the first two cross-section observations (i.e., individuals) may consider Coca-Cola and Pepsi-Cola as perfect substitutes, and therefore, since $p_1 > p_2$, all the income spent on soft drinks goes to Pepsi-Cola, the individuals 3 and 4 prefer to consume a quantity of Coca-Cola X^* different from zero, even though the price of Coca-Cola is higher (see Fig. 2).

In other words, since Agent 1 and Agent 3 have different preferences, their optimization process is different

Fig. 1 X^* is the consumption of Coca, while Q^* is the consumption of Pepsi

Observations	Coca-Cola	Pepsi-Cola
1	0	Q_1^*
2	0	Q_2^*
3	X_3^*	Q_3^*
4	X_4^*	Q_4^*
$\vdots$	$\vdots$	$\vdots$

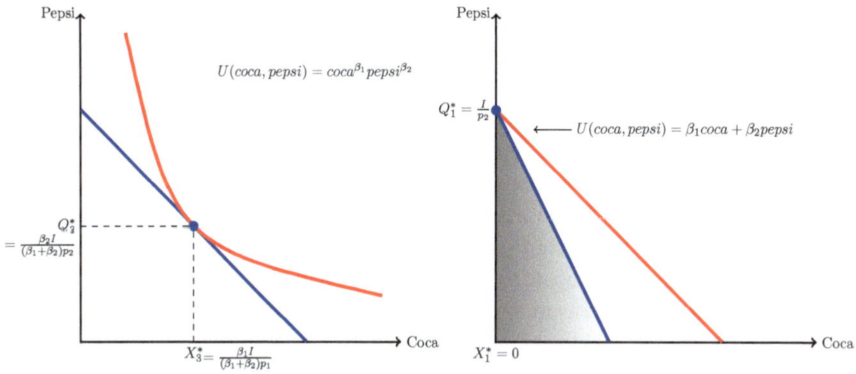

Fig. 2 The (individual) demand functions for a given budget constraint $p_1 X + p_2 Q = I$ varies accordingly to individual preferences. Agent 3 and 4 do not consider Coca-Cola and Pepsi to be equally good (*left graph*), while Agent 1 and 2 do (*right graph*)

$$\text{Agent 1 max process} \qquad \begin{cases} \max\limits_{X,Q} U(X, Q) = \beta_1 X + \beta_2 Q \\ \text{s.t.} I = p_1 X + p_2 Q \leq I_{ub} \end{cases}$$

$$\text{Agent 3 max process} \qquad \begin{cases} \max\limits_{X,Q} U(X, Q) = X^{\beta_1} Q^{\beta_2} \\ \text{s.t.} I = p_1 X + p_2 Q \leq I_{ub} \end{cases}$$

with I_{ub} being the budget constraint. In order to check whether the previous conjecture is true or not, a structural model that enables us to empirically validate the hypothesized choice structure must be specified. If the utility functions are not cardinal, the results of the two maximization processes cannot be compared directly. To the contrary, the study of the expenditure functions allows to monetize the otherwise incommensurable trade-offs between the benefits of the consumptions and their costs. In particular, an expenditure function indicates the minimum amount of money that an individual would need to spend in order to achieve a certain level of utility (given an utility function and a set of prices). If the conjectured choice models are correct, then for those agents that consider Coca-Cola and Pepsi-Cola as perfect substitutes (like Agent 1), the expenditure function should be

$$I(p_1, p_2, \bar{v}(X, Q)) = min\left(p_1 \frac{\beta_1 X + \beta_2 Q}{\beta_1}, p_2 \frac{\beta_1 X + \beta_2 Q}{\beta_2} \right),$$

where $\bar{v}(X, Q)$ is the level of utility for the observed consumption levels $(0, Q)$. For individuals that do not consider the two soft drinks as perfect substitutes (like Agent 3), the amount of expenditures for the given level (X, Q) should be

$$I(p_1, p_2, \bar{v}(X, Q)) = \left[p_1 \left(\frac{\beta_1 p_2}{\beta_2 p_1} \right)^{\frac{\beta_2}{\beta_1+\beta_2}} + p_2 \left(\frac{\beta_2 p_1}{\beta_1 p_2} \right)^{\frac{\beta_1}{\beta_1+\beta_2}} \right] X^{\frac{\beta_1}{\beta_1+\beta_2}} Q^{\frac{\beta_2}{\beta_1+\beta_2}}.$$

In this case both Definition 1 and/or assumption (1) are useless, because the required factorization for the vector $W = [X, Q]^T$, given a set of parameters $\lambda_1 = [\beta_1, \beta_2]$, can be true for the perfect substitute case or for the imperfect one, but not for both.

This simple introductory example shows that when micro-data exhibit holes (non-participation in the activity of interest), kinks (switching behaviors), and corners (nonconsumption or nonparticipation at specific points in time), then relations like (1) become meaningless [2]. There are at least three solutions to deal with an assumption like (1) in a context where heterogeneity among individuals is a major concern.

A first solution is to aggregate the data and study a much smoother problem (being smoother due to the compensations of the movements in opposite directions) typical for macro-data. Consider, for example, a relation between two variables which at a micro level may be piecewise linear with many nodes. After the aggregation is done, the relationship can probably be approximated by a smooth function that can satisfy Eq. (1) [3]. However, if an econometrician is interested in the analysis of individual-level data, in order to describe the economic behavior of individuals or firms, this option does not help.

A second possibility is to accept the heterogeneous nature of the parameters at a micro-level, but to ignore it, and use a parametric (possibly linear) specification with constant coefficients. Let us now abstract from the above example and denote the response by Y and the two explanatory variables X and Q such that

$$Y_i = t(X_i, Q_i) + e_i = \beta_0 + \beta_1 X_i + \beta_2 Q_i + \varepsilon_i \qquad E[\varepsilon_i | X_i, Q_i] = 0\,. \tag{2}$$

In this case all the heterogeneity is absorbed by the regression disturbance ε. Even if many applied economists recognize the limits of a standard parametric specification that most likely suffers from a functional form misspecification because $t(X_i, Q_i) \neq \beta_0 + \beta_1 X_i + \beta_2 Q_i$, which means that $e \neq \epsilon$, they still use it as an approximation because their least squares estimates (like OLS) converge to the value of β that tries to minimize (we say *try* because its success depends also on other factors like the scedasticity function) the mean-squared prediction error $E[t(X_i, Q_i) - \beta_0 - \beta_1 X_i - \beta_2 Q_i]^2$. It is well known that under homoscedasticity OLS gives the best linear predictor of the nonlinear regression function (the mean-squared error (MSE) being the loss function), and this even when there is a functional form misspecification [4]. However, this property is not useful if the objective of the researcher is to interpret the regression coefficients as a true micro-relationship in the form of $E[Y_i | X_i, Q_i]$, because the standard OLS would typically be inconsistent when estimating the marginal effect of the variables,

$$\hat{\beta}_1^{OLS} \rightarrow \beta_1 \neq \frac{\partial t(X_i, Q_i)}{\partial X_i} =: \beta_{1i} \qquad \hat{\beta}_2^{OLS} \rightarrow \beta_2 \neq \frac{\partial t(X_i, Q_i)}{\partial Q_i} =: \beta_{2i}\,.$$

In particular, if the returns are heterogeneous in the data generating process (DGP), a modeling strategy like (2) might not be able to derive consistent estimates. For example, if $Y_i = \beta_i^T W_i + \varepsilon_i$ with $W = [1, X, Q]^T$, for $\beta_i = [\beta_{0i}, \beta_{1i}, \beta_{2i}]$ is modeled as

$$Y_i = \beta^T W_i + e_i \quad \text{with} \quad \beta = E[\beta_i] \quad \text{and} \quad e_i = W_i^T[\beta_i - \beta] + \varepsilon_i \ ,$$

then the standard OLS estimators would give

$$\begin{aligned}
\hat{\beta}^{OLS} &= \left[\sum_{i=1}^{n} W_i W_i^T\right]^{-1} \sum_{i=1}^{n} W_i Y_i = \left[\sum_{i=1}^{n} W_i W_i^T\right]^{-1} \sum_{i=1}^{n} W_i \left[W_i^T E(\beta_i) + e_i\right] \\
&\xrightarrow[n\to\infty]{p} E(\beta_i) + E(W_i W_i)^{-1} E(W_i W_i^T[\beta_i - E(\beta_i)]) + E(W_i W_i^T)^{-1} E(W_i \varepsilon_i) \\
&= E(W_i W_i^T)^{-1} E(W_i W_i^T \beta_i).
\end{aligned}$$

From the last equality it follows that $\hat{\beta}^{OLS} \nrightarrow E[\beta_i|W_i]$ unless $E[\beta_i|W_i] = E[\beta_i]$.

A third solution is to conclude that the discreteness and nonlinearity typical for micro-data requires to model heterogeneity directly. But how? A first option is to transform the density requirement of Definition 1 into an individual-level factorization like

$$f(y_i, w_i|\lambda_{1i}, \lambda_{2i}) = f(y_i|w_i; \lambda_{1i}) f(w_i|\lambda_{2i}) \ , \quad i = 1, ..., n \tag{3}$$

(here w_i needs not to include a 1 for the intercept), where every cross-sectional observation is characterized by a set of individual parameters $(\lambda_{1i}, \lambda_{2i})$. This creates the complication that the parameters (λ_1, λ_2) are no longer variation free, which is not *stricto sensu* a problem because it is possible to transform λ_{1i} into a random coefficient to which it is possible to associate an invariant hyperparameter θ that characterizes the prior density $f(\lambda_{1i}|w_i, \theta)$. In this specification, the invariance assumption can be reproduced in the form $f(y_i|w_i, g(w_i, \theta))$, where θ is estimated globally by a maximum likelihood or in a neighborhood, e.g., by a kernel-based local likelihood. This Bayesian solution allows to have variation-free hyperparameters and, at the same time, random coefficients that capture individual heterogeneity due to the randomness of λ_{1i}.

No matter how elegant the solution might look like, it presents many and interdependent problems. The main one is the low degree of robustness of the estimates $\hat{\theta}$. One may use shrinking priors to overcome this, but in order to make sure that the prior decays quickly enough (to produce robust estimates), it is necessary to impose stringent conditions both on the priors' tails and on the decay rates of the tails. This kind of assumptions are very hard to understand in practice and even harder to relate to economic theory.

A less controversial way to directly model heterogeneity is to allow the value of the coefficients to change when and observable variable F, called here 'effect modifier(s),' allows to write Eq. (2) as

$$Y_i = \beta_i^T W_i + \varepsilon_i \quad \text{with} \quad \beta_i = g(F_i) + \delta_i \ . \tag{4}$$

This is the well-known varying coefficient model (VCM), cf. Hastie and Tibshirani [5]. In this specification, Y_i is the dependent variable, W_i is a $d_W \times 1$ vector of explanatory variables, and the coefficient β_i is allowed to vary across i. In particular, it is a function of a $d_F \times 1$ vector of observable variables F_i (which might include also elements of W), while $g(.)$ is a vector of functions of the effect modifier, and δ_i is a stochastic mean-zero disturbance with finite variance. The exogeneity assumption is centered on the idea of correctly estimating the causal impact of W, not the one of F, on Y, therefore it is possible to imagine $\hat{g}$ as the best nonparametric predictor of β_i for a given F_i. This implicates that the expected value of δ given Q would be equal to zero by construction: $E[\delta_i|F_i] = 0$. The new structure of the model produces a very flexible and yet interpretable semiparametric specification.

The hybrid nature of the VCMs has several advantages. First, it reduces the level of complexity of a pure nonparametric model allowing to interpret the coefficients like in a parametric specification. Second, it enables to incorporate insights that come from economic theory into the modeling process. Third, it produces a good trade-off between the loss in fitting ability, which is (hopefully) small compared to the nonparametric specification, and the increased facility of the estimation process, which is almost as easy as in a parametric model.

The empirical potentials of the VCM modeling can be understood reconsidering the soft drink example. In this case, depending on whether the agent considers the goods as perfect substitutes or not, the coefficients resulting from the optimal allocations are different. However, in both cases, they are functions of the level of expenditure I, the prices (p_1, p_2) and the quantity consumed (X, Q) by individuals with some characteristics also included in F.

The previous consideration suggests that a VCM, in which the returns are functions of the prices and of the quantities of the goods, allows to keep a linear specification for the expenditure function (or expenditure shares) in the form of

$$Y_i = \beta_{0i} + \beta_{1i} X_i + \beta_{2i} Q_i + \varepsilon_i \tag{5}$$

with $\beta_j = g_j(F_i) + \delta_{ji}$, $j = 0, 1, 2$. In other words, a VCM allows us to transform the structural specification of (2) into a model able to take into account heterogeneity *sive natura*, making an assumption like (1) meaningful and often also plausible. Of course, the presence of numerous effect modifiers makes an equation like (5) hard to compute. To the contrary, a function with few effect modifiers is more easily interpretable and, at the same time, reduces the course of dimensionality of the nonparametric regression $\beta_j = g_j(F_i) + \delta_{ji}$, $j = 0, 1, 2$. Therefore it makes sense to reduce the number of effect modifiers for each j separately (e.g., by canceling those that present a low level of nonlinearity with respect to the regressors).

The introduction of a second, more complex, economic example helps explaining the potentials of a VCM, even when the conjectures about the individual decision-making process behind the observed covariates is less easy to deduce than in a simple demand analysis environment. Let us suppose that an applied economist wants to study the impact of education and experiences on wages in a cross-sectional dataset. The concerns about the disaggregated nature of the data might induce the researcher

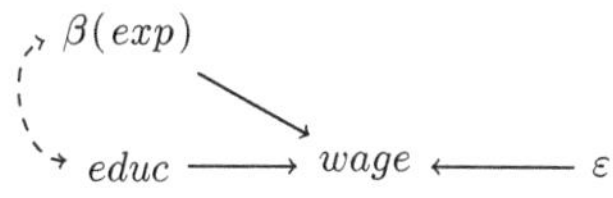

Fig. 3 A causal graph of (6) highlighting the causal links among the variables

to do an a priori analysis that most likely reveals that marginal returns to education vary for different levels of working experience, see e.g., Schultz [6]. Merging the insights that come from the economic theory with the intuitions resulting from the scrutiny of the data we end up with a VCM of the form

$$wage_i = \beta_{0i} + \beta_{1i} educ_i + \varepsilon_i \ , \tag{6}$$

where the intercept and the slope are functions of the level of experience, $\beta_i = g(exp_i) + \delta_i$, with $g(\cdot)$ and δ belonging to $\mathbb{R}^2$. A structural specification like (6) is very appealing because it corrects the (downward) bias that would emerge using a linear modeling that ignores the interaction between experience and education and therefore systematically underestimates the returns on schooling [7]. In this new formulation, it is important to discuss the role of δ. As indicated above, the nature of δ is not the one of an isotonic deviation from the mean but rather the one of a stochastic disturbance in a nonparametric equation. Therefore the role that δ plays in Eq. (4) is related to its disturbance nature.

Unlike in the soft drinks example, where the choices' structure was easy to *reverse engineer*,[2] the relationships among the three variables $(wage, educ, exp)$ are more complex. The lack of knowledge about the objectives that individuals have, and the ignorance about the means that they may use to achieve them, does not allow to have an insight in the choice structure based uniquely on the observed covariates. In other words, in the analysis of the wage-education-experience relationship, even assuming that the only objective of (all) individuals is to obtain a high salary, there is no perfect insight about which actions an individual would take in order to achieve this result. For example, in order to have a higher wage, agent i could start working immediately after high school and accumulate experience, which is valued in the labor market. In this case the decision to go to university would be postponed. At the same time, agent j could do the opposite having the same objective function. This means that it is highly probable that the nonlinear nature of the discrete data that describe the individual choices can be largely absorbed by $g(exp_i)educ_i$, but there could still be a local deviation from the mean of the level of education, here denoted as $\delta_i educ_i$, due to the uncertainty about the individual decision-making process. This is reflected in the causal graph given in Fig. 3.

Since each coefficient is not an average given a particular sample realization (but a function), the parameters are allowed to have different degrees of heterogeneity

[2]Reverse engineering, also called *back engineering*, is the process of extracting knowledge or design information from anything man-made, and reproducing it. In economics, the reverse engineering process consists of extracting the structure of individual preferences from observed outcomes and then reproduce the outcomes using the conjectured informations.

even for the same levels of the effect modifier, reflected in the presence of δ. So, here heterogeneity can explicitly imply deviations from the slopes. In the semiparametric framework, the quantity of local deviation δ is a function of the degree of smoothness of $g(exp_i)$. At the same time, since the primary objective of the research is not to estimate correctly the causal impact of F on Y but rather the one of W, it is sufficient to think of $\hat{g}(\cdot)$ as the best nonparametric predictor of β_i *such that* $E[\delta_i|F_i] = 0$ becomes true by construction. As a result, the average returns to education are equal for all the cross-section observations that have the same level of exp.

2 Triangular Varying Coefficient Models with Instruments

The previous section highlighted the necessity to model heterogeneity directly in order to make the assumptions of Definition 1 plausible. But still, exogeneity can be violated by the nature of the observed variables irrespectively of the semiparametric characteristics of the VCM. In particular, a regressors could be endogenous in the sense that in the structural equation $Y_i = \beta_i^T W_i + \varepsilon_i$ one has $E[\varepsilon_i|W_i, F_i] \neq 0$. The three usual sources of endogeneity typically mentioned are: the omission of explanatory variables correlated with the included covariates, a measurement error, and reversed causality. All the three sources of endogeneity cannot be solved using the varying coefficient approach alone.[3] A popular solution is to introduce some additional variables called instruments.

Definition 2 A variable Z is called an instrumental variable (IV) for W if

1. is partially correlated with the endogenous variable W once the other explanatory variables have been netted out.
2. is mean independent with respect to the stochastic error ε.

This definition suggests that the addition of a second structural equation to the VCM creates a triangular model able to exogenize W while modeling heterogeneity directly. For simplification, let us set for a moment $dim(W) = dim(Z) = 1$. Keeping a specification like (4), it is sufficient to add a selection equation that relates the endogenous W with the instrument(s) Z, namely

$$W_i = m(Z_i) + \eta_i \qquad E[\eta_i|Z_i] = 0 \tag{7}$$

and assume a finite variance for η_i. In this formulation the vector of explanatory variables is allowed to contain endogenous components, while Z is a vector of IVs, which may have F as one of its arguments. Furthermore, $m(\cdot)$ is a smooth function, or a vector of smooth functions if $dim(W) > 1$, while ε and η are, respectively, the endogenous error and a stochastic disturbance that has expected value equal to zero and finite variance.

[3]However, the most typical, though in economics rarely mentioned, endogeneity problem, i.e., the functional misspecification, can be largely diminished by the VCM.

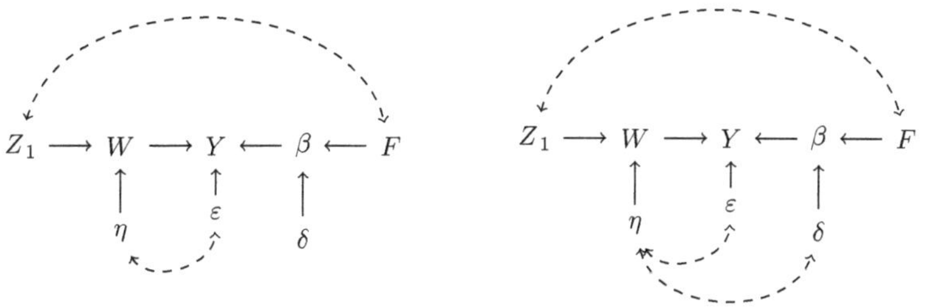

Fig. 4 The mechanism of the endogeneity process changes depending on the assumptions about the relationship between the error ε and the stochastic disturbances (η, δ). The *left* picture is the only possibility in a world of homogeneous coefficients, while the *right* specification (with $cov(\eta, \delta) \neq 0$) is the situation resulting from introducing a varying coefficient structure. The direct connection between δ and ε is not taken into account because the interest is about the causal link between Y and W for a given level of F

The triangular nature of Eqs. (4) and (7) implies a simple endogeneity mechanism. In order for the error term ε to be correlated with at least one of the explanatory variables W, it must be $cov(\eta, \varepsilon) \neq 0$. To see how the mechanism of the model works in practice, it is useful to consider the simplest possible specification, namely a model that would include only one heterogeneous intercept, one heterogeneous slope, and one endogenous explanatory variable. The latter is instrumented by one exogenous variable Z_1, which is correlated with W even if the impact of the (exogenous) effect modifier has been netted out, namely

$$\begin{aligned} Y_i &= \beta_{0i} + \beta_{1i} W_i + \varepsilon_i & \qquad E[\varepsilon_i | F_i, W_i] &\neq 0 \\ W_i &= m(F_i, Z_{1i}) + \eta_i & \qquad E[\eta_i | F_i, Z_{1i}] &= 0 \, . \end{aligned}$$

In this specification, irrespectively of the relation between the error ε and the two disturbances δ and η, endogeneity comes only through $cov(\varepsilon, \eta)$, see causal graph Fig. 4.

The considerations about the mechanisms of the endogeneity problem combined with the observation that a VCM is a special case of a semiparametric linear specification, suggest that the model can be identified and later estimated using the *control function* approach [8]. The control function, say $h(\cdot)$, handles the relation between η and ε (irrespectively of the behavior of δ) in the following form

$$\epsilon_i = \delta_i W_i + \varepsilon_i = h(\eta_i) + \vartheta_i \qquad E[\vartheta_i | \eta_i] = 0 \, . \tag{8}$$

This added to (4) eliminates the endogeneity problem giving unbiased estimates for $g(.)$

$$E[Y_i | Z_i, \eta_i] = g(F_i) W_i + h(\eta_i) \qquad Z_i = (F_i, Z_{1i}) \, . \tag{9}$$

It is important to notice that the higher complexity of a VCM increases the chance to successfully eliminate the endogeneity problem via the control function approach.

Specifically, even if a set of valid instruments $(Z_i)_{i=1}^n$ is available, a linear IV estimator would generally be biased. For example, if the equation $Y_i = \beta_{01} + \beta_{1i} W_i + \varepsilon_i$ (with $\varepsilon_i \perp\!\!\!\perp Z_i$) is modeled using homogeneous coefficients $Y_i = \beta_0 + \beta_1 W_i + e_i$ with $e_i = [\beta_{0i} - \beta_0] + W_i[\beta_{1i} - \beta_1] + \varepsilon_i$ and $\beta_j = E(\beta_{ji})$, $j = 0, 1$, then the instrumentation using Z_i does not produce consistent estimates. Consider for example the case where $dim(Z) = dim(W) \geq 1$. In this setting the estimated returns are

$$\hat{\beta}^{IV} = \left[\sum_{i=1}^{n} Z_i W_i^T\right]^{-1} \sum_{i=1}^{n} Z_i Y_i = \left[\sum_{i=1}^{n} Z_i W_i^T\right]^{-1} \sum_{i=1}^{n} Z_i \left[W_i^T E(\beta_i) + e_i\right]$$
$$\xrightarrow[n\to\infty]{p} E(\beta_i) + E(Z_i W_i^T)^{-1} E(Z_i W_i^T[\beta_i - E(\beta_i)]) + E(Z_i W_i^T)^{-1} E(Z_i \varepsilon_i)$$
$$= E(Z_i W_i^T)^{-1} E(Z_i W_i^T \beta_i) .$$

The last equality cannot be simplified further unless a new assumption, namely $\beta_i \perp\!\!\!\perp (W_i, Z_i)$, is made—which is clearly in contradiction with the spirit of the model, cf. the causal graphs in Fig. 4. Basically, the heterogeneous nature of the returns transforms Z_1 into a 'poor' instrument if the simple linear structure is used.

In order to proceed and correctly estimate the unknown terms in Eq. (9), it is necessary to impose additional identification conditions. Identification can be obtained imposing a conditional mean independence in the form of

$$E[\epsilon_i | Z_i, \eta_i] = E[\epsilon_i | \eta_i] \qquad \text{CMI}, \tag{10}$$

or a conditional moment restriction

$$E[\epsilon_i | Z_i] = 0 \qquad \text{CMR}. \tag{11}$$

The CMI and the CMR are not equivalent (Kim and Petrin [9]). The CMI requires Z and η to be additively separable in W, which often is not the case. To the contrary, the CMR can be easily justified by the use of economic primitives that describe the structural specification [10]. The use of the CMR, however, requires to include the instrument(s) in the control function, such that the relation between ϵ and η becomes

$$\epsilon_i = h(Z_i, \eta_i) + \vartheta_i \quad E[\vartheta_i | Z_i, \eta_i] = 0 . \tag{12}$$

In any case, if the amplitude of the control function increases, a less precise estimate $\hat{g}(.)$ might be produced (multi-functionality). This is the statistical counterpart of the econometric problem called 'weak' instruments, i.e., instruments that are weakly correlated with the endogenous regressors.

The estimation of VCM in its simplest specification has been proposed in different forms. Hastie and Tibshirani [5] used a smoothing spline based on a penalized least squares minimization, while Fan and Zhang [11] proposed a kernel weighted polynomials. However, this last method and its surrogates are designed for a single effect modifier for all coefficients, which is a strong limitation in the context we discussed so far.

Estimating an equation like (9) is a more complicated procedure than the one required for a simple VCM. The presence of a control function, which depends upon η, requires the use of specific tools that are designed for additive models. The two most common alternatives are the marginal integration method [12] and the smooth backfitting [13]. The latter method suffers less from the curse of dimensionality and can be applied as part of a 2-step procedure. The first step consists in the estimation of $m(Z)$ in Eq. (7) using a standard nonparametric technique. The second step consists in the substitution of the estimated residuals $\hat{\eta}$ into (9), which creates an equation characterized by a finite sample disturbance whose impact can be mitigated asymptotically [14]. For an exhaustive survey on the VCM estimation techniques see Park et al. [15]. For a comparison of implementations of these methods in R [16], including the control function approach, see Sperlich and Theler [17].

All the previous considerations are particularly important in the treatment effect literature. For a discrete W with finite support, Imbens and Angrist ([18], [19]) named the impact of a treatment (i.e., a change in W) local average treatment effect (LATE). By construction, the LATE can only compute the average of the β_i for the individuals that choose to switch their w because of an instrument's change. In other words, in the LATE environment, the parameter of interest can only be estimated for people responding to the selection equation and is therefore an instrument (or selection) specific parameter. They imposed a conditional independence assumption in the form of $Y_i(w) \perp\!\!\!\perp Z_i \; \forall w$, as well as the request of independence of the (so called) compliers sub-population to an instrument's change. Reconsidering these assumptions in the presence of heterogeneous returns shows that the LATE is not defined if $cov(\beta, Z) \neq 0$. In the case of a VCM this means that, unless the effect modifier is indeed a constant, the standard independence assumption used to define and identify the LATE is not fulfilled.

The model we outlined above suggests that, if some effect modifiers F_i are observed, they should be used to construct a VCM that makes the LATE conditions more credible. For example, in the case of a binary endogenous treatment W which is instrumented by a binary instrument Z, the varying LATE becomes

$$LATE(q) = \frac{E[Y|F = q, Z = 1] - E[Y|F = q, Z = 0]}{E[W|F = q, Z = 1] - E[W|F = q, Z = 0]}.$$

Integrating over q gives the value of the LATE. In this case, the more heterogeneity of returns to W is captured by $g(F)$ the less the LATE will vary over the IVs' choice. In other words, a VCM reduces the typical LATE problem to a minimum because it controls for the correlation between the effect modifier and the instrument. Therefore, the VCM enables to identify a LATE type parameter that can be estimated nonparametrically regressing Y and W on F and Z. The interesting point here is that the parameter of interest depends on both, the instruments' choice and the values taken by F. An interesting next step would be to find a meaningful model specification that merges the effect modifier and the instrument.

3 An Example

In order to see all the potentials of the triangular VCM specification in practice, it is useful to reconsider the wages-experience-education relationship. Experience and education are crucial variables in the determination of a worker's wage. Yet, labor economists have argued for many years that cognitive and noncognitive abilities are also critical in order to determine labor market outcomes. A large empirical literature has confirmed the positive connection between cognitive test scores and high wages [20]. Unfortunately, many datasets do not provide ability's measures. The lack of information about the skills misleads the researcher to mistake the data generating process (DGP). Even if a VCM modeling strategy is used, if the ability of the individual is not included, such that

$$\begin{aligned} wage_i &= t(educ_i, exp_i, ability_i) + \zeta_i \quad \text{is modeled as} \\ wage_i &= g_0(exp_i) + g_1(exp_i)educ_i + \epsilon_i, \end{aligned}$$

then the exogeneity assumption $E[\varepsilon_i|educ_i, exp_i] = 0$ does not hold, because of an omitted variable bias. This problem can be solved using an instrument.

There exist at least two classical errors that arise when searching for an IV. The first one is the selection of a variable that is clearly correlated with the endogenous regressor but hardly independent from the error ε. For example, the level of education (of one) of the parents would be hardly independent from the omitted variable $ability$. A second wrong choice would be the selection of a variable that has the opposite characteristics, namely a variable that is exogenous but that is hardly correlated with the endogenous regressor. For example, the last digit of the person social security number. The choice of good instruments must come from both, a deep knowledge of the origin of the IV and the source of endogeneity.

Take instead the example proposed by [21]. In most American states education legislation requires students to enter school in the calender year when they turn six. Therefore the age at which students start school is a function of the date of birth of the pupils. For example, if the 31st of December is the legal cutoff point, children born in the fourth quarter enter school shortly before turning six, while those born in the first quarter enter school when they are around six years and an half. Furthermore, because compulsory schooling laws require students to remain in school only till they turn 16, these groups of students will be in different grades, or through a given grade to a different degree, when they reach the legal drop out age. The combination of the school start-age policies and the school attendance laws creates a situation where children attend school for different times depending upon their birthdays. Assuming that the day of birth of a person is not correlated with his abilities seems to make the quarter of birth (qob) a valid IV. The typical mistake made here is to conclude from no-causality to no-correlation. But first, there is clearly the possibility that the IV is correlated with the education of the parents, and second, being the youngest could mean to be the smallest and physically weakest in the class resulting in maturity disadvantages. All these facts could change the wage-path invalidating the IV.

Nonetheless, let us consider a VC triangular model with the same instrument proposed by Angrist and Krueger

$$wage_i = g_0(exp_i) + g_1(exp_i)educ_i + h(\eta_i) + \vartheta_i \tag{13}$$
$$educ_i = m(exp_i, qob_i) + \eta_i \; . \tag{14}$$

In this specification, they identify the LATE of education on wages for those who do not drop out in spite of the ones that could have thanks to their birth date. Note that, if this is not the parameter of interest, it might have been much better and easier to use a proxy approach instead of an IV one. In order to reverse engineer the preferences' structure it is necessary to model a situation where a rational individual has to decide, when turning 16, to stay for the rest of the academic year or leave school. In this context, the agent's wage is a function of the years of education $educ$, but also of his unobserved ability, ε. The agent's ability is not observed, but the information set that the student can consult before the decision to stay or not is made includes a signal of his individual ability η, for example his past grades. The cost to stay until the end of the year is a function of an exogenous cost shifter, namely the quoter of birth qob, if a student turns 16 in January the cost to stay till the end of the year is higher than if he turns 16 in May, so it makes sense to consider the quoter of birth an argument of the cost function. At the same time, the agent's utility has to be function of the education's choice, the cost-shifters, and the unobserved ability, $U(educ, qob, \varepsilon) = p(educ, \varepsilon) - c(educ, qob)$, where $p(.)$ is the education production function and $c(.)$ is the cost function. The optimal choice problem becomes

$$educ = \underset{\tilde{educ}}{\text{argmax}}\{E[U(\tilde{educ}, qob, \varepsilon)|qob, \eta]\} \; . \tag{15}$$

The specification of the utility function is crucial. The functional form $U(educ, qob, \varepsilon) = p(educ, \varepsilon) - c(educ, qob)$ is not chosen for convenience. The quarter of birth must be part of the cost function, otherwise qob would not be valid instruments—but at the same time it cannot be part of the educational production function because otherwise the causal effect of $educ$ cannot be excluded from the joint effect of $(educ, qob)$. The costs can depend among the ability's signal, η, if for example a staying-based financial aid is available. This possibility, however, is not taken into account. The decision problem just described is illustrated in Fig. 5.

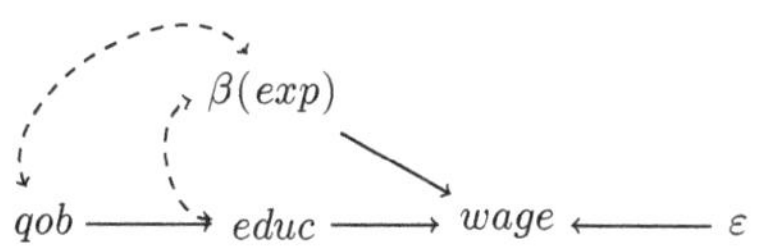

Fig. 5 When endogeneity is an issue due to the presence of a model misspecification, the use of VCM is not enough to guarantee causal analysis, and the introduction of IVs becomes necessary to ensure the exogeneity

In this context the exclusion restriction requires the choice variable $educ$ to be separable in (ε, qob). This depends upon the assumptions that the researcher is willing to make about the educational production function $p(\cdot)$ and the cost function $c(\cdot)$.

All the previous considerations show how a model like (13)–(14) is able to: 1. make individual returns heterogeneous, 2. solve the endogeneity problems that are due to the functional form misspecification using the VCM nature of the model, 3. solve the endogeneity problems that are due to the nature of the regressors using IVs, and 4. relate the structural specification to the economic theory providing a rigorous microfoundation of the outcome equation.

References

1. Engle, R.F., Hendry, D.F., Richard, J.-F.: Exogeneity. Econometrica **51**(2), 277–304 (1983)
2. Pudney, S.: Modelling Individual Choice: The Econometrics of Corners, Kinks, and Holes. Blackwell, Oxford (1989)
3. Eilers, P., Marx, B.: Flexible smoothing with B-splines and penalties. Stat. Sci. **11**(2), 89–121 (1996)
4. White, H.L.: A heteroskedasticity-consistent covariance matrix estimator and a direct test for heteroskedasticity. Econometrica **48**(4), 817–838 (1980)
5. Hastie, T., Tibshirani, R.: Varying-coefficient models. J. Roy. Stat. Soc. Ser. B (Methodological) **55**(4), 757–796 (1993)
6. Schultz, T.P.: Human capital, schooling and health returns. Econ. Hum. Biol. **1**(2), 207–221 (2003)
7. Card, D.: Estimating the return to schooling: progress on some persistent econometric problems. Econometrica **69**, 1127–1160 (2001)
8. Tesler, L.: Iterative estimation of a set of linear regression equations. J. Am. Stat. Assoc. **59**, 845–862 (1964)
9. Kim, K., Petrin, A.: A New Control Function Approach for Non-Parametric Regressions with Endogenous Variables. NBER Working Paper, No. 16679 (2013)
10. Benini, G., Sperlich, S.: Modeling Heterogeneity by Structural Varying Coefficient Models Paper presented at the 2015 IAAE Annual Conference (2016)
11. Fan, J., Zhang, W.: Statistical methods with varying coefficient models. Stat. Interface **1**(1), 179–195 (2008)
12. Linton, O., Nielsen, J.P.: A kernel method of estimating structured nonparametric regression based on marginal integration. Biometrika **82**(2), 93–100 (1995)
13. Roca-Pardinas, J., Sperlich, S.A.: Feasible estimation in generalized structured models. Stat. Comput. **20**, 367–379 (2010)
14. Sperlich, S.: A note on nonparametric estimation with predicted variables. Econ. J. **12**, 382–395 (2009)
15. Park, B.U., Mammen, E., Lee, Y.K., Lee, E.R.: Varying coefficient regression models: a review and new developments. Int. Stat. Rev. 1–29 (2013)
16. R Core Team: R: A Language and Environment for Statistical Computing, R Foundation for Statistical Computing, Vienna, Austria (2014). http://www.R-project.org/
17. Sperlich, S., Theler, R.: Modeling heterogeneity: a praise for varying-coefficient models in causal analysis. Comput. Stat. **30**, 693–718 (2015)
18. Imbens, G. W., Angrist, J. D.: Identification and Estimation of Local Average Treatment Effects. Econometrica. **62**(2), 467–475 (1994)
19. Angrist, J. D., Imbens, G. W.: Two-Stage Least Squares Estimation of Average Causal Effects in Models with Variable Treatment Intensity. J. Am. Stat. Assoc. **90**(430), 431–442 (1995)

20. Murnane, R., Willett, J., Levy, F.: The growing importance of cognitive skills in wage determination. Rev. Econ. Stat. **xxxvii**(2), pp. 251–266 (1995)
21. Angrist, J.D., Krueger, A.B.: Does compulsory school attendance affect schooling and earnings? Q. J. Econ. **106**(4), 979–1014 (1991)

Kalman Filtering and Forecasting Algorithms with Use of Nonparametric Functional Estimators

Gennady Koshkin and Valery Smagin

Abstract The paper deals with the Kalman filtering and forecasting algorithms for a class of time-varying systems with unknown additive inputs. Such classes include object models with possible failures and also with models of controlled processes with unknown disturbances. The designed algorithms are based on combining the Kalman filter and nonparametric estimator. Examples are given to illustrate the usefulness of the proposed approach.

Keywords Kalman filtering and forecasting · Unknown disturbances · Nonparametric estimator

1 Introduction

The Kalman filtering [1] provides the synthesis of algorithms for the class of systems with unknown additive perturbations. Such systems are used as the models of real physical systems, for example, as the models of objects with unknown errors, and also, in control problems for stochastic systems with unknown disturbances.

The known methods to calculate estimates of a state vector are based on the estimators of unknown perturbations [2–10]. In the survey [2], the extension algorithms of the states space requiring complete information on the model of this input are considered. In [4–7] is considered the case when additional information for the models with an unknown input is not required.

The problem with making use of compensation methods for linear stochastic systems with an unknown constant input is solved in papers [11, 12].

In this paper, for a discrete object with an unknown input, we propose the modification of the Kalman filtering algorithm using nonparametric estimators of the observed process according to [13]. The suggested approach allows improving the

G. Koshkin (✉) · V. Smagin
National Research Tomsk State University, 36 Lenin Avenue, 634050 Tomsk, Russia
e-mail: kgm@mail.tsu.ru

V. Smagin
e-mail: vsm@mail.tsu.ru

R. Cao et al. (eds.), *Nonparametric Statistics*, Springer Proceedings in Mathematics & Statistics 175, DOI 10.1007/978-3-319-41582-6_6

estimation accuracy of state vector and unknown input. Also, the problem, considered in [13], is extended to the time-varying system.

An example illustrating the advantages of the proposed algorithm in comparison with the known algorithms from [5, 6] is given.

2 Problem Formulation

Consider, the mathematical model of the linear discrete-time stochastic system with unknown input in the form:

$$x_{k+1} = A_k x_k + B_k r_k + q_k, \quad x_{k=0} = x_0, \tag{1}$$

$$y_k = H_k x_k + \nu_k, \tag{2}$$

where x_k is a state of an object, r_k is an unknown input, y_k is an observation vector, A_k, B_k, and H_k are matrices of the appropriate dimensions. It is assumed that random perturbations q_k and noise measurements ν_k are not correlated between themselves and are subject to the Gaussian distribution with zero mean and the corresponding covariance: $\mathbf{E}\left[q_k q_t^{\mathrm{T}}\right] = Q_k \delta(k,t)$, $\mathbf{E}\left[\nu_k \nu_t^{\mathrm{T}}\right] = V_k \delta(k,t)$, where $\delta(k,t)$ is the Kronecker symbol, i.e., $\delta(k,t) = 1$ if $k = t$, and $\delta(k,t) = 0$ if $k \neq t$, $\mathbf{E}[\cdot]$ denotes the expectation of a random variable, ^{T} denotes matrix transposition. It is assumed also that the vector of initial conditions is uncorrelated with values q_k and ν_k. This vector has the following characteristics:

$$\mathbf{E}\left[x_0\right] = \bar{x}_0, \quad \mathbf{E}\left[(x_0 - \bar{x}_0)(x_0 - \bar{x}_0)^{\mathrm{T}}\right] = P_0.$$

In the simple case, when r_k is a zero-mean white random vector with the known variance, the optimal filtering problem for the model (1), (2) reduces to the Kalman filtering algorithm [1]. If input r_k is a deterministic component and its evolution in time is described by the known linear system, the optimal estimates of r_k and x_k can be obtained by making use of the extended state Kalman filter [2]. In this paper, the case is considered when prior knowledge about the time evolution of r_k is not available. Vector r_k is assumed to be completely unknown.

3 Filtering Algorithm for Systems with Unknown Input

The optimal filter considered in this paper is defined by the full-order Kalman filter. Therefore, filter equations have the form:

$$\hat{x}_{k+1} = A_k \hat{x}_k + B_k \hat{r}_k + L_k[y_{k+1} - H_k(A_k \hat{x}_k + B_k \hat{r}_k)], \quad \hat{x}_{k=0} = \bar{x}_0, \tag{3}$$

$$L_k = P_{k+1/k} H_k^{\mathrm{T}} [H_k P_k H_k^{\mathrm{T}} + V_k]^{-1}, \tag{4}$$

$$P_{k+1/k} = A_k P_k A_k^{\mathrm{T}} + Q_k, \tag{5}$$

$$P_{k+1} = (I - L_k H_k) P_{k+1/k}, \quad P_{k=0} = P_0, \tag{6}$$

where $\hat{x}_k$ and $\hat{r}_k$ are estimators, $P_k = \mathbf{E}\left[(x_k - \hat{x}_k)(x_k - \hat{x}_k)^{\mathrm{T}}\right]$.

However, formulas (3)–(6) cannot be applied immediately because $\hat{r}_k$ is unknown. Obtain estimator $\hat{r}_k$ using the criteria

$$J(r_{k-1}) = \mathbf{E}\left[\sum_{i=1}^{k} \|u_i\|_{C_k}^2 + \|r_{i-1}\|_{D_k}^2\right], \tag{7}$$

where $u_i = y_i - H_i \hat{x}_i$ is an innovation process, $\|\cdot\|_{C_k}^2$ is the Euclidian norm, C_k and D_k are symmetric positive definite weight matrices.

The optimal estimator of unknown input at the moment $k = 1$ is found by minimization of the following criteria:

$$J(r_0) = \min_{r_0} \mathbf{E}\left[\|y_1 - H_1 \hat{x}_1\|_{C_0}^2 + \|r_0\|_{D_0}^2\right]. \tag{8}$$

Substituting $\hat{x}_1 = A_0 \hat{x}_0 + B_0 r_0$ into (8), we have

$$J(r_0) = \min_{r_0} \mathbf{E}\left[\|y_1 - H_0 A_0 \hat{x}_0 - H_0 B_0 r_0\|_{C_0}^2 + \|r_0\|_{D_0}^2\right]. \tag{9}$$

Transform the norms in (9) and obtain

$$J(r_0) = \min_{r_0} \mathbf{E}\left[\alpha_0 - 2 r_0^{\mathrm{T}} B_0^{\mathrm{T}} H_0^{\mathrm{T}} C_0 (y_1 - H_0 A_0 \hat{x}_0) + \|r_0\|_{B_0^{\mathrm{T}} H_0^{\mathrm{T}} C_0 H_0 B_0 + D_0}^2\right]. \tag{10}$$

Here, the parameter α_0 is independent of r_0. First, we differentiate (10) w.r.t. r_0, and then, find the optimal estimator of unknown input from the equation

$$\frac{dJ(r_0)}{dr_0} = 2(B_0^{\mathrm{T}} H_0^{\mathrm{T}} C_0 H_0 B_0 + D_0) r_0 - 2 B_0^{\mathrm{T}} H_0^{\mathrm{T}} C_0 \mathbf{E}[y_1 - H_0 A_0 \hat{x}_0]. \tag{11}$$

So, at the moment $k = 1$, we obtain the optimal estimator of unknown input:

$$\hat{r}_0 = (B_0^{\mathrm{T}} H_0^{\mathrm{T}} C_0 H_0 B_0 + D_0)^{-1} B_0^{\mathrm{T}} H_0^{\mathrm{T}} C_0 \mathbf{E}[y_1 - H_0 A_0 \hat{x}_0]. \tag{12}$$

Analogously, at the moment $k = 2$, the optimal estimator of unknown input is found from the criteria:

$$J(r_1) = \min_{r_1} \mathbf{E}\left[\|y_2 - H_2 \hat{x}_2\|_{C_1}^2 + \|r_1\|_{D_1}^2\right] + J(\hat{r}_0). \tag{13}$$

Taking into account (13) and the expression $\hat{x}_2 = A_1\hat{x}_1 + B_1 r_1$ at the moment $k = 2$, we have

$$J(r_1) = \min_{r_1} \mathbf{E}\left[\|y_2 - H_1 A_1 \hat{x}_1 - H_1 B_1 r_1\|^2_{C_1} + \|r_1\|^2_{D_1}\right] + J(\hat{r}_0).$$

As in the case of (10), we obtain:

$$J(r_1) = \min_{r_1} \mathbf{E}\left[\alpha_1 - 2r_1^{\mathrm{T}} B_1^{\mathrm{T}} H_1^{\mathrm{T}} C_1 (y_2 - H_1 A_1 \hat{x}_1) + \|r_1\|^2_{B_1^{\mathrm{T}} H_1^{\mathrm{T}} C_1 H_1 B_1 + D_1}\right], \quad (14)$$

where the value α_1 is independent of r_1. Differentiating (14) w.r.t. r_1, as in the first step, we obtain the optimal estimator

$$\hat{r}_1 = (B_1^{\mathrm{T}} H_1^{\mathrm{T}} C_1 H_1 B_1 + D_1)^{-1} B_1^{\mathrm{T}} H_1^{\mathrm{T}} C_1 \mathbf{E}[y_2 - H_1 A_1 \hat{x}_1]. \quad (15)$$

By the mathematical induction, the estimators for the next steps take the form

$$\hat{r}_k = (B_k^{\mathrm{T}} H_k^{\mathrm{T}} C_k H_k B_k + D_k)^{-1} B_k^{\mathrm{T}} H_k^{\mathrm{T}} C_k \mathbf{E}[w_k], \quad (16)$$

here $w_k = y_{k+1} - H_k A_k \hat{x}_k$.

Now, let us calculate value $\mathbf{E}[w_k]$ using nonparametric estimators [14]. Applying the well-known kernel estimates, we have

$$\hat{r}_k = (B_k^{\mathrm{T}} H_k^{\mathrm{T}} C_k H_k B_k + D_k)^{-1} \hat{w}_k = S_k \hat{w}_k, \quad (17)$$

where $S_k = (B_k^{\mathrm{T}} H_k^{\mathrm{T}} C_k H_k B_k + D_k)^{-1}$, and the jth component of the vector takes the form:

$$\hat{w}_{k,j} = \frac{\sum\limits_{i=1}^{k} w_{i,j} K\left(\dfrac{k-i+1}{h_j}\right)}{\sum\limits_{i=1}^{k} K\left(\dfrac{k-i+1}{h_j}\right)}. \quad (18)$$

In the ratio (18), $K(\cdot)$ is a kernel function, h_j is a bandwidth parameter. We use the Gaussian kernels, and the bandwidths are calculated by the cross-validation method [15]. Note that ratio (18) is similar to the estimate of the simple conditional functional, namely, the regression function (see [14]). The conditional central moments, for example, the residual conditional variance [14, p. 27], the conditional coefficient of asymmetry and excess [16] are the conditional functionals also and give the important information on properties of random variables. We can obtain nonparametric estimators of such functionals by the methods from [14].

As an example, take the following analogue of the residual conditional variance on the base of ratio (18):

$$\hat{d}_{k,j}^{2} = \frac{\sum_{i=1}^{k} w_{i,j}^{2} K\left(\frac{k-i+1}{h_j}\right)}{\sum_{i=1}^{k} K\left(\frac{k-i+1}{h_j}\right)} - \hat{w}_{k,j}^{2}. \tag{19}$$

4 Prediction Algorithm for Systems with Unknown Input

The equations of the optimal prediction with estimate $\hat{r}_k$ are of the form:

$$\hat{x}_{k+1} = A_k\hat{x}_k + B_k\hat{r}_k + K_k[y_k - H_k\hat{x}_k], \quad \hat{x}_{k=0} = \bar{x}_0, \tag{20}$$

$$K_k = A_k P_k H_k^{\mathrm{T}}[H_k P_k H_k^{\mathrm{T}} + V_k]^{-1}, \tag{21}$$

$$P_{k+1} = (A_k - K_k H_k)P_k(A_k - K_k H_k)^{\mathrm{T}} + K_k V_k K_k^{\mathrm{T}} + Q_k, \quad P_{k=0} = P_0, \tag{22}$$

where $P_k = \mathbf{E}\left[(x_k - \hat{x}_k)(x_k - \hat{x}_k)^{\mathrm{T}}\right] = P_0$, $\hat{r}_k$ is the estimate of unknown input defined by formulas (17) and (18).

5 An Illusrative Example

Apply the filtering algorithm (3)–(6), combined with nonparametric estimates (17) and (18), to the model (1) and to the observations (2) with the following parameters:

$$A = \begin{pmatrix} 0 & 1 \\ 0.05 & 0.9 + 0.1\sin(0.05k) \end{pmatrix}, \quad B = \begin{pmatrix} 1.0 & 0 \\ 0 & 1.0 \end{pmatrix}, \quad Q = \begin{pmatrix} 0.01 & 0 \\ 0 & 0.02 \end{pmatrix},$$

$$V = \begin{pmatrix} 1.8 & 0 \\ 0 & 2.2 \end{pmatrix}, \quad H = \begin{pmatrix} 1.0 & 0 \\ 0 & 1.0 \end{pmatrix}, \quad P_0 = \begin{pmatrix} 1.0 & 0 \\ 0 & 1.0 \end{pmatrix}, \quad C = \begin{pmatrix} 1.0 & 0 \\ 0 & 1.0 \end{pmatrix},$$

$$D = \begin{pmatrix} 0 & 0 \\ 0 & 0 \end{pmatrix}, \quad \bar{x}_0 = \begin{pmatrix} 0 \\ 0 \end{pmatrix}, \quad r_{1,k} = \begin{cases} -1 + \sin(0.1k), & \text{if } 0 \le k < 50, \\ 1 + \sin(0.1k), & \text{if } 50 \le k < 130, \\ \sin(0.2k), & \text{if } 130 \le k \le 200, \end{cases}$$

$$r_{2,k} = \begin{cases} \sin(0.1k), & \text{if } 0 \le k < 50, \\ -1 + \sin(0.2k), & \text{if } 50 \le k < 130, \\ 0.5 + \sin(0.1k), & \text{if } 130 \le k \le 200. \end{cases}$$

As the kernel function, we take the Gaussian density, i.e., $K(u) = \frac{1}{\sqrt{2\pi}}\exp(-u^2/2)$.

By simulations, the proposed filtering algorithms are compared with the algorithms combined with the least mean square (LMS) estimates from [3, 4]. These comparisons are given in Figs. 1, 2, 3 and 4.

The proposed extrapolation algorithms (18), (20)–(22) are compared with the algorithms using the LMS estimates. These comparisons are given in Figs. 5, 6, 7 and 8.

Below, in Tables 1, 2, 3 and 4 the errors of estimation

$$\sigma_{x,i} = \sqrt{\frac{\sum_{k=1}^{N}(x_{ik} - \hat{x}_{ik})^2}{N-1}}, \quad \sigma_{r,i} = \sqrt{\frac{\sum_{k=1}^{N}(r_{ik} - \hat{r}_{ik})^2}{N-1}}, \quad i = 1, 2,$$

are given for two filtering algorithms (N = 200) and by averaging 50 realizations. The results of modeling prediction algorithms for the same data are presented in Tables 3 and 4.

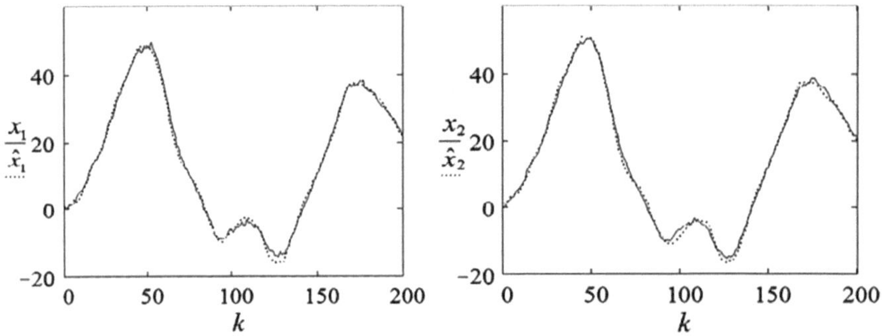

Fig. 1 The dependence of the components and their estimates combined with nonparametric algorithm (3)–(6), (18)

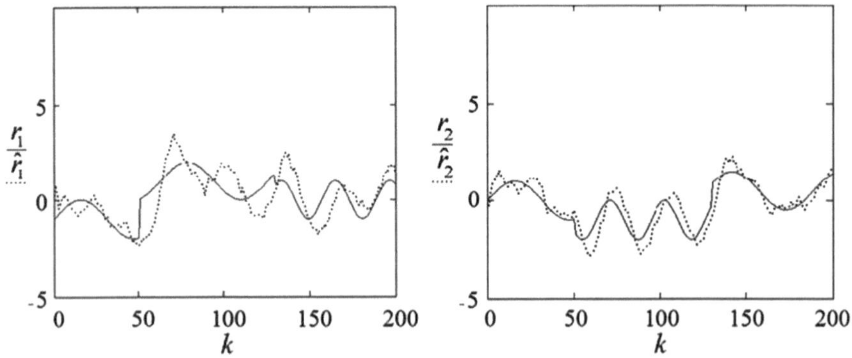

Fig. 2 The estimation of unknown inputs by nonparametric algorithm (3)–(6), (18)

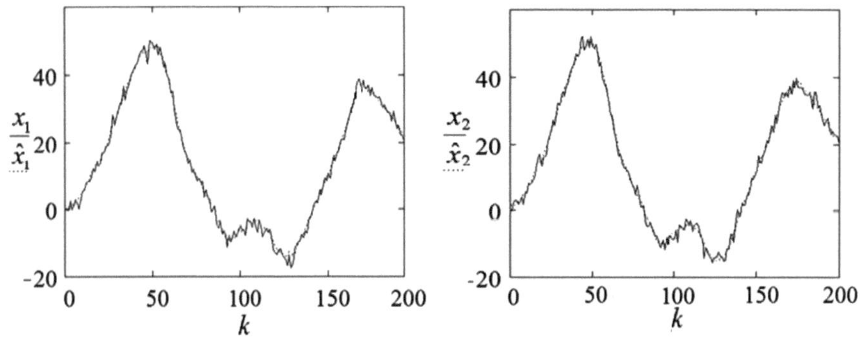

Fig. 3 The dependence of the components x_1, x_2 and their filtering estimates using the LMS estimates from [3, 4]

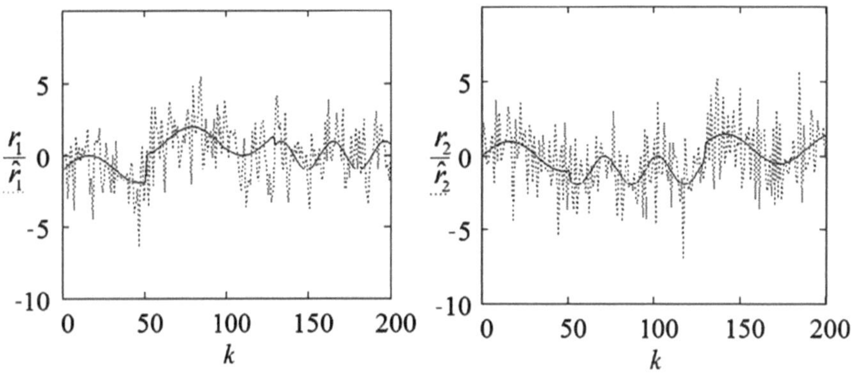

Fig. 4 The estimation of unknown inputs by the LMS estimates

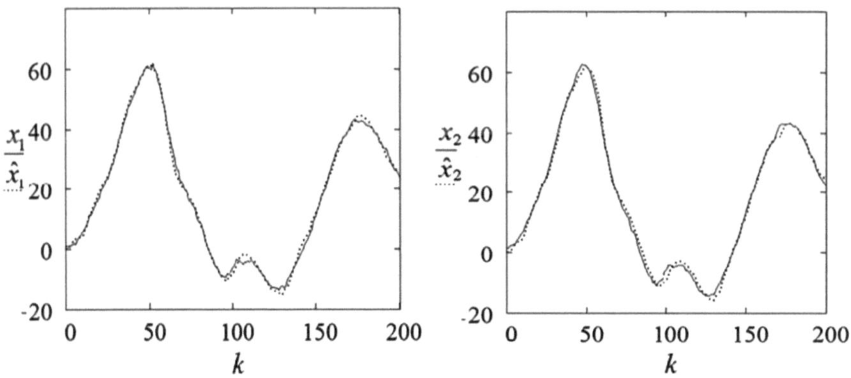

Fig. 5 The dependence of the components and their extrapolation estimates combined with non-parametric algorithm (18), (20)–(22)

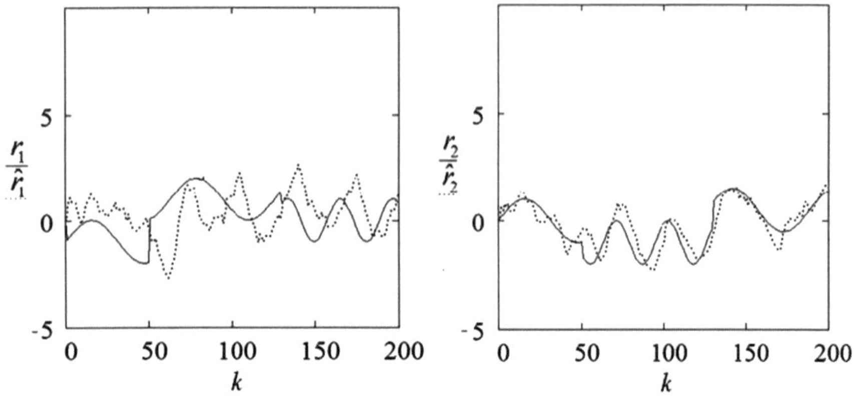

Fig. 6 The estimation of unknown inputs by nonparametric algorithm (18), (20)–(22)

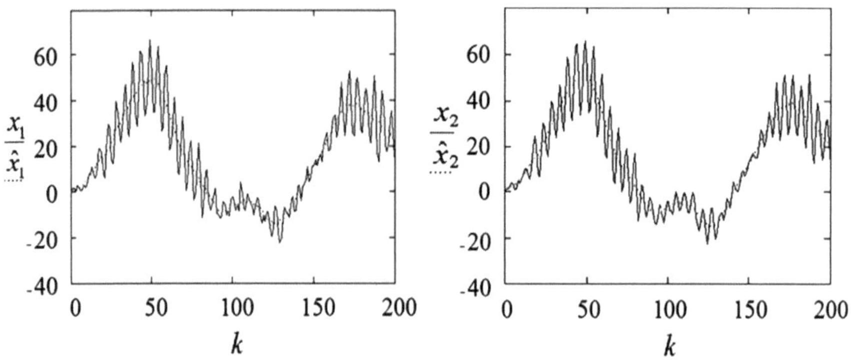

Fig. 7 The dependence of the components and their extrapolation estimates with using the LMS estimates

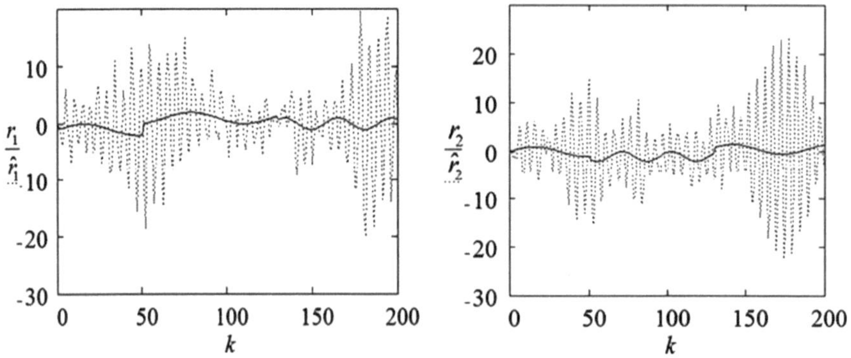

Fig. 8 The estimate extrapolation of unknown inputs by the LMS estimates

Table 1 Errors for filtering algorithm combined with nonparametric estimates

$\sigma_{x,1}$	$\sigma_{x,2}$	$\sigma_{r,1}$	$\sigma_{r,2}$
1.198	1.269	1.112	0.535

Table 2 Errors for filtering algorithm using the LMS estimates

$\sigma_{x,1}$	$\sigma_{x,2}$	$\sigma_{r,1}$	$\sigma_{r,2}$
1.324	1.466	2.657	2.621

Table 3 Errors for prediction algorithm combined with nonparametric estimates

$\sigma_{x,1}$	$\sigma_{x,2}$	$\sigma_{r,1}$	$\sigma_{r,2}$
1.343	1.503	1.398	0.532

Table 4 Errors for prediction algorithm using the LMS estimates

$\sigma_{x,1}$	$\sigma_{x,2}$	$\sigma_{r,1}$	$\sigma_{r,2}$
5.139	4.892	6.936	6.921

6 Conclusion

The paper deals with the algorithms of the Kalman filtering and prediction for systems with unknown input. The proposed method has been verified by simulations. Figures and Tables show that the procedures with nonparametric estimates have the advantages in accuracy in comparison with the known algorithms using the LMS estimates. It is seen that the presented nonparametric technique may be used in solving the general filtering and prediction problems with unknown input.

Acknowledgments Work supported by Russian Foundation for Basic Research (projects 13-08-00744, 13-08-01015), Program for Improving the Competitiveness of TSU (Tomsk State University) among the Worlds Leading Scientific, and Laboratory of Geological Informatics of Computer Science Department of TSU.

References

1. Kalman, R.E., Busy, R.: A new results in linear filtering and prediction theory. Trans. ASME J. Basic Engr. **83**, 95–108 (1961)
2. Astrom, K., Eykhoff, P.: System identification—a survey. Automatica **7**, 123–162 (1971)
3. Anderson, B.D.O., Moore, J.B.: Optimal Filtering. Prentice-Hall (1979)
4. Kitanidis, P.K.: Unbiased minimum-variance linear state estimation. Automatica **23**(6), 775–778 (1987)
5. Janczak, D., Grishin, Y.: State estimation of linear dynamic system with unknown input and uncertain observation using dynamic programming. Control Cybern. **35**(4), 851–862 (2006)
6. Gillijns, S., Moor, B.: Unbiased minimum-variance input and state estimation for linear discrete-time systems. Automatica **43**, 111–116 (2007)
7. Hsien, C.-S.: On the optimal of two-stage Kalman filter for systems whis unknown input. Asian J. Control **12**(4), 510–523 (2010)
8. Darouach, M., Zasadzinski, M., Xu, S.J.: Full-order observers for linear systems with unknown inputs. IEEE Trans. Automat. Control **39**, 606–609 (1994)

9. Hou, M., Patton, R.: Optimal filtering for systems with unknown inputs. IEEE Trans. Automat. Control **43**, 445–449 (1998)
10. Witczak, M.: Fault diagnosis and fault-tolerant control strategies for non-linear systems. Chapter 2. Unknown input observers and filters. Lecture Notes in Electrical Engineering, pp. 19–56, Springer International Publishing. Switzerland (2014)
11. Smagin, S.V.: Filtering in linear discrete systems with unknown perturbartion. Optoelectron. Instrum. Data Process. **45**(6), 513–519 (2009)
12. Smagin, V.I.: State estimation for linear discrete systems with unknown input using compensations. Russ. Phys. J. **57**(5), 682–690 (2014)
13. Koshkin, G.M., Smagin, V.I.: Filtering and prediction for discrete systems with unknown input using nonparametric algorithms. In: Proceedings of 10th International Conference on Digital Technologies, pp. 120–124. Zilina, Slovakia, 9–11 July 2014
14. Dobrovidov, A., Koshkin, G., Vasiliev, V.: Non-parametric state space models. Heber, UT 84032, USA. Kendrick Press Inc. (2012)
15. Leung, D.: Cross-validation in nonparametric regression with outliers. Ann. Stat. **33**, 2291–2310 (2005)
16. Kitaeva, A.V., Koshkin, G.M.: Recurrent nonparametric estimation of functions from functionals of multidimensional density and their derivatives. Autom. Remote Control **70**(3), 389–407 (2009)

Regularization of Positive Signal Nonparametric Filtering in Multiplicative Observation Model

Alexander V. Dobrovidov

Abstract A solution to the problem of useful random signal extraction from a mixture with a noise in the multiplicative observation model is proposed. Unlike conventional filtering tasks, in the problem under consideration it is supposed that the distribution (and the model) of the useful signal is unknown. Therefore, in this case one cannot apply such well-known techniques like Kalman filter or posterior Stratonovich-Kushner evolution equation. The new paper is a continuation and development of the author's article, reported at the First ISNPS Conference (Halkidiki'2012), where the filtering problem of positive signal with the unknown distribution had been solved using the generalized filtering equation and nonparametric kernel techniques. In the present study, new findings are added concerning the construction of stable procedures for filtering, the search for optimal smoothing parameter in the multidimensional case and some of the convergence results of the proposed techniques. The main feature of the problem is the positive distribution support. In this case, the classical methods of nonparametric estimation with symmetric kernels are not applicable because of large estimator bias at the support boundary. To overcome this drawback, we use asymmetric gamma kernel functions. To have stable estimators, we propose a regularization procedure with a data-driven optimal regularization parameter. Similar filtering algorithms can be used, for instance, in the problems of volatility estimation in statistical models of financial and actuarial mathematics.

Keywords Multiplicative observation model · Nonparametric filtering · Incomplete statistical information

A.V. Dobrovidov (✉)
V.A. Trapeznikov Institute of Control Sciences of Russian Academy of Sciences, Moscow 117997, Russia
e-mail: dobrovidov@gmail.com

R. Cao et al. (eds.), *Nonparametric Statistics*, Springer Proceedings in Mathematics & Statistics 175, DOI 10.1007/978-3-319-41582-6_7

1 Introduction

Consider a stationary two-component hidden Markov process $(X_n, S_n)_{n\geqslant 1}$ where (X_n) is an observable component and (S_n) is an unobservable one. The problem of filtering the random signal $(S_n)_{n\geqslant 1}$ or known one-to-one function $\vartheta_n = Q(S_n)$ from the mixture (observation model)

$$X_n = \varphi(S_n, \eta_n), \tag{1}$$

with the noise η_n, where $\varphi(\cdot)$ is a given function, is solved under a *nonparametric uncertainty* about the distribution of the useful signal S_n. Note that the notion "unobservable signal" presupposes that signal realizations are never observed without a noise and, therefore, it is impossible to collect statistics about it and to solve the problem of identification, i.e., construction of mathematical model of the useful signal. Without knowledge of characteristics or the state equation of the unobserved useful signal S_n one cannot take advantage of the rich baggage of optimal filtering techniques such as, for instance, the Kalman's filter or Stratonovich-Kushner evolution equation. In many practically important problems of underwater acoustics, radar, financial and actuarial mathematics accurate models of useful signals are unknown. Therefore, the approach to solving the problems of signal processing with unknown distributions is relevant.

In general, when the characteristics of the unobservable signal S_n are unknown, it is impossible to build an optimal filtering procedure. Let a conditional density of observations $g(x_n|s_n)$ under fixed signal $S_n = s_n$ belong to the exponential family [1, 2] (univariate case)

$$g(x_n|s_n) = \widetilde{C}(s_n)h(x_n)\exp\{T(x_n)Q(s_n)\}, \quad x_n \in \mathbb{R}, s_n \in \mathbb{R}, \tag{2}$$

where $h(x_n), \theta_n = Q(s_n), T(x_n)$ are given Borel functions and $\widetilde{C}(s_n)$ is a normalizing factor. Then the mean-square Bayes estimator $\hat{\theta}_n^{opt} = \mathsf{E}(\vartheta_n|x_1^n) = \mathsf{E}_{\mathsf{x}}(\vartheta_n) = \mathsf{E}(Q(S_n)|x_1^n)$ satisfies the generalized optimal filtering equation [1]

$$T'(x_n)\hat{\theta}_n^{opt} = \frac{\partial}{\partial x_n}\left(\ln \frac{f_c(x_n|x_1^{n-1})}{h(x_n)}\right), \tag{3}$$

where $T' = dT/dx_n$, $f_c(\cdot|\cdot)$ is a conditional density of observation x_n given fixed observations $x_1^{n-1} = (x_1, \ldots, x_{n-1})$. To our best knowledge, this equation was firstly obtained in [3]. The main feature of the equation is that it depends explicitly on probabilistic characteristics of the observable process (X_n) only [1]. Such property of the estimator equation is the embodiment of the well-known empirical Bayesian approach of G. Robbins to the problems of signal processing. Empirical Bayesian approach was used in some papers [4, 5] to find a Bayesian estimators of the unknown constant parameters of probability densities from the exponential family in the case of i.i.d. observations. For the time being, the author has not met any papers of other researches concerning the application of empirical Bayesian approach to signal processing where observations have been statistically dependent.

If the compound process $(X_n, S_n)_{n\geqslant 1}$ is Gaussian, $Q(S_n) = S_n$ and, consequently, state and observation equations are linear then the Kalman filter follows from the generalized Eq. (3) [6].

If the observation Eq. (1) is nonlinear multiplicative

$$X_n = S_n\eta_n, \quad S_n, X_n, \eta_n \geqslant 0, \tag{4}$$

where η_n is the noise distributed as χ^2-similar density

$$p_{\eta_n}(y) = C(k,\sigma)\cdot y^{k/2-1}\exp\left(-\frac{ky}{2\sigma^2}\right), \quad C(k,\sigma) = \left(\frac{k}{2\sigma^2}\right)^{k/2}\Gamma^{-1}(k/2),$$

with k degree of freedom and known parameter σ^2, then the conditional density $g(x_n|s_n)$, corresponding to the Eq. (4), has the form (2), where $T(x_n) = -kx_n/(2\sigma^2)$, $h(x_n) = x_n^{k/2-1}$ and $Q(s_n) = 1/s_n$. Then the Eq. (3) becomes an equation with respect to the optimal estimator $\hat{\theta}_n^{opt}$ of the form [7]

$$\hat{\theta}_n^{opt} = \frac{\sigma^2(k-2)}{kx_n} - \frac{2\sigma^2}{k}\frac{\partial}{\partial x_n}\ln f_c(x_n|x_1^{n-1}). \tag{5}$$

Here s_n is a natural parameter of the exponent family (2) and θ_n is a canonical parameter, linearly entering under exponent of (2). Generalized Eq. (3) is written with respect to estimator $\hat{\theta}_n$ of canonical parameter θ_n. Since $\theta_n = Q(s_n)$ and Q is one-to-one function by condition, i.e., $s_n = Q^{-1}(\theta_n)$, then it is natural to set $\hat{s}_n = Q^{-1}(\hat{\theta}_n)$. This means that $\hat{S}_n$ is also optimal estimator but only with the risk function $\mathsf{E}(Q(S_n) - Q(\hat{S}_n))^2$. One can see that the Eq. (5) does not depend explicitly on functionals of distribution of signal (S_n).

In the Eq. (5) the conditional density of observations $f_c(x_n|x_1^{n-1})$ appears via logarithmic density derivative

$$\frac{\partial}{\partial x_n}\ln f_c(x_n|x_1^{n-1}) = \frac{\partial/\partial x_n f_c(x_n|x_1^{n-1})}{f_c(x_n|x_1^{n-1})}. \tag{6}$$

We will estimate it using the nonparametric kernel techniques adapted to statistically dependent sequences with the strong mixing property (α-mixing). This assumption allows one to replace the expression (6) by closely related expression

$$\frac{\partial}{\partial x_n}\ln \bar{f}_c(x_n|x_{n-\tau}^{n-1}) = \frac{\partial/\partial x_n f(x_{n-\tau}^n)}{f(x_{n-\tau}^n)} \doteq \psi(x_{n-\tau}^n) \tag{7}$$

with truncated marginal density $f(\cdot)$ where $\tau \in \mathbb{N}$ is a *dependence zone*. One of the ways to evaluate τ by observations X_1^n is given in the book [1] (p. 402–416). Thus to solve the Eq. (5) our goal is to restore $(\tau+1)$-dimensional density $f(\cdot)$ and its partial derivative by weakly dependent vector observations $\mathbf{x}_i = x_{n-\tau-i}^{n-i}$, $i=\overline{1,N}$, $N = n-\tau+1$, $\tau+1=d$.

Since by condition, the functional form of the density f is unknown, then to construct the density estimator $\hat{f}$ the method of nonparametric kernel estimation is chosen. In contrast to the classical method of Rosenblatt-Parzen with symmetric kernel functions, in this paper we develop an approach for estimating densities and their derivatives with nonnegative support $[0,\infty)$. Such supports appear in nonlinear signal processing problems, an example of which is the multiplicative model (4).

From (7), it follows that for evaluation of the density logarithm it is necessary to estimate the density and its derivative separately. Obtained estimator in the form of the relationship is unstable in some points x when the denominator in (7) is close to zero. Therefore we propose some regularized procedure with a regularization parameter which has to be found by sample. The feature of the approach consists in application of multidimensional asymmetric gamma kernel functions for evaluation unknown densities and its derivatives by dependent observations and investigation of its asymptotical properties.

Section 2 discusses the asymptotic properties of multivariate density and its derivative estimators including bandwidth selection. Calculation of the data-based regularization parameter is fulfilled in Sect. 3. Next Sect. 4 is devoted to numerical comparison of the nonparametric filter and the optimal nonlinear Stratonovich's filter.

2 Multivariate Gamma Kernel Estimator

Let $\mathbf{X}_1,\ldots,\mathbf{X}_N$ be a sample of α-mixing random d-vectors with unknown probability density function (pdf) $f(\mathbf{x})$, $\mathbf{x}\in\mathbb{R}^{+d}$. To estimate $f(\mathbf{x})$ we use a product density estimator [8]

$$\hat{f}(x_1,\ldots,x_d) = \frac{1}{N}\sum_{i=1}^{N}\prod_{s=1}^{d} K^s_{\rho_{b(x_s)},b}(X_{is}), \tag{8}$$

where K^s is a gamma kernel for a variable x_s, and X_{is} is a sth element of the vector sample $\mathbf{X}_i$ at the moment i. Gamma kernel has the form [9]

$$K_{\rho_b(x),b}(t) = \frac{t^{\rho_b(x)-1}\exp(-t/b)}{b^{\rho_b(x)}\Gamma(\rho_b(x))}.$$

Here b is a smoothing parameter (bandwidth) with a property $b\to 0$ as $N\to 0$, $\Gamma(\cdot)$ is a standard gamma function and

$$\rho_b(x) = \begin{cases} \rho_1(x) = \frac{x}{b}, & \text{if} \quad x\geq 2b,\\ \rho_2(x) = \left(\frac{x}{2b}\right)^2+1, & \text{if} \quad x\in[0,2b) \end{cases}$$

is a shape parameter. Since the process (X_n) is strictly stationary, we assume that the bandwidths b_s for each component x_s of the vector $\mathbf{x}$ coincide and are equal to b.

Since the logarithmic derivative (7) contains the partial derivative $f'_{x_n}(x^n_{n-\tau}) = \partial f(x^n_{n-\tau})/\partial x_n$ at the point x_n, then it is necessary to construct a nonparametric estimator of the partial derivative of a density on the support $\mathbb{R}^{+d}$, $d = \tau + 1$. By differentiating (8) in x_d, we have

$$\hat{f'}_{x_d}(\mathbf{x}) = \hat{f'}_{x_d}(x_1^d) = N^{-1}\sum_{i=1}^{N}\tilde{\tilde{K}}(\mathbf{x},\mathbf{X}_i) = N^{-1}\sum_{i=1}^{N} b_d^{-1}L(x_d, X_{id})\prod_{s=1}^{d}K^s_{\rho_{b(x_s)},b}(X_{is}), \quad (9)$$

where $\hat{f'}_{x_d}(\mathbf{x}) = \dfrac{\partial}{\partial x_d}\hat{f}(\mathbf{x})$, $\tilde{\tilde{K}}(\mathbf{x},\mathbf{X}_i) = \dfrac{\partial}{\partial x_d}\tilde{K}(\mathbf{x},\mathbf{X}_i) = b_d^{-1}L(x_d, X_{id})\tilde{K}(\mathbf{x},\mathbf{X}_i) = b_d^{-1}L(x_d, X_{id})\prod_{s=1}^{d}K^s_{\rho_{b(x_s)},b}(X_{is})$, $L(x_d, X_{id}) = \ln(X_{id}) - \ln b_d - \Psi(x_d/b_d)$ and $\Psi(\cdot)$ is a Digamma function (derivative from gamma function logarithm).

Here and after we will install all propositions and proofs for the case when $x \geqslant 2b$ because for the selected criterion in the form of mean integrated square error (MISE)

$$MISE(\hat{f}^{(r)}(\mathbf{x})) = \int_0^{\infty}\mathsf{E}(\hat{f}^{(r)}(\mathbf{x}) - f^{(r)}(\mathbf{x}))^2 d\mathbf{x} = \int_0^{2b} + \int_{2b}^{\infty}, \quad r = 0, 1, \quad (10)$$

$$f^{(0)}(\cdot) = f(\cdot), \quad f^{(1)}(\cdot) = f'(\cdot)$$

the integral over $[0, 2b)$ converges to zero when $b \to 0$ (see [8]).

2.1 Properties of the Gamma Kernel Density Estimator by Dependent Data

Convergence and convergence rate at a point $\mathbf{x}$ of the multivariate pdf estimator (8) are evaluated with the help of $MSE(\hat{f}(\mathbf{x})) = \mathsf{E}(f(\mathbf{x}) - \hat{f}(\mathbf{x}))^2$. By definition,

$$MSE(\hat{f}(\mathbf{x})) = Bias(\hat{f}(\mathbf{x}))^2 + Var(\hat{f}(\mathbf{x})), \quad (11)$$

where $Bias(\hat{f}(\mathbf{x})) = \mathsf{E}(\hat{f}(\mathbf{x})) - f(\mathbf{x})$ is the same for dependent and independent random variables.

Lemma 2.1.1 (Bias of $\hat{f}(\mathbf{x})$) *If the pdf $f(\mathbf{x})$ is twice continuously differentiable then for $x_j \geqslant 2b$, $j = \overline{1, d}$ and $b \to 0$*

$$Bias(\hat{f}(\mathbf{x})) = \frac{b}{2}\sum_{j=1}^{d} x_j \frac{\partial^2 f(\mathbf{x})}{\partial x_j^2} + d\,o(b). \quad (12)$$

The proof can be found in [8].

For dependent random vectors the variance of the sum (8) consists of two terms

$$Var(\hat{f}(\mathbf{x})) = Var(N^{-1}\sum_{i=1}^{N}\prod_{j=1}^{d} K(\mathbf{X}_i)) = N^{-1}var(K(\mathbf{X}_1))$$

$$+2N^{-1}\sum_{i=1}^{N-1}\left(1-\frac{1}{N}\right) Cov\,(K(\mathbf{X}_1), K(\mathbf{X}_{1+i})) = V + C, \tag{13}$$

where V is a kernel variance and C is a kernel covariance.

Lemma 2.1.2 (Variance of $\hat{f}(\mathbf{x})$) *If* $b \to 0$ *and* $Nb^{d/2} \to \infty$ *when* $N \to \infty$ *then*

$$V = O\left(N^{-1}b^{-d/2}\right) \to 0.$$

Conditions of convergence in mean-square sense of the density estimator (8) one can find in [10].

Lemma 2.1.3 (Covariance of $\hat{f}(\mathbf{x})$) *1) If* $b \to 0, Nb^{d(\upsilon+1)/2} \to \infty$ *as* $N \to \infty, f(\mathbf{x}_1, \mathbf{x}_k) < M < \infty\ \forall k, \int_1^\infty \alpha(\tau)^\upsilon d\tau < \infty, 0 < \upsilon < 1$, *where* $\alpha(\tau)$ *is a coefficient of strong mixing then*

$$C = O\left(N^{-1}b^{-d(\upsilon+1)/2}\right).$$

2) If in addition $b \to 0$, $Nb^{d/2} \to \infty$, $\int_1^\infty \alpha(\tau)^{\upsilon/2} d\tau < \infty$, *then*

$$C = o\left(N^{-1}b^{-d/2}\right).$$

From Lemma 2.1.3 one can see that under the condition $\int_1^\infty \alpha(\tau)^{\upsilon/2} d\tau < \infty$ the convergence rate of the estimator covariance is higher than the similar rate for the estimator variance. Therefore, in the calculation of the asymptotic behavior of *MSE* we can neglect the contribution of covariance compared with the contribution of variance. Then $MSE(\hat{f}(\mathbf{x}))$ for the weakly dependent and independent observations will be the same as $N \to \infty$.

Theorem 2.1.1 (MSE of $\hat{f}(\mathbf{x})$) *If* $b \to 0$, $Nb^{d/2} \to \infty$ *as* $N \to \infty$, *and* $\int_1^\infty \alpha(\tau)^{\upsilon/2}\, d\tau < \infty, 0 < \upsilon < 1$ *then*

$$MSE(\hat{f}(\mathbf{x})) = O\left(N^{-1}b^{-d/2}\right). \tag{14}$$

The results of Lemmas 2.1.1, 2.1.2 and Theorem 2.1.1 coincide with the corresponding results of [8] in the case of independent observations. They are listed here in order to understand the conditions under which the properties of the estimators for independent and weakly dependent observations coincide asymptotically. The proof of Lemma 2.1.3 is given in Appendix.

2.2 Properties of the Gamma Kernel Density Derivative Estimator by Dependent Data

For the bias of estimator (9) the following Lemma holds.

Lemma 2.2.1 (Bias of $\hat{f'}_{x_d}(\mathbf{x})$) *Let partial derivative $\hat{f'}_{x_d}(\mathbf{x})$ of multivariate density be twice continuously differentiable. Then for $x \geqslant 2b$, $b_j = b\ \forall j, j = \overline{1,d}$, and $b \to 0$*

$$Bias(\hat{f}'_{x_d}(\mathbf{x})) = bB_1(\mathbf{x}) + b^2 B_2(\mathbf{x}) + o(b),\ B_1(\mathbf{x}) = \frac{f(\mathbf{x})}{12x_d^2} + \frac{1}{4x_d}\sum_{j=1}^{d} x_j \frac{\partial^2 f(\mathbf{x})}{\partial x_j^2},$$

$$B_2(\mathbf{x}) = \frac{1}{24x_d^2}\sum_{j=1}^{d} x_j \frac{\partial^2 f(\mathbf{x})}{\partial x_j^2}.$$

For the density derivative estimator (9) the same formulae as (13) hold

$$MSE(\hat{f'}_{x_d}(\mathbf{x})) = Bias(\hat{f'}_{x_d}(\mathbf{x}))^2 + Var(\hat{f'}_{x_d}(\mathbf{x})), \tag{15}$$

where $Var(\hat{f'}_{x_d}(\mathbf{x})) = N^{-1}Var(\tilde{\tilde{K}}(\mathbf{x}, \mathbf{X}_i)) + 2N^{-1}\sum_{i=1}^{N-1}\left(1 - \frac{1}{N}\right) Cov\left(\tilde{\tilde{K}}(\mathbf{x}, \mathbf{X}_1), \tilde{\tilde{K}}(\mathbf{x}, \mathbf{X}_{1+i})\right) = V' + C'$.

Lemma 2.2.2 (Variance of $\hat{f'}_{x_d}(\mathbf{x})$) *Under the conditions of Lemma 2.2.1 and $b \to 0$ as $Nb^{(d+2)/2} \to \infty$,*

$$V' = O\left(N^{-1}b^{-(d+2)/2}\right).$$

Lemma 2.2.3 (Covariance of $\hat{f'}_{x_d}(\mathbf{x})$) *If $b \to 0$, $Nb^{d(\upsilon+1)/2} \to \infty$, $f'(\mathbf{x}_1, \mathbf{x}_k) << M < \infty\ \forall k$, $\int_1^\infty \alpha(\tau)^\upsilon d\tau < \infty$, $0 < \upsilon < 2/d$, then*

$$C' = O\left(N^{-1}b^{-d(\upsilon+1)/2}\right).$$

From Lemma 2.2.3 it follows that provided the condition $\int_1^\infty \alpha(\tau)^{\upsilon/2}\, d\tau < \infty$, $0 < \upsilon < 2/d$, the rate of convergence of the estimator covariance is higher than the rate of convergence of the estimator variance.

The last three Lemmas define the convergence rate of *MSE* of partial derivative of multivariate density [10].

Theorem 2.2.1 (MSE of $\hat{f}'(\mathbf{x})$) *If* $b \to 0$, $Nb^{(d+2)/2} \to \infty$ *when* $N \to \infty$, *and* $\int_1^\infty \alpha(\tau)^{\upsilon/2} d\tau < \infty, 0 \leqslant \upsilon \leq 2/d$, *then*

$$MSE(\hat{f}'(\mathbf{x})) = O\left(N^{-1}b^{-(d+2)/2}\right) \to 0. \tag{16}$$

The proofs of these assertions one can find in [10, 11].

2.3 Optimal Bandwidths for Multivariate Density and Its Partial Derivative

It is well-known that convergence rates of estimators (8) and (9) are strongly dependent on a selection of bandwidths. Therefore, to apply nonparametric estimators in practice one is trying to find the optimal bandwidth b which is usually determined by minimizing the *MISE* (10) or its asymptotic equivalent as $b \to 0$.

Theorem 2.3.1 (Optimal bandwidth for density estimator) *Under the conditions of Lemmas 2.1.1, 2.1.2 and 2.1.3, the optimal bandwidth minimizing asymptotic* $MISE(\hat{f}(\mathbf{x}))$ *equals*

$$b_{opt} = \left(\frac{d\int V(\mathbf{x})d\mathbf{x}}{\int U^2(\mathbf{x})d\mathbf{x}}\right)^{2/(d+4)} N^{-2/(d+4)}. \tag{17}$$

Here

$$V(\mathbf{x}) = \frac{f(\mathbf{x})}{\left(2\sqrt{\pi}\right)^d}\prod_{j=1}^{d} x_j^{-1/2}, \quad U(\mathbf{x}) = \sum_{j=1}^{d} x_j \frac{\partial^2 f(\mathbf{x})}{\partial x_j^2}. \tag{18}$$

Substitution (17) *in* (10) *leads to the optimal convergence rate of* $MISE(\hat{f}(\mathbf{x})) \sim N^{-4/(d+4)}$.

This result coincides with the result of [8], where a sequence of only independent observations were considered.

Theorem 2.3.2 (Optimal bandwidth for density derivative estimator) *Under the conditions of Lemmas 2.2.1, 2.2.2, 2.2.3, the optimal bandwidth minimizing asymptotic* $MISE(\hat{f}'(\mathbf{x}))$ (16) *equals*

$$b'_{opt} = \left(\frac{(d+2)\int_0^\infty V_1(\mathbf{x})d\mathbf{x}}{4\int_0^\infty U_1^2(\mathbf{x})d\mathbf{x}}\right)^{2/(d+6)} N^{-2/(d+6)}, \tag{19}$$

where

$$V_1(\mathbf{x}) = \frac{f(\mathbf{x})}{2x_d\left(2\sqrt{\pi}\right)^d}\prod_{j=1}^{d} x_j^{-1/2}, \quad U_1(\mathbf{x}) = \frac{f(\mathbf{x})}{12x_d^2} + \frac{1}{4x_d^2}U(\mathbf{x}). \tag{20}$$

Substitution of (19) *in* (10) *leads to the optimal convergence rate of* $MISE(\hat{f}'(\mathbf{x})) \sim N^{-4/(d+6)}$.

Proofs of Theorem 2.3.2 are mostly borrowed from [10].

From formulas (17) and (19) one can see that the optimum bandwidths depend on the unknown density and its derivatives. Therefore, it is impossible to calculate their exact values. To obtain the numerical values of the parameters b_{opt} and b'_{opt} it is necessary to evaluate the integrals $v = \int V(\mathbf{x})d\mathbf{x}$, $u = \int U^2(\mathbf{x})d\mathbf{x}$ $v_1 = \int V_1(\mathbf{x})d\mathbf{x}$, $u_1 = \int U_1^2(\mathbf{x})d\mathbf{x}$ in (17) and (19) by observations which are drawn from the unknown density $f(\mathbf{x})$. This can be done using the method of cross-validation [12]. For this we have to substitute instead of the unknown density f (x) its nonparametric estimator (8), which once again at the second stage will depend on other bandwidth b_g. This procedure can be continued, but, as is noted in [13], the effect of each subsequent step onto the initial evaluation procedure is rapidly waning. Therefore, we restrict ourselves to estimating the bandwidth parameter b_g in the second stage. This procedure is discussed in the next subsection.

Then, taking into account (18) and (19), for estimates $\hat{v}$, $\hat{u}$ of integrals v, u we have the following expressions:

$$\hat{v} = \frac{1}{\left(2\sqrt{\pi}\right)^d}\frac{1}{N}\sum_{k=1}^{N}\prod_{j=1}^{d} X_{kj}^{-1/2}, \quad \hat{u} = \frac{1}{N}\sum_{k=1}^{N}\left(\hat{U}(\mathbf{X}_k)\right)^2 / \hat{f}(\mathbf{X}_k), \tag{21}$$

$$\hat{U}(\mathbf{X}_k) = \sum_{\substack{i=1\\ i\neq k}}^{N}\sum_{j=1}^{d} X_{kj}\left[L^2(X_{kj}, X_{ij}) - \Psi\left(1, \frac{X_{kj}}{\hat{b}_g}\right)\right]\prod_{s=1}^{d} K^s_{\rho_{\hat{b}(X_{ks}),\hat{b}}}(X_{is}), \tag{22}$$

$$\hat{f}(\mathbf{X}_k) = \frac{1}{N-1}\sum_{j\neq k}^{N}\tilde{K}(\mathbf{X}_k, \mathbf{X}_j), \tag{23}$$

where $\Psi(1, \cdot)$ is trigamma function. Using cross-validation (23), for the estimators $\hat{v}_1$, $\hat{u}_1$ of integrals v_1, u_1 one can obtain the following expressions:

$$\hat{v}_1 = \frac{1}{2\left(2\sqrt{\pi}\right)^d}\frac{1}{N}\sum_{k=1}^{N}\frac{1}{X_{kd}}\prod_{j=1}^{d} X_{kj}^{-1/2}, \quad \hat{u} = \frac{1}{N}\sum_{k=1}^{N}\left(\hat{U}_1(\mathbf{X}_k)\right)^2 / \hat{f}(\mathbf{X}_k), \tag{24}$$

$$\hat{U}_1(\mathbf{X}_k) = \frac{1}{12X_{kd}^2} + \frac{1}{X_{kd}^2}\hat{U}(\mathbf{X}_k). \tag{25}$$

Thus, the formulae (21)–(25) allow us to calculate the bandwidth parameter at a finite sample length N.

2.4 Gamma Rule of Thumb

In the case of gamma kernel estimator, the univariate bandwidth b_g at the second stage is again calculated by the formula (17), where instead of the unknown $f(\mathbf{x})$ is substituted the density of multivariate gamma distribution $Gamma(\mathbf{x}, \tilde{b}) = \prod_{j=1}^{d} Gamma(x_j, \tilde{b})$ with the parameter $x_1 = x_2 = \cdots = x$. This distribution is called a reference distribution and has the same support and the first two moments with the true distribution. Under these conditions, the expression for $b_g = b_g(d)$ takes the form

$$b_g(d) = A(d)\big(H(d)/B(d)\big)^{2/(d+4)}, \tag{26}$$

$$A(d) = (0.25641)^{2d/(d+4)}256^{1/(d+4)}, \quad H(d) = \pi^{d/2}\tilde{b}^{(d-4)/2}$$

$$B(d) = (2x - \tilde{b})\big(4x^3 + 59x^2\tilde{b} + 92x\tilde{b}^2 + 16\tilde{b}^3(1-d) + d(2x^3 - 17x^2\tilde{b} + 40x\tilde{b}^2)\big).$$

Unknown parameters $x, \tilde{b}$ of univariate gamma distribution are replaced by sampled value $x = \bar{m}$, $\bar{\tilde{b}} = \bar{D}/\bar{m}$ according to the method of moments, where $\bar{m}$ and $\bar{D}$ are the sample mean and the sample variance of observations.

3 Asymptotically Optimal Regularization Parameter and Its Estimation

3.1 The Equation for the Asymptotically Optimal Regularization Parameter

Now we can write the nonparametric counterpart of the optimal Eq. (5) taking into account the expression (7):

$${}^{\tau}\hat{\theta}_n = \frac{\sigma^2(k-2)}{kx_n} - \frac{2\sigma^2}{k}\,{}^{\tau}\hat{\psi}(x_{n-\tau}^n), \tag{27}$$

where ${}^{\tau}\hat{\psi}(x_{n-\tau}^n) = \hat{f'}_{x_n}(x_{n-\tau}^n)/\hat{f}(x_{n-\tau}^n)$ is a nonparametric estimator of logarithmic density derivative. This statistics is unstable when denominator is near zero. Therefore, we introduce a regularization procedure for obtaining the stable estimator of the form

$$\widetilde{\psi}(x_{n-\tau}^{n}) \doteq \widetilde{\psi}_{n}(x_{n-\tau}^{n};\delta_{n}) = \frac{{}^{\tau}\hat{\psi}_{n}(x_{n-\tau}^{n})}{1+\delta_{n}^{\tau}\hat{\psi}_{n}(x_{n-\tau}^{n})^{2}}, \tag{28}$$

where the regularization parameter $\delta_n \to 0$ has to be evaluated for a finite n.

Estimator $\hat{\psi}(x_{n-\tau}^{n})$ of logarithmic derivative (7) is a special case of the general substitution estimator of complex functions $G(t_n(\mathbf{x}))$, where $\mathbf{x} \in \mathbb{R}^d, t_n : \mathbb{R}^d \to \mathbb{R}^m$, $G : \mathbb{R}^m \to \mathbb{R}$. In this case $m = 2$, $t_n = (t_{n1}, t_{n2})$, $t_{n1}(x) = f_n(x)$, $t_{n2} = f_n'(x)$ and $G(t_n) = t_{n2}/t_{n1}$. If the sequence (t_n) converges to t with $n \to \infty$, then under certain regularity conditions, $G(t_n)$ tends to $G(t)$ in some probabilistic sense. The most important of these conditions are as follows:

(1) the existence and boundedness of a number of derivatives of $G(t)$;
(2) the sequence $(|G(t_n)|)$ is dominated by a numerical sequence $\left(C_0 d_n^{\gamma}\right)$, where C_0 is a constant; a sequence of numbers $d_n \to \infty$, as $n \to \infty$ and $0 \leqslant \gamma \leqslant \infty$.

The meaning of the second condition is that the sequence $(G(t_n))$ with $n \to \infty$ can not grow faster than a power function of n. This condition, along with other less important conditions (see. Corollary 1.8.1. in [1]) provides the mean-square convergence $G(t_n) \to G(t)$.

If $\mathsf{E} \parallel t_n - t \parallel \to 0$ then by Theorem 1.9.1 in [1]

$$|\mathsf{E}[G(t_n) - G(t)]^2 - \mathsf{E}[\nabla_t G(t)(t_n - t)^{\mathrm{T}}]^2| \to 0, \quad n \to \infty, \tag{29}$$

i.e. mean-square convergence of complex functions $G(t_n)$ to $G(t)$ is replaced by the mean-square convergence of simpler statistics t_n to t.

There are a number of examples where the conditions (1) and (2) are not satisfied. For the logarithmic derivative $t = (t_1, t_2)^{\mathrm{T}}, G(t) = t_2/t_1$, where $t_1 = f(x)$ and $t_2 = f'(x)$. For example, for a Gaussian density $f(x)$, we have $G = -x$. This function is unbounded on $\mathbb{R}$ and it results that Eq. (28) is not applicable. So, instead of $G(t_n)$ it is proposed to use regularization procedure (called in [1] as *a piecewise-smooth approximation*)

$$\widetilde{G}(t_n, \delta_n) = \frac{G(t_n)}{1+\delta_n|G(t_n)|^2}, \tag{30}$$

where $\delta_n > 0$ is a regularization parameter tending to zero as $n \to \infty$. As shown in [1], the function $\widetilde{G}(t_n, \delta_n)$ satisfies the conditions (1) and (2) and therefore it is dominated by a power function of n. Due to this property, the result of (29) can be used to obtain the mean-square convergence

$$\lim_{n\to\infty} \mathsf{E}(\widetilde{G}(t_n, \delta_n) - G(t))^2 = 0 \tag{31}$$

when $\mathsf{E} \parallel t_n - t \parallel \to 0$ and $\delta_n \to 0$.

Since the statistics $\hat{\psi}_n(\mathbf{x})$ in expression (7) for the logarithmic derivative is unstable when the denominator is close to zero, we will use the regularization procedure to obtain stable estimates of the form

$$\widetilde{\psi}(\mathbf{x}) \doteq \widetilde{\psi}_n(\mathbf{x};\delta_n) = \frac{\hat{\psi}_n(\mathbf{x})}{1+\delta_n(\hat{\psi}_n(\mathbf{x}))^2}, \quad \mathbf{x} = x_1^d, \tag{32}$$

where the regularization parameter δ_n have to be found. For a fixed sample size, N the optimal value of this parameter may be found by minimizing the standard deviation of $\widetilde{\psi}(\mathbf{x})$ from $\psi(\mathbf{x})$ at each point $\mathbf{x}$. However, such criterion is not very convenient in practice because of a long enough numerical minimization procedure must be repeated for each time step. The alternative to it serves an integral criterion,[1]

$$J = \int_0^\infty \mathsf{E}(\widetilde{\psi} - \psi)^2 \omega(x)dx = \int_0^\infty \mathsf{E}\left(\frac{\hat{\psi}}{1+\delta(\hat{\psi})^2} - \psi\right)^2 \omega(\mathbf{x})d\mathbf{x}, \tag{33}$$

where $\omega(\mathbf{x})$ is some weighting function. For the existence of this criterion, the weighting function can be selected, for example, in the form of $\omega(\mathbf{x}) = f(\mathbf{x})$. Differentiation of the (33) in δ leads to the equation

$$\int_0^\infty \mathsf{E}\frac{\widehat{\psi}^2(\hat{\psi})^2}{(1+\delta(\hat{\psi})^2)^3}\omega dx = \int_0^\infty \psi \mathsf{E}\frac{(\hat{\psi})^2}{(1+\delta(\hat{\psi})^2)^2}\omega d\mathbf{x}, \tag{34}$$

which cannot be solved analytically with respect to δ. Then, taking into account that δ is infinitesimal, we expand the $\widetilde{\psi}$ in (28) in a series from δ up to the first order including

$$\widetilde{\psi} = \frac{\hat{\psi}}{1+\delta(\hat{\psi})^2} = \hat{\psi} - \delta\hat{\psi}(\hat{\psi})^2 + o(\delta). \tag{35}$$

Thus, an approximate criterion will be represented as follows

$$J_1 = \int_0^\infty \mathsf{E}(\hat{\psi}(1-\delta(\hat{\psi})^2) - \psi)^2\,\omega(\mathbf{x})d\mathbf{x}. \tag{36}$$

Minimizing it in δ leads to the equation for the asymptotically optimal regularization parameter

$$\delta_{opt} = \frac{\int_0^\infty \mathsf{E}(\hat{\psi})^4\,\omega d\mathbf{x} - \int_0^\infty \psi \mathsf{E}(\hat{\psi})^3\,\omega d\mathbf{x}}{\int_0^\infty \mathsf{E}(\hat{\psi})^6\,\omega d\mathbf{x}}. \tag{37}$$

---

[1] For brevity, in the future the arguments of some functions are omitted.

Thereby, the problem reduces to finding $\mathsf{E}\hat{\psi}^k$ at $k = 3, 4, 6$. To calculate the mathematical expectations in (37) the formula (29) is used twice: once for the function $G_1 = t^k$ and the second for the function $G_2(t) = t_2/t_1$, where $t_1 = f$, $t_2 = f'$. Thus, for the first time

$$\mathsf{E}\hat{\psi}^k \approx \psi^k + k\psi^{k-1}[\mathsf{E}\hat{\psi} - \psi]. \tag{38}$$

For the second time

$$\mathsf{E}\hat{\psi} - \psi = \left(\frac{\partial G_2}{\partial t_1}\ \frac{\partial G_2}{\partial t_2}\right)\begin{pmatrix}\mathsf{E}\hat{t}_1 - t_1\\ \mathsf{E}\hat{t}_2 - t_2\end{pmatrix} = \left(-\frac{t_2}{t_1^2}\ \frac{1}{t_1}\right)\begin{pmatrix}\mathsf{E}\hat{t}_1 - t_1\\ \mathsf{E}\hat{t}_2 - t_2\end{pmatrix}$$

$$= -\frac{t_2}{t_1^2}(\mathsf{E}\hat{t}_1 - t_1) + \frac{1}{t_1}(\mathsf{E}\hat{t}_2 - t_2) = -\frac{f'}{f^2}Bias(\hat{f}) + \frac{1}{f}Bias(\hat{f}')$$

$$= (b/2)\frac{\upsilon(\mathbf{x})}{f(\mathbf{x})}, \quad \upsilon(\mathbf{x}) = \frac{f}{6x_n^2} + \left(\frac{1}{2x_n} - \psi(\mathbf{x})\right)u(\mathbf{x}), \quad u(\mathbf{x}) = \sum_{j=1}^{d} x_j \frac{\partial^2 f(\mathbf{x})}{\partial x_j^2}.$$

Substituting (38) in (37), we get

$$\mathsf{E}\hat{\psi}^k = \psi^k + \frac{1}{2}k\psi^{k-1}\frac{b}{2f}\upsilon(\mathbf{x}) + o(b). \tag{39}$$

3.2 *Cross-Validation Estimate*

Returning to the formula (37), calculate its numerator using (39):

$$\int_0^\infty \mathsf{E}(\hat{\psi})^4\,\omega d\mathbf{x} - \int_0^\infty \psi\mathsf{E}(\hat{\psi})^3\,\omega d\mathbf{x} \approx \frac{b}{4}\int_0^\infty \psi^3(\mathbf{x})\frac{\upsilon(\mathbf{x})}{f(\mathbf{x})}\omega(\mathbf{x})d\mathbf{x}. \tag{40}$$

Denominator of (37) is

$$\int_0^\infty \mathsf{E}\hat{\psi}^6\,\omega dx \approx \int_0^\infty \left(\psi^6 + \frac{5b}{4}\psi^4\frac{\upsilon(\mathbf{x})}{f(\mathbf{x})}\right)\omega(\mathbf{x})d\mathbf{x}.$$

Under unknown density f, these integrals can be evaluated from a sample using the method of cross-validation. For example, if $\omega(\mathbf{x}) = f(\mathbf{x})$ the numerator of (40) can be represented as

$$\frac{b}{4}\int_0^\infty \psi^3(\mathbf{x})\frac{\upsilon(\mathbf{x})}{f(\mathbf{x})}\omega(\mathbf{x})d\mathbf{x} \approx \frac{\hat{b}}{4}\sum_{k=1}^{N}\hat{\psi}^3(\mathbf{X}_k)\frac{\hat{\upsilon}(\mathbf{X}_k)}{\hat{f}(\mathbf{X}_k)},$$

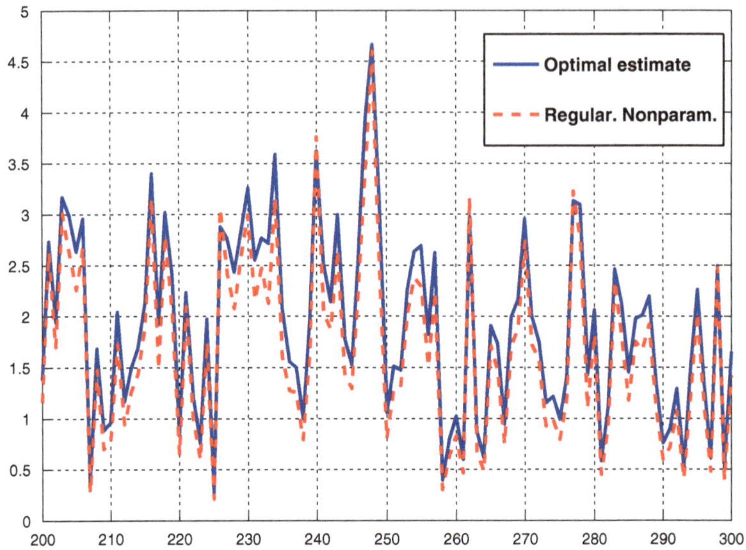

Fig. 1 Optimal and regularized nonparametric filtering

where

$$\hat{\psi}(\mathbf{X}_k) = \frac{\hat{f}'(\mathbf{X}_k)}{\hat{f}(\mathbf{X}_k)}, \quad \hat{\upsilon}(\mathbf{X}_k) = \frac{\hat{f}(\mathbf{X}_k)}{6x_n^2} + \left(\frac{1}{2x_n} - \hat{\psi}(\mathbf{X}_k)\right)\hat{U}(\mathbf{X}_k),$$

$$\hat{f}(\mathbf{X}_k) = \frac{1}{N-1}\sum_{j\neq k}^{N} \tilde{K}(\mathbf{X}_k, \mathbf{X}_j), \quad \hat{f}'(\mathbf{X}_k) = \frac{1}{N-1}\sum_{j\neq k}^{N} \tilde{\tilde{K}}(\mathbf{X}_k, \mathbf{X}_j),$$

and $\hat{U}(\mathbf{X}_k)$ is determined by (22). Here, as at the second step in the evaluation of density and its derivative, the bandwidth is calculated by the gamma rule of thumb, cited above. Evaluation of the denominator (37) is performed similarly.

Now everything is ready to build the unsupervised algorithm of nonlinear nonparametric filtering of the nonnegative desired signal.

The main steps of the algorithm

1. Selection of the conditional density from the exponent family, which is adequate to observation model.
2. Derivation of the generalized equation of optimal filtering.
3. Generation of observations based on the observation model.
4. The data-based bandwidth calculation.
5. Calculation of the regularization parameter.
6. Nonparametric filtering of nonnegative random signal.

4 Numerical Comparison of Filters

In Fig. 1, two filtering estimators of the desired signal S_n in the multiplicative model (4) were represented: optimal mean-square estimator $\hat{s}_n^{opt}$ and adaptive nonparametric estimator ${}^{\tau}\hat{s}_n = 1/{}^{\tau}\hat{\theta}_n$ (5) obtained under unknown distribution of the signal S_n.

It is easy to notice the closeness of filtering estimators with full (solid line) and incomplete (dashed line) statistical information.

5 Conclusion

This paper presents a nonparametric unsupervised filter to extract the useful nonnegative signal with unknown distribution from a mixture with a noise. The solution is based on the generalized filtering equation and nonparametric estimators with asymmetric kernel functions. The convergence properties of multivariate kernel estimators on $\mathbb{R}^{+d}$-support are studied. Numerical simulation shows that quality of the unsupervised filter is close to the Stratonovitch's optimal nonlinear filter constructed from complete statistical information. It is of interest to develop a proposed nonparametric estimation methods for signals with limited support in the form of a bounded bar $\mathbb{B}^d = \{a_j \leqslant x_j \leqslant b_j, j = 1, \ldots, d\}$.

Appendix

Proof of Lemma 2.1.3. We give here only outline of the proof. Provided the condition (1), from formula (13) it follows that for a sequence of statistically dependent random variables the variance of their sum is expressed through the covariance

$$C = 2N^{-1} \sum_{i=1}^{N-1} \left(1 - \frac{1}{N}\right) Cov\left(\tilde{K}(\mathbf{x}, \mathbf{X}_1), \tilde{K}(\mathbf{x}, \mathbf{X}_{1+i})\right). \tag{41}$$

This covariance is estimated from above by using the Davydov's inequality [14]

$$|Cov\left(\tilde{K}(\mathbf{x}, \mathbf{X}_1), \tilde{K}(\mathbf{x}, \mathbf{X}_{1+i})\right)| \leq 2\pi\alpha(i)^{1/r} \parallel \tilde{K}(\mathbf{x}, \mathbf{X}_1) \parallel_q \parallel \tilde{K}(\mathbf{x}, \mathbf{X}_{1+i}) \parallel_p,$$

where $p^{-1} + q^{-1} + r^{-1} = 1$, $\alpha(i)$ is a strong mixing coefficient of the process (X_n), and $\| \cdot \|_q$ is a norm in the space L_q. This norm can be estimated by the following expression

$$\| \tilde{K}(\mathbf{x}, \mathbf{X}_1) \|_q = \left(\int \left(\prod_{j=1}^{d} K_{\rho_1(x_j),b_j}(t_j) \right)^q f(t_1^d) dt_1 \ldots dt_d \right)^{1/q}$$
$$= \left(\mathsf{E} \left(\prod_{j=1}^{d} K^{q-1}_{\rho_1(x_j),b_j}(\xi_j) f(\xi_1^d) \right) \right)^{1/q}, \tag{42}$$

where a kernel $\prod_{j=1}^{d} K_{\rho_1(x_j),b_j}(\xi_j)$ is used as density function and random variables ξ_j is distributed like $Gamma(\rho_1(x_j), b_j)$-distribution with expectation $\mu_j = x_j$ and variance $\sigma_\xi^2 = x_j b_j$. Expectation in parentheses of (42) is calculated by expanding the function $f(\xi_1^d)$ in a Taylor series at the point $\mu = \mu_1^d$. After some algebra we get

$$|C| = |C(\hat{f}(\mathbf{x})| \leqslant \frac{D(\upsilon, \mathbf{x})}{N} b^{-d\frac{1+\upsilon}{2}} \int_1^\infty \alpha(\tau)^\upsilon d\tau, \quad 0 < \upsilon < 1, \quad D(\upsilon, \mathbf{x}) < \infty.$$

Provided the condition (2), we use the technique of the proof from [15]. To do this, we divide (41) into two terms $|C| = (2/N)\sum_{i=1}^{N-1}(\cdot) = (2/N)\big(\sum_{i=1}^{c(N)}(\cdot) + \sum_{i=c(N)+1}^{N-1}(\cdot)\big) = I_1 + I_2$ and estimate each at $N \to \infty$. As a result we get

$$I_1 = O\left(\frac{c(N)b^{d/2}}{nb^{d/2}}\right), \quad I_2 = O\left(\frac{1}{nb^{d/2}c(N)b^{d\upsilon/2}}\right).$$

It remains to find $c(N)$ satisfying conditions $c(N)b^{d/2} \to 0$ and $c(N)b^{d\upsilon/2} \to \infty$. Let $c(N) = b^{-\varepsilon}$. Then, these conditions are met simultaneously if $\frac{d}{2} > \varepsilon > \frac{d}{2}\upsilon > 0, \quad 0 < \upsilon < 1$, and $|C| = o\left(Nb^{d/2}\right)^{-1}$. □

Proof of Lemma 2.2.3. Proof of this Lemma is carried out, in principle, in the same way as Lemma 2.1.3, but it occupies a lot of space because of a complex estimator expression containing a special functions (see [10]). □

References

1. Dobrovidov Alexander V., Koshkin Gennady M., Vasiliev Vyacheslav A.: Non-Parametric State Spase Models. Kendrick Press, USA (2012)
2. Lehmann, E.L.: Testing Statistical Hypotheses. Wiley, N.Y. (1959)
3. Dobrovidov, A.V.: Nonparametric methods of nonlinear filtering of stationary random sequences. Automat. Remote Control **44**(6), 757–768 (1983)
4. Pensky, M.: A general approach to nonparametric empirical Bayes estimation. Statistics **29**, 61–80 (1997)
5. Pensky, M., Singh, R.S.: Empirical Bayes estimation of reliability characteristics for an exponential family. Can. J. Stat. **27**, 127–136 (1999)
6. Markovich, L.A.: The equation of optimal filtering, Kalman's filter and theorem on normal correlation. In: Proceedings of the 11th International Vilnius Conference on Probability and Mathematical Statistics. Vilnius, Lithuania, 182, 30 June–4 July 2014. ISBN: 978-609-433-220-3

7. Dobrovidov, A.V.: Stable nonparametric signal filtration in nonlinear models. In: Topics in Nonparametric Statistics: Proceedings of the First Conference of the International Society for Nonparametric Statistics, vol. XVI, pp. 61–74. Springer, New York (2014)
8. Taoufik, B., Rambouts, J.: Nonparametric density estimation for multivariate bounded data. J. Stat. Plann. Infer. **140**(1), 139–152 (2007)
9. Chen, S.X.: Probability density function estimation using gamma kernels. Ann. Inst. Statist. Math. **52**(3), 471–480 (2000)
10. Markovich, L.A.: Gamma kernel estimation of multivariate density and its derivative on the nonnegative semi-axis by dependent data (2015). arXiv:1410.2507v2
11. Dobrovidov, A.V., Markovich, L.A.: Nonparametric gamma kernel estimators of density derivatives on positive semi-axis. In: Proceedings of IFAC MIM 2013, pp. 1–6. Petersburg, Russia, 19–21 June 2013
12. Hall, P., Marron, J.S., Park, B.U.: Smoothed cross-validation. Probab. Theory Relat. Fields **92**, 1–20 (1992)
13. Devroye, L., Gyorfi, L.: Nonparametric Density Estimation. The L1 View. John Wiley, New York, N.Y. (1985)
14. Davydov, Y.A.: On Convergence of distributions induced by stationary random processes. Probab. Theory Appl. V. **XIII**(4), 730–737 (1968)
15. Masry, E.: Probability density estimation from sampled data. IEEE Trans. Inf. Theory. V. **IT-29**(5), 696–709 (1983)

Nonparametric Estimation of Heavy-Tailed Density by the Discrepancy Method

Natalia Markovich

Abstract The nonparametric estimation of the probability density function (pdf) requires smoothing parameters like bandwidths of kernel estimates. We consider the so-called discrepancy method proposed in [13, 14, 21] as a data-driven smoothing tool and alternative to cross-validation. It is based on the von Mises-Smirnov's (M-S) and the Kolmogorov–Smirnov's (K-S) nonparametric statistics as measures in the space of distribution functions (cdfs). The unknown smoothing parameter is found as a solution of the discrepancy equation. On its left-hand side stands the measure between the empirical distribution function and the nonparametric estimate of the cdf. The latter is obtained as a corresponding integral of the pdf estimator. The right-hand side is equal to a quantile of the asymptotic distribution of the M-S or K-S statistic. The discrepancy method considered earlier for light-tailed pdfs is investigated now for heavy-tailed pdfs.

Keywords Heavy-tailed density · Kernel estimator · Bandwidth · Discrepancy method

1 Introduction

Let $X^n = \{X_1, X_2, \ldots, X_n\}$ be a sample of independent identically distributed (iid) random variables (r.v.s) with marginal cumulative distribution function (cdf) $F(x)$ and probability density function (pdf) $f(x)$. Estimation of the pdf is one of the basic concepts which appears in different contexts of applied statistics. Most of the known estimators are oriented on light-tailed distributions. Those may include distributions located on bounded and unbounded intervals. Among these estimators are a histogram, projection, and kernel estimators.

We focus on nonparametric estimation of heavy-tailed pdfs. The estimation of the latter requires special procedures due to their specific features. The latter are characterized by slower decay to zero of heavy tails at infinity than that of an exponen-

N. Markovich (✉)
V.A. Trapeznikov Institute of Control Sciences of Russian Academy of Sciences, Moscow 117997, Russia
e-mail: nat.markovich@gmail.com

R. Cao et al. (eds.), *Nonparametric Statistics*, Springer Proceedings in Mathematics & Statistics 175, DOI 10.1007/978-3-319-41582-6_8

tial rate, the lack of some or all moments of the distribution, and sparse observations at the tail domain of the distribution.

Nonparametric estimators require an appropriate selection of tuning (smoothing) parameters, e.g., a bandwidth of the kernel estimators and the bin width of the histogram. The bandwidth selection impacts on the accuracy of the pdf estimates more than the form of the kernel function [19]. The exception are gamma kernels which are free from boundary bias for positively defined pdfs, [6]. A comprehensive survey of methods to compute the smoothing parameter is given in [22]. A cross-validation is the most popular method among data-driven tools [20]. However, for heavy-tailed pdfs the estimates with the bandwidth h selected by cross-validation do not converge in the space L_1, since $h \to \infty$ as $n \to \infty$, [7]. Such methods like plug-in and rule of thumb are based on the minimization of the mean integrated squared error of the pdf estimate, where some pilot estimates of the pdf and its derivatives as well as a specification of bandwidths for pilot estimates are required, [10, 20]. The popular plug-in method is proposed in [18]. A direct formula for the plug-in bandwidth calculation that requires pilot second and third pdf derivatives can be found in [9]. Due to the arbitrary specification of the pilot estimators the plug-in methods tend to larger bandwidths than the cross-validation and to over-smoothing of the pdf estimates. Particularly, the standard normal pdf is usually used as a pilot pdf that may not be appropriate for the estimation of heavy-tailed pdfs with infinite second moments. The rule of thumb with gamma reference function for positively defined pdfs and for univariate iid and dependent data is proposed in [8, 11], respectively, and for multivariate dependent data in [12].

In [5] a minimum distance (MD) approach for parametric pdf estimation based on the minimizing of the distance between a pilot nonparametric density estimator and the assumed parametric pdf model, namely, $\widehat{\theta}_n = \arg\min D(\widehat{f_h}, f_\theta)$ is proposed. Here, D can be selected as a metric L_1, L_2 or L_∞. The problem of such MD-estimators arises from the choice of h.

We consider the discrepancy method of the bandwidth selection that determines an approach different from the mentioned methods. It is based on the solution of a specific discrepancy equation. On its left-hand side stands a measure between the empirical distribution function and the nonparametric estimate of the cdf. The latter is obtained as a corresponding integral of the pdf estimator. Well-known nonparametric statistics like the von Mises-Smirnov's (M-S) and the Kolmogorov–Smirnov's (K-S) statistics are used as such measures in the space of cdfs. The right-hand side of the discrepancy equation is equal to a quantile of the asymptotic distribution of the M-S or K-S statistics.

It is remarkable that the asymptotic distributions of the M-S and the K-S statistics are invariant regarding the cdf, i.e., they are the same for any cdf. However, if one substitutes the unknown cdf by its estimate, such invariancy is violated. One may use the asymptotic distributions of the M-S and K-S statistics due to an expected consistency of the kernel estimates. The discrepancy method is general. It can be applied to any pdf estimator, but not necessarily to kernel estimators.

The discrepancy method demonstrates better results for non-smooth distributions at a compact support like a triangular and a uniform one than cross-validation. The discrepancy method is free from drawbacks of cross-validation like falling into local extremes and of plug-in methods like the specification of pilot estimators of the pdf derivatives. In [21] it was proved that a projection estimator with a regularization parameter selected by a ω^2-discrepancy method based on the M-S statistic has the rate of convergence in L_2 not worse than $n^{-(k+1/2)/(2k+3)}$, which is close to the best rate $n^{-(k+1/2)/(2k+2)}$ for pdfs with a bounded variation of the kth derivative. In [14], Sect. 4.8.1 it is derived that the mean-squared error of a variable bandwidth kernel estimator [1] may attain the fastest achievable order $n^{-8/9}$ if the bandwidth h is selected by the discrepancy method D based on the K-S statistic and the pdf has four continuous derivatives.

In this paper, we will not construct another refinement to existing bandwidth selectors. We aim to extend the discrepancy method to heavy-tailed pdfs. As it was shown in [14] the discrepancy method may have no solutions for the heavy-tailed distributions. To overcome this problem we proposed in [15] a modified discrepancy method based only on the k largest order statistics of the sample X^n. In this paper, we investigate the impact of the selection of k on the heavy-tailed pdf estimation by a simulation study.

The paper is organized as follows. In Sect. 2 the discrepancy method and its modification for heavy-tailed distributions are described. In Sect. 3 the simulation study of the kernel estimates with the bandwidth selected by the discrepancy method is presented. The exposition is finalized by some conclusions.

2 Discrepancy Method and Heavy-Tailed Densities

The discrepancy ω^2 and D methods were proposed and investigated in [13, 14, 21] to estimate smoothing parameters (bandwidths h) of nonparametric pdf estimators $\widehat{f}_h(x)$, e.g., kernel and projection estimators by samples. The discrepancy methods are based on nonparametric statistics like Kolmogorov–Smirnov D_n and von Mises-Smirnov ω_n^2. The idea of the method is to solve the discrepancy equation

$$\rho(\widehat{F}_h(x), F_n(x)) = \delta \tag{1}$$

regarding the bandwidth h, where δ is the unknown uncertainty, $\rho(\cdot,\cdot)$ is a metric in the space of cdfs, $\widehat{F}_h(x) = \int_{-\infty}^{x} \widehat{f}_h(t)dt$ is the estimate of the df $F(x)$, and $F_n(x) = 1/n \sum_{i=1}^{n} \Theta(x - X_i)$ is the empirical distribution function. Here, $\Theta(x) = 1$ if $x \ge 0$ and $\Theta(x) = 0$ if $x < 0$. The statistics ω_n^2 and D_n are used as metric $\rho(\cdot,\cdot)$. One may use other nonparametric statistics, e.g., Anderson–Darling or Rényi [2, 16] as $\rho(\cdot,\cdot)$, since the Eq. (1) is general.

The ω_n^2 is determined by

$$\omega_n^2 = n \int_{-\infty}^{\infty} (F_n(x) - F(x))^2 \, dF(x)$$

or equivalently by

$$\omega_n^2 = n \int_0^1 (F_n(t) - t)^2 \, dt$$

for a transformed sample $t_i = F(X_i), i = 1, \ldots, n$. The D_n is determined by

$$\sqrt{n} D_n = \sqrt{n} \sup_{-\infty < x < \infty} |F(x) - F_n(x)|.$$

Limit distributions of both statistics are invariant regarding the cdf $F(x)$, [2].

For practice one can use the simple expressions

$$\hat{\omega}_n^2(h) = \sum_{i=1}^{n} \left(\widehat{F}_h(X_{(i)}) - \frac{i - 0.5}{n} \right)^2 + \frac{1}{12n}, \tag{2}$$

and

$$\sqrt{n} \hat{D}_n(h) = \sqrt{n} \max(\hat{D}_n^+, \hat{D}_n^-),$$

$$\hat{D}_n^+ = \max_{1 \le i \le n} \left(\frac{i}{n} - \widehat{F}_h(X_{(i)}) \right), \qquad \hat{D}_n^- = \max_{1 \le i \le n} \left(\widehat{F}_h(X_{(i)}) - \frac{i-1}{n} \right),$$

calculated by the order statistics $X_{(1)} \le \cdots \le X_{(n)}$ corresponding to the sample X^n. Using the tables of statistics ω_n^2 and D_n [2] it was found in [13] that quantiles 0.05 and 0.5 correspond to pdf modes of the statistics ω_n^2 and D_n, respectively. Since such quantiles provide the most likely values of the corresponding statistics, it was proposed to use them as the discrepancy values δ. Hence, one may find the bandwidth h as solutions of the equations

$$\hat{\omega}_n^2(h) = 0.05 \tag{3}$$

and

$$\sqrt{n} \hat{D}_n(h) = 0.5. \tag{4}$$

In our discrepancy method, we have to substitute the unknown $F(x)$ by its estimate $\widehat{F}_h(x)$. Relying on consistent estimates of $\widehat{F}_h(x)$ one can use the quantiles of the asymptotic distributions of statistics ω_n^2 and D_n as δ.

For heavy-tailed distributions (e.g., Cauchy, Pareto, Weibull with shape parameter less than 1) it was noted in [14] that the discrepancy Eqs. (3) and (4), where $\widehat{F}_h(x)$ and $F_n(x)$ are calculated by the entire sample X^n, may have no solutions, i.e., the statistics $\hat{\omega}_n^2(h)$, $\sqrt{n}\hat{D}_n(h)$ may never reach values 0.05 and 0.5 for any h, or the solution is provided by too small values of h. Such h may be not satisfactory to fit the pdf at the tail, where over-smoothed kernels are required.

In [15] it was observed that this can be overcome by means of using the largest order statistics $X_{(n-k+1)} \le \cdots \le X_{(n)}$ instead of the entire sample, namely,

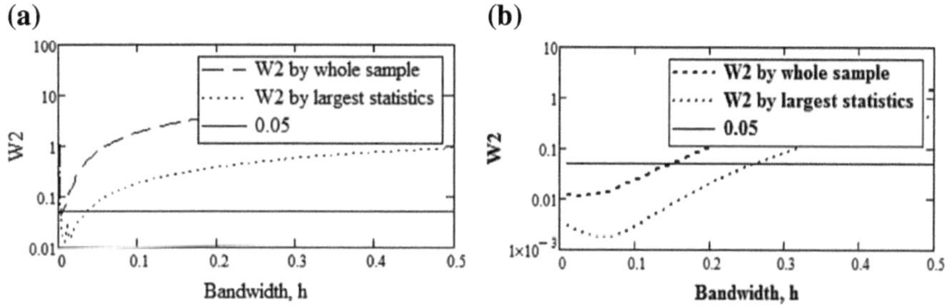

Fig. 1 The $\widehat{\omega^2}_n$ statistic based on the entire sample $X_{(1)}, \ldots, X_{(n)}$ of the size $n = 100$ and on the largest order statistics $X_{(n-k)}, \ldots, X_{(n)}$, $k = 20$, simulated from the heavy- and light-tailed Weibull distributions with shape parameter equal to 0.5 (**a**) and 1.5 (**b**), respectively, against h. For the light-tailed Weibull distribution the quantile 0.05 is reached for larger h than for the heavy-tailed case

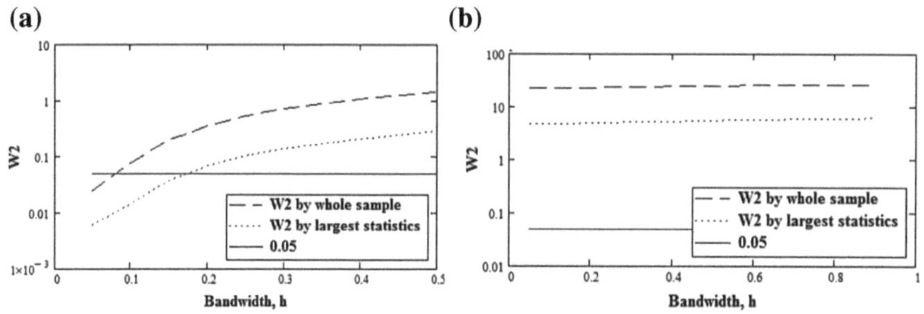

Fig. 2 The $\widehat{\omega^2}_n$ statistic based on the entire sample $X_{(1)}, \ldots, X_{(n)}$ of the size $n = 100$ and on the largest order statistics $X_{(n-k)}, \ldots, X_{(n)}$, $k = 20$, simulated from a Pareto distribution with $\gamma = 0.8$ (Fig. 2a) and a standard Cauchy distribution (Fig. 2b) against h

$$\hat{\omega}_n^2(h) = \sum_{i=n-k+1}^{n} \left(\widehat{F}_h(X_{(i)}) - \frac{i-0.5}{n} \right)^2 + \frac{1}{12n}, \tag{5}$$

and

$$\sqrt{n}\hat{D}_n^+ = \sqrt{n} \max_{n-k+1 \le i \le n} \left(\frac{i}{n} - \widehat{F}_h(X_{(i)}) \right), \tag{6}$$

$$\sqrt{n}\hat{D}_n^- = \sqrt{n} \max_{n-k+1 \le i \le n} \left(\widehat{F}_h(X_{(i)}) - \frac{i-1}{n} \right)$$

in (3) and (4). This approach is natural since for heavy-tailed distributions only largest order statistics are significant.[1] Traditional methods use the entire samples.

In Figs. 1 and 2 the effect of excluding the lowest order statistics in the discrepancy equation is demonstrated, where the kernel estimator

[1] For example, the Hill's estimator of the tail index is based on the k largest statistics.

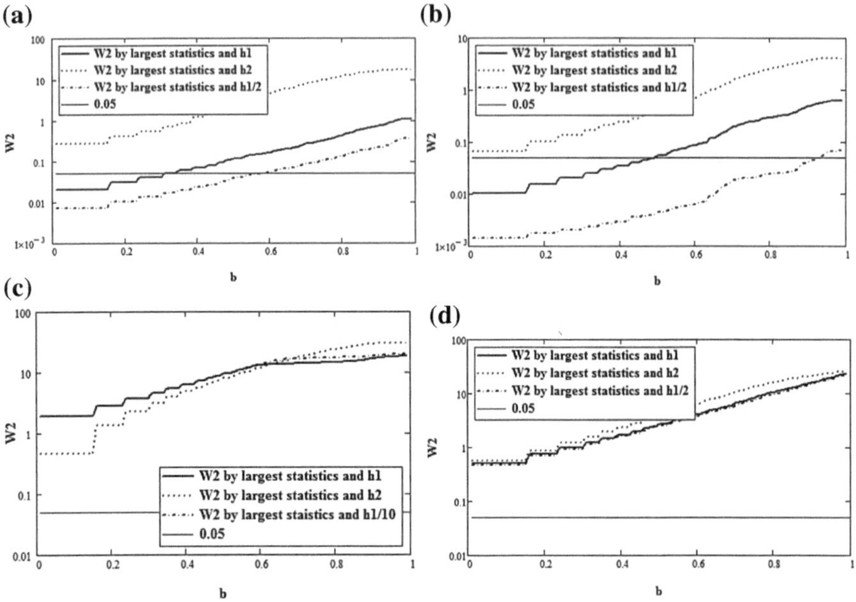

Fig. 3 The ω_n^2 statistic based on the largest order statistics $X_{(n-k)}, \ldots, X_{(n)}$ simulated from the Pareto distribution with $\gamma = 0.8$ (Fig. 3a), from the standard lognormal distribution (Fig. 3b), from the lognormal distribution (11) with parameters $\mu = 0$, $\sigma = 10$ (Fig. 3c) and from a standard Cauchy distribution (Fig. 3d) against $b \in [0, 1]$ such that $k = n^b$, where $h_1 = 0.398$ and $h_2 \in \{5.577, 1.006, 3.708 \times 10^9, 2.604\}$ in Fig. 3a–d, respectively

$$\widehat{f}_h(x) = 1/(nh) \sum_{i=1}^{n} K\left(\frac{x - X_i}{h}\right) \tag{7}$$

with the Gaussian kernel

$$K(x) = (1/\sqrt{2\pi}) \exp(-x^2/2) \tag{8}$$

is used. The discrepancy equation has no solution for the standard Cauchy distribution.

One can get similar examples for the D-method.

In case, that only the $k = [n^b]$,[2] $0 \le b \le 1$ largest order statistics are used, the solution of the discrepancy equation may not exist for sufficiently large values of h, see Fig. 3. The bandwidth h is calculated as $h_1 = n^{-1/5}$, $h_1/2$ and

$$h_2 = 1.06\widehat{\sigma} n^{-1/5}, \tag{9}$$

[2] $[x]$ denotes the integer part of x.

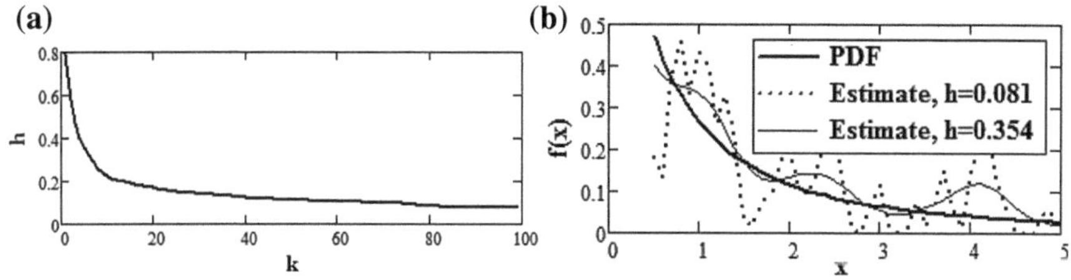

Fig. 4 The bandwidths h obtained by the ω^2 method against the number of the largest order statistics k for the Pareto distribution with $\gamma = 0.8$ (Fig. 4a) and the kernel estimate (7) with the Gaussian kernel (8), where $h \in \{0.081, 0.354\}$ are obtained using the ω^2 method by the entire sample and by only $k = 5$ largest order statistics, respectively (Fig. 4b)

where $\widehat{\sigma}$ is an empirical standard deviation.[3] The case $b = 1$ implies that the $\hat{\omega}_n^2(h)$-statistic is calculated by the entire sample.

The heavier the tail, the smaller k is required to have a solution of the discrepancy equation for the same h. One can see this both for standard lognormal and Pareto distributions that have lighter and heavier tails, respectively, when the same bandwidth $h_1/2$ was taken for the kernel estimate, Fig. 3a, b. Remarkable is that the Eq. (3) has no solution for any b and h both for the lognormal distribution (Fig. 3c) with a heavier tail than the standard lognormal one and for the Cauchy distribution (Fig. 3d). The latter distribution belongs to the class of super heavy-tailed distributions arising rarely in practice. One can transform a super heavy-tailed distributed sample to ease the heaviness of the tail and to get a heavy-tailed distribution that is easier to estimate.

In Fig. 4a it is shown that an increasing the k does not impact significantly on value of h. The smaller k corresponds to the larger h providing the better estimation of the pdf, Fig. 4b.

3 A Simulation Study of Heavy-Tailed Density Estimation

For our simulation study we select the kernel estimator (7) with the Gaussian kernel (8). Let the underlying samples have sizes $n \in \{100, 500\}$. We generate $B = 500$ samples of each size n. We evaluate the accuracy of kernel estimate $\widehat{f_h}(x)$ at an interval $[a, X_{max}], a > 0, X_{max} = \max\{X_1, \ldots, X_n\}$ in order to separate the estimate from zero where the underlying pdfs may have infinite values. The a is some constant that may be larger than a location parameter. Our simulation is not affected by the choice of a since the accuracy of all examined bandwidth selectors is calculated the same way. The bandwidth h is selected by the ω^2 and D methods.

[3]The rule-of-thumb selected h_2 is recommended in ([20], p. 45) as an optimal value for the Gaussian kernel. This method is however very sensitive to outliers due to possibly large σ.

Table 1 The bandwidth selection for the standard lognormal distribution with parameters $\mu = 0$, $\sigma = 1$ at $[1, X_{max}]$

	D-method				$\overline{\rho}_1$	σ_1^2	$\overline{\rho}_2$	σ_2^2
$n = 100$								
k	$\overline{\rho}_1$	σ_1^2	$\overline{\rho}_2$	σ_2^2	KS method			
n^1	0.00259	$5e-06$	0.0904	0.0011	0.00178	$5e-06$	0.0745	0.0014
$[n^{0.8}]$	0.00132	$3e-06$	0.0552	0.0017	Rule of thumb (9)			
$[n^{0.6}]$	0.00249	$3e-06$	0.109	0.0020	0.9351	0.2578	1.9903	0.3863
$[n^{0.4}]$	0.00817	$2e-05$	0.201	0.0022	LSCV method			
$[n^{0.2}]$	0.01650	$5e-05$	0.268	0.0018	0.0026	$3e-6$	0.0787	$8e-4$
					JS method			
					0.0013	$2.1569e-6$	0.0544	$5.6261e-4$
$n = 500$								
k	$\overline{\rho}_1$	σ_1^2	$\overline{\rho}_2$	σ_2^2	KS method			
n^1	$9e-04$	$4.7e-07$	0.0612	$4e-04$	$6.35e-04$	$3e-07$	0.0525	$4e-04$
$[n^{0.8}]$	$4.2e-04$	$2e-07$	0.0311	$1.6e-04$	Rule of thumb (9)			
$[n^{0.6}]$	0.00297	$2e-06$	0.138	$9e-04$	0.9745	0.0881	2.3481	0.2394
$[n^{0.4}]$	0.01290	$1e-05$	0.254	$6e-04$	LSCV method			
$[n^{0.2}]$	0.02320	$2e-05$	0.314	$3e-04$	$6.4e-4$	$2e-7$	0.0529	$3e-4$
					JS method			
					$4.0160e-4$	$1.1771e-7$	0.0376	$1.9921e-4$

To evaluate the accuracy of the pdf estimates, we calculate the following statistics for each sample:

$$\nu^1 = \frac{1}{n - i^* + 1} \sum_{i=i^*}^{n} \left(\widehat{f_h}(X_{(i)}) - f(X_{(i)})\right)^2, \qquad \nu^2 = \sup_{i=i^*,\ldots,n} |\widehat{f_h}(X_{(i)}) - f(X_{(i)})|.$$

Here, i^* is selected as a maximal i such that $X_{(i)}$ is less or equal to a. Note that ν^1 and ν^2 are not sensitive to rare observations at the tails where the pdf values could be relatively small. Furthermore, we compare the accuracy of the pdf estimates by calculating the statistics

$$\overline{\rho}_j = \frac{1}{B} \sum_{i=1}^{B} \nu_i^j, \qquad \sigma_j^2 = \frac{1}{B} \sum_{i=1}^{B} (\nu_i^j - \overline{\rho}_j)^2, \qquad j = 1, 2. \tag{10}$$

We generate heavy-tailed distributions, namely, the lognormal, the Fréchet, the Weibull and Pareto distributions with pdfs

$$f(x) = \frac{1}{\sqrt{2\pi} x \sigma} \exp(-\frac{(\ln x - \mu)^2}{2\sigma^2}), \qquad x > 0, \qquad \mu \in R, \qquad \sigma > 0, \tag{11}$$

$$f(x) = \alpha x^{-\alpha-1} \exp(-x^{-\alpha}) \mathbf{1}(x > 0), \qquad \alpha > 0,$$

Table 2 The bandwidth selection for the Weibull distribution with parameters $\lambda = 2$, $b = 0.2$ at $[0.4, X_{max}]$

	D-method				$\overline{\rho}_1$	σ_1^2	$\overline{\rho}_2$	σ_2^2
$n = 100$								
k	$\overline{\rho}_1$	σ_1^2	$\overline{\rho}_2$	σ_2^2	KS method			
n^1	$4e+03$	$9e+05$	63	10	0.220	0.328	0.428	0.137
$[n^{0.8}]$	0.12100	0.0563	0.377	0.0873	Rule of thumb (9)			
$[n^{0.6}]$	0.01590	0.0002	0.256	0.0129	0.1418	0.1301	0.3163	0.1095
$[n^{0.4}]$	0.00657	$9e-05$	0.118	0.0047	LSCV method			
$[n^{0.2}]$	0.00232	$6e-06$	0.100	0.0009	$2e+2$	$1e+6$	3.865	$2e+2$
					JS method			
					0.0115	$1.3068e-4$	0.1911	0.0117
$n = 500$								
k	$\overline{\rho}_1$	σ_1^2	$\overline{\rho}_2$	σ_2^2	KS method			
n^1	$2e+02$	$2e+02$	14.237	12.5331	0.0615	0.00584	0.467	0.0683
$[n^{0.8}]$	0.0008	$3e-07$	0.0691	0.0006	Rule of thumb (9)			
$[n^{0.6}]$	0.00968	$3e-05$	0.170	0.0028	0.0159	$5.6536e-04$	0.1940	0.0163
$[n^{0.4}]$	0.00145	$2e-07$	0.111	0.0003	LSCV method			
$[n^{0.2}]$	0.00246	$4e-07$	0.142	$6e-05$	2.273	$3e+2$	0.857	4
					JS method			
					0.0118	$5.8259e-5$	0.1989	0.0093

$$f(x) = (b/\lambda)(x/\lambda)^{b-1}\exp(-(x/\lambda)^b), \qquad x > 0, \qquad 0 < b < 1, \qquad \lambda > 0,$$

$$f(x) = (1+\gamma x)^{-1/\gamma-1}\mathbf{1}(x \geq 0), \qquad \gamma > 0,$$

respectively.

In Tables 1, 2, 3, 4 and 5 we compare the accuracy of the discrepancy method D with different values of $k = [n^b]$, the least-squares cross-validation method (LSCV), [3, 17], the method KS[4] which is theoretically optimal for estimating pdfs for normal distribution [4], the Sheather and Jones bandwidth selector (the SJ-method[5]) [18] and the Silverman's rule of thumb (9). The latter four methods are calculated by entire samples. The LSCV selects such h that provides the minimum of the sum

$$LSCV(h) = n^{-1}\sum_{i=1}^{n}\int \hat{f}_{-i}(x;h)^2 dx - 2n^{-1}\sum_{i=1}^{n}\hat{f}_{-i}(X_i;h), \tag{12}$$

where

[4]This is ksdensity-procedure in Matlab.

[5]It was calculated by Dynara Team Matlab code $mh - optimal - bandwidth$.

Table 3 The bandwidth selection for the Fréchet distribution with parameter $\alpha = 0.1$ at $[1, X_{max}]$

	D-method				$\overline{\rho}_1$	σ_1^2	$\overline{\rho}_2$	σ_2^2
$n = 100$								
k	$\overline{\rho}_1$	σ_1^2	$\overline{\rho}_2$	σ_2^2	KS method			
n^1	$4.7925e-5$	$6.5176e-10$	0.0284	$4.0102e-5$	$2.6258e-4$	$2.7650e-6$	0.0413	$7.6768e-4$
$[n^{0.8}]$	$4.3359e-5$	$3.4315e-9$	0.0246	$7.7639e-5$	Rule of thumb (9)			
$[n^{0.6}]$	$4.7925e-5$	$6.5176e-10$	0.0284	$4.0102e-5$	$5.0627e-5$	$7.0372e-10$	0.0289	$4.0136e-5$
$[n^{0.4}]$	$5.0627e-5$	$7.0372e-10$	0.0289	$4.0136e-5$	LSCV method			
$[n^{0.2}]$	$4.7925e-5$	$6.5176e-10$	0.0284	$4.0102e-5$	$1.8591e-5$	0.0135	$7.9145e-11$	$3.2397e-6$
					JS method			
					$4.6146e-5$	$6.4771e-10$	0.0287	$4.5047e-5$
$n = 500$								
k	$\overline{\rho}_1$	σ_1^2	$\overline{\rho}_2$	σ_2^2	KS method			
n^1	$3.9522e-5$	$9.3177e-11$	0.0344	$3.5088e-6$	$1.5233e-4$	$7.0980e-9$	0.0605	$5.9227e-4$
$[n^{0.8}]$	$3.9425e-5$	$9.2474e-11$	0.0343	$3.4896e-6$	Rule of thumb (9)			
$[n^{0.6}]$	$3.9522e-5$	$9.3177e-11$	0.0344	$3.5088e-6$	$4.1755e-5$	$1.0053e-10$	0.0349	$3.5124e-6$
$[n^{0.4}]$	$4.1755e-5$	$1.0053e-10$	0.0349	$3.5124e-6$	LSCV method			
$[n^{0.2}]$	$3.9522e-5$	$9.3177e-11$	0.0344	$3.5088e-6$	$1.7256e-5$	$1.2914e-11$	0.0138	$5.5496e-7$
					JS method			
					$4.4284e-5$	$1.1308e-10$	0.0348	$3.4797e-6$

Table 4 The bandwidth selection for the Fréchet distribution with parameter $\alpha = 1$ at $[1, X_{max}]$

	D-method				$\overline{p}_1$	σ_1^2	$\overline{p}_2$	σ_2^2
$n = 100$								
k	$\overline{p}_1$	σ_1^2	$\overline{p}_2$	σ_2^2	KS method			
n^1	0.0187	$2.6069e-5$	0.3044	$6.1792e-4$	$9.1117e-4$	$7.9479e-7$	0.0626	$8.8957e-4$
$[n^{0.8}]$	0.0015	$3.3110e-6$	0.0879	0.0017	Rule of thumb (9)			
$[n^{0.6}]$	0.0890	$1.6817e-5$	0.2313	0.0017	0.0392	0.0032	0.3438	0.0637
$[n^{0.4}]$	0.0303	$2.9317e-5$	0.3579	$9.6628e-5$	LSCV method			
$[n^{0.2}]$	0.0250	$2.9470e-5$	0.3361	$2.7394e-4$	0.0192	$1.3284e-5$	0.3181	$1.6205e-4$
					JS method			
					0.0095	$5.8863e-5$	0.2189	0.0062
$n = 500$								
k	$\overline{p}_1$	σ_1^2	$\overline{p}_2$	σ_2^2	KS method			
n^1	0.0237	$5.9958e-6$	0.3432	$4.9610e-5$	$2.3846e-4$	$4.2647e-8$	0.0360	$1.3020e-4$
$[n^{0.8}]$	$9.7296e-4$	$2.7294e-7$	0.0970	$9.4370e-4$	Rule of thumb (9)			
$[n^{0.6}]$	0.0139	$6.6393e-6$	0.2881	$2.5530e-4$	0.0094	$1.0738e-4$	0.2484	0.0235
$[n^{0.4}]$	0.0291	$5.7607e-6$	0.3658	$3.8527e-6$	LSCV method			
$[n^{0.2}]$	0.0275	$5.7902e-6$	0.3595	$1.4676e-5$	0.0195	$2.9142e-6$	0.3292	$6.9918e-6$
					JS method			
					0.0087	$5.3185e-5$	0.2250	0.0058

$$\hat{f}_{-i}(x; h) = \frac{1}{(n-1)h} \sum_{j=1,\ j \neq i}^{n} K\left(\frac{x - X_j}{h}\right), \tag{13}$$

For the Gaussian kernel $N(x, h^2) = (1/(h\sqrt{2\pi})) \exp\left(-x^2/(2h^2)\right)$ the integral in (12) can be calculated analytically, that is

$$\begin{aligned} LSCV(h) = {} & \frac{1}{n-1} N(0, 2h^2) + \frac{n-2}{n(n-1)^2} \sum_{i \neq j} N(X_i - X_j, 2h^2) \\ & - \frac{2}{n(n-1)} \sum_{i \neq j} N(X_i - X_j, h^2). \end{aligned}$$

The main objective is to study how the selection of k impacts the accuracy of the kernel estimate when h is selected by the D-method. Only heavy-tailed distributions are considered. The Fréchet distribution considered in Table 3 has a heavier tail than the one in Table 4 since the tail index $\alpha = 0.1$ is smaller. The Pareto and the Fréchet distributions in Tables 3 and 5 provide special cases. $\gamma = 10$ corresponds to the tail index $\alpha = 1/\gamma = 0.1$ which implies that all moments are infinite beginning from the first according to Breiman's theorem, [14]. This complicates the pdf estimation.

From the tables one may conclude that using the largest order statistics may improve the accuracy of the D-method significantly. Regarding $k = [n^{0.8}]$ the accuracy is the best. This also provides much better accuracy than other methods for

Table 5 The bandwidth selection for the Pareto distribution with parameter $\gamma = 10$ at $[1, X_{max}]$

	D-method				$\overline{\rho}_1$	σ_1^2	$\overline{\rho}_2$	σ_2^2
$n = 100$								
k	$\overline{\rho}_1$	σ_1^2	$\overline{\rho}_2$	σ_2^2	KS method			
n^1	$2.1967e-4$	$1.4729e-9$	0.0692	$2.4081e-6$	$1.8062e-04$	$1.1129e-09$	0.0655	$7.0771e-06$
$[n^{0.8}]$	$2.1962e-4$	$1.4735e-9$	0.0692	$2.4079e-6$	Rule of thumb (9)			
$[n^{0.6}]$	$2.1967e-4$	$1.4729e-9$	0.0692	$3.4081e-6$	$2.2510e-04$	$1.5255e-09$	0.0697	$2.4081e-06$
$[n^{0.4}]$	$2.2510e-4$	$1.5255e-9$	0.0697	$3.4081e-6$	LSCV method			
$[n^{0.2}]$	$2.1967e-4$	$1.4729e-9$	0.0692	$3.4081e-6$	$1.0037e-04$	$4.1824e^{-}10$	0.0546	$4.0967e-06$
					JS method			
					$2.2042e-4$	$5.9902e-9$	0.0629	$5.9688e-5$
$n = 500$								
k	$\overline{\rho}_1$	σ_1^2	$\overline{\rho}_2$	σ_2^2	KS method			
n^1	$2.1526e-4$	$7.1284e-09$	0.0617	$6.6867e-05$	$1.7835e-04$	$5.9744e-9$	0.0580	$7.7639e-5$
$[n^{0.8}]$	$2.1180e-4$	$7.0013e-09$	0.0614	$6.6900e-05$	Rule of thumb (9)			
$[n^{0.6}]$	$2.1525e-4$	$7.1284e-09$	0.0617	$6.6867e-05$	$2.2068e-04$	$7.3908e-9$	0.0622	$6.6923e-5$
$[n^{0.4}]$	$2.2068e-4$	$7.3908e-09$	0.0622	$6.6923e-05$	LSCV method			
$[n^{0.2}]$	$2.1525e-4$	$7.1284e-09$	0.0617	$6.6867e-05$	$9.7794e-05$	$2.0441e-09$	0.0471	$6.6308e-05$
					JS method			
					$2.2994e-4$	$1.4144e-9$	0.0695	$4.1975e-6$

lognormal and Weibull distributions. The D-method is better than the SJ-method for all distributions. The LSCV method is better than the D-method for the Fréchet distribution with tail index $\alpha = 0.1$. For Pareto distribution the LSCV method demonstrates non-consistency. For the Fréchet distribution with $\alpha = 1$ the KS method has the best accuracy among all examined methods. The rule of thumb is generally worse than other methods.

4 Conclusions

Nonparametric estimation of heavy-tailed pdfs is considered. In earlier papers of the author [13, 21] the selection of the bandwidth h in nonparametric density estimates by means of discrepancy methods based on the von Mises-Smirnov and the Kolmogorov–Smirnov statistics has been proposed. It was investigated for light-tailed pdfs. In this paper, we adapt the discrepancy methods to heavy-tailed pdfs. The adaptation implies that the discrepancy statistics are calculated only by the k largest order statistics belonging to the distribution tail. We investigated by a simulation study for several heavy-tailed distributions how to improve the selection of k better and found that $k = n^{0.8}$ where n is the sample size could be an appropriate value. The theoretical value of k providing an optimal accuracy of the kernel estimate is the subject of our future work. For distributions with very heavy tails when no finite moments exist, it may be recommended to transform the data to a distribution with lighter tail.

Acknowledgments This work was supported in part by the Russian Foundation for Basic Research, grant 13-08-00744.

References

1. Abramson, I.S.: On bandwidth estimation in kernel estimators—a square root law. Ann. Stat. **10**, 1217–1223 (1982)
2. Bolshev, L.N., Smirnov, N.V.: Tables of Mathematical Statistics. Nauka, Moscow (1965) (in Russian)
3. Bowman, A.W.: An alternative method of cross-validation for the smoothing of density estimates. Biometrika **71**(2), 353–360 (1984)
4. Bowman, A.W., Azzalini, A.: Applied Smoothing Techniques for Data Analysis. Oxford University Press (1997)
5. Cao, R., Cuevas, A., Fraiman, R.: Minimum distance density-based estimation. Comput. Stat. Data Anal. **20**, 611–631 (1995)
6. Chen, S.X.: Probability density function estimation using gamma kernels. Ann. Inst. Stat. Math. **52**(3), 471–480 (2000)
7. Devroye, L., Györfi, L.: Nonparametric Density Estimation. The L_1 View. Wiley, New York (1985)
8. Dobrovidov A.V., Markovich L.A.: Data-driven bandwidth choice for gamma kernel estimates of density derivatives on the positive semi-axis. In: Proceedings of IFAC International Workshop

on Adaptation and Learning in Control and Signal Processing, pp. 500–505. Caen, France, 3–5 July (2013). doi:10.3182/20130703-3-FR-4038.00086, arXiv:1401.6801
9. Hall, P., Sheather, S.J., Jones, M.C., Marron, J.S.: On optimal data-based bandwidth selection in kernel density estimation. Biometrika **78**, 263–269 (1991)
10. Loader, C.R.: Bandwidth selection: classical or plug-in? Ann. Stat. **27**(2), 415–438 (1999)
11. Markovich, L.A.: Nonparametric gamma kernel estimators of density derivatives on positive semiaxis by dependent data. RevStat Stat. J. (2016). arXiv:1401.6783 (in appear)
12. Markovich, L.A.: Nonparametric estimation of multivariate density and its derivative by dependent data using gamma kernels. Submitted J. Nonparametric Stat. (2016). arXiv:1410.2507
13. Markovich, N.M.: Experimental analysis of nonparametric probability density estimates and of methods for smoothing them. Autom. Remote Control **50**, 941–948 (1989)
14. Markovich, N.M.: Nonparametric Analysis of Univariate Heavy-Tailed Data. Wiley, Chichester (2007)
15. Markovich, N.M.: Nonparametric estimation of extremal index using discrepancy method. In: Proceedings of the X International Conference System Identification and Control Problems SICPRO-2015, pp. 160–168. V.A. Trapeznikov Institute of Control Sciences, Moscow, 26–29 Jan 2015. ISBN: 978-5-91450-162-1
16. Martynov, G.V.: Omega-Square Criteria. Nauka, Moscow (1978) (in Russian)
17. Rudemo, M.: Empirical choice of histogram and kernel density estimators. Scand. J. Stat. **9**, 65–78 (1982)
18. Sheather, S.J., Jones, M.C.: A reliable data-based bandwidth selection method for kernel density estimation. J. Roy. Stat. Soc. Ser. B **53**(3), 683–690 (1991)
19. Scott, D.W.: Multivariate Density Estimation. Theory, Practice, and Visualization. Wiley, New York, Chichester (1992)
20. Silverman, B.W.: Density Estimation for Statistics and Data Analysis. Chapman and Hall, London, New York (1986)
21. Vapnik, V.N., Markovich, N.M., Stefanyuk, A.R.: Rate of convergence in L_2 of the projection estimator of the distribution density. Autom. Remote Control **53**, 677–686 (1992)
22. Zambom, A.Z., Dias, R.: A Review of Kernel Density Estimation with Applications to Econometrics (2012). arXiv:1212.2812

Robust Estimation in AFT Models and a Covariate Adjusted Mann–Whitney Statistic for Comparing Two Sojourn Times

Sutirtha Chakraborty and Somnath Datta

Abstract A Mann–Whitney statistic may be used to compare two sets of state waiting times; however, wrong conclusions may be reached if there are confounding covariates. We develop a Mann–Whitney type test statistic based on the residuals from an accelerated failure time model fitted to two groups of sojourn times with a common set of covariates. This covariate adjusted test statistics handles right censoring via the inverse probability of censoring weights. These weights were devised to improve efficiency in the sense that certain pairs in which at least one state entry time is uncensored could be compared. Extensive simulation studies were undertaken to evaluate the performance of this test. A real data illustration of our methodology is also provided.

Keywords Confounding · U-statistic · Two sample · Sojourn time · AFT model

1 Introduction

Comparison of sojourn time distributions corresponding to a given transient state between two populations (groups) of individuals in a multistate model is an important but relatively unexplored area of research. The presence of censoring (with respect to both the entry and exit times) in the multistate network makes this problem of extending standard statistical testing procedures particularly difficult.

A nonparametric approach to this problem in absence of subject-level covariates was developed by Fan and Datta [7], where they used certain product weights based on the principle of inverse probability of censoring weighting (IPCW) [4, 5, 13] with an indicator kernel to generalize a Mann–Whitney type test. Although quite effective as a testing procedure for comparing the overall waiting time distributions,

S. Chakraborty
National Institute of Biomedical Genomics, Kalyani, India
e-mail: sc4@nibmg.ac.in

S. Datta (✉)
University of Florida, Gainesville, FL, USA
e-mail: somnath.datta@ufl.edu

R. Cao et al. (eds.), *Nonparametric Statistics*, Springer Proceedings
in Mathematics & Statistics 175, DOI 10.1007/978-3-319-41582-6_9

this methodology may fail to perform an appropriate comparison between the two distributions when confounding covariates are present. In this article, we seek to develop a solution to this extended problem of comparing the sojourn time distributions for the two populations after necessary adjustment for the available covariate information.

We develop an IPCW modified U-statistic [10, 12] based on the residuals from the fitting of two separate accelerated failure time (AFT) models [11] to the sojourn times from the two populations. This U-statistic serves as a Mann–Whitney test statistic [12] for comparing the covariate adjusted sojourn time distributions corresponding to the two groups of individuals. The performance of this novel statistic has been thoroughly studied using simulation. The methodology is illustrated using a data set on motor activities of patients suffering a spinal cord injury [8, 9].

2 Methodology

2.1 Data Structure and Notations

Let us envisage a scenario where we have data on the right censored entry and exit times of individuals from two independent populations (groups). Let $X^*_{i,j}$ and $V^*_{i,j}$ denote the original uncensored entry and exit times for the ith individual in the jth group, $C_{i,j}$ be a common censoring variable which affects both of them and is assumed to be independent of the pair $(X^*_{i,j}, V^*_{i,j})$; $j = 1, 2$. In addition we have information on a p-dimensional covariate Z for all the individuals in the model. To that end, let $Z_{i,j}$ denote the observed value of Z for the ith individual in the jth group. Thus overall, our entire observed dataset is composed of the 5-tuples $(X_{i,j}, \delta_{i,j}, V_{i,j}, \eta_{i,j}, Z_{i,j})$, $(i = 1, 2, \ldots, n_j;\ j = 1, 2;\ n_1 + n_2 = n)$, where $X_{i,j} = \min(X^*_{i,j}, C_{i,j})$ and $V_{i,j} = \min(V^*_{i,j}, C_{i,j})$ are the observed right censored entry and exit times for the ith individual in the jth group and $\eta_{i,j} = I(C_{i,j} \geq X^*_{i,j})$ and $\delta_{i,j} = I(C_{i,j} \geq V^*_{i,j})$ are the corresponding censoring indicators.

Define $W^*_{i,j} = V^*_{i,j} - X^*_{i,j}$ as the actual uncensored waiting time for the ith individual in the jth group and $W_{i,j}$ be its corresponding observed version in the right censored data. Clearly, $W_{i,j}$ is uncensored and equals $W^*_{i,j}$ if and only if $\eta_{i,j} = 1$. For methodological and theoretical developments, we assume that for each group j, the data vectors $(X_{i,j}, \delta_{i,j}, V_{i,j}, \eta_{i,j}, Z_{i,j})$ are for $1 \leq i \leq n_j$ and also the censoring C_{ij} is independent of (X_{ij}, V_{ij}, Z_{ij}).

2.2 A Covariate Adjusted Mann–Whitney U-Statistic

In the absence of censoring the Mann–Whitney U-statistic to be used for comparing the marginal waiting time distributions between the two different groups is given by

$$U = \frac{1}{n_1 n_2} \sum_{i_1=1}^{n_1} \sum_{i_2=1}^{n_2} I(W^*_{i_1,1} \le W^*_{i_2,2}). \tag{1}$$

However, in the presence of right censoring not all waiting times can be observed in both the two groups and hence they need to be replaced by their corresponding right censored values. Fan and Datta [7] proposed a modified Mann–Whitney U-statistic that compensates for this selection bias using the IPCW reweighting principle [4, 5]:

$$\widehat{U}_{\mathrm{FD}} = \frac{1}{n_1 n_2} \sum_{i_1=1}^{n_1} \sum_{i_2=1}^{n_2} \frac{I(W_{i_1,1} \le W_{i_2,2})\delta_{i_1,1}\eta_{i_2,2}}{\widehat{K}_1(V_{i_1,1}-)\widehat{K}_2(W_{i_1,1}+X_{i_2,2}-)}. \tag{2}$$

The IPCW principle adjusts for the underlying censoring in the data by generating a set of weights corresponding to the observed (uncensored) data values. Each of these weights is a one over the estimated probability of an individual not being censored up to a certain observed event. Estimation of a certain population parameter with these weights compensates for the underlying selection bias in the data used in constructing the estimator. As for example, the weight $\widehat{K}_1(V_{i_1,1}-)$ is the estimated conditional probability of the event $\{\delta_{i_1,1} = 1\}$ given the value of $V^*_{i_1,1}$ (which equals $V_{i_1,1}$ when $\delta_{i_1,1} = 1$) under independent censoring and therefore it compensates for the factor $\delta_{i_1,1}$ in the numerator.

Although, this statistic can eliminate the selection bias due to underlying right censoring in the data using the IPCW principle, it only provides a marginal comparison of two sets of waiting times. In the presence of subject level covariates, one may be interested in a comparison of the sojourn times after adjusting for such covariates. In the case of a discrete (or categorical) covariate, one option will be to carry out subgroup analysis by restricting the Fan-Datta analysis to a subgroup of waiting times corresponding to a particular value of this covariate. However such an approach will suffer from low power due to smaller sample size which will be even more problematic when one has more than one covariates. In this work, we propose a residual based extension of the Datta-Fan statistic to incorporate the effects of a collection of baseline covariates, both discrete and continuous, in differentiating the sojourn time patterns in the two groups of individuals that can be applied to the entire data set.

As mention before, we propose an extension of the classical Mann–Whitney U-statistic [12] that can be used to build a test for comparing the waiting time distributions between the two groups, after adjusting for subject level covariates Z. For this purpose we pursue a regression approach to build a set of model residuals, which can in turn be used to build such a modified U-statistic. Although other types of regression models (both semi-parametric such as the Cox regression model [2], as well as, nonparametric such as Beran's estimator [1]) can be used for the purpose of covariate adjustment, we choose a transformation model for the waiting times in order to calculate the residuals due to its relative simplicity. To this end, we define the following two accelerated failure time (AFT) models [17] corresponding to the waiting times of the individuals from the two groups:

$$\log(W^*_{i_1,1}) = \alpha_1 + Z^T_{i_1,1}\beta_1 + \epsilon_1, 1 \le i_1 \le n_1,$$

$$\log(W^*_{i_2,2}) = \alpha_2 + Z^T_{i_2,2}\beta_2 + \epsilon_2, 1 \le i_2 \le n_2,$$

where α_1, α_2 are the intercepts, β_1, β_2 are the vectors of regression coefficients in the two groups; ϵ_1, ϵ_2 denote the zero mean random error terms for the two regression models.

First consider the hypothetical setup, when the regression coefficients β_j in the two models are known and also suppose censoring is absent. In that case, the log-transformed waiting times from the two groups after eliminating the covariate effects would have been given by $R^*_{i_j,j} = \log(W^*_{i_j,j}) - Z^T_{i_j,j}\beta_j$; $j = 1, 2$, and it would be possible to apply standard Mann–Whitney U-statistic to the $R^*_{i_j,j}$ to draw inference on $\theta = P(R^*_{i_1,1} \le R^*_{i_2,2})$. Note that $R^*_{i_j,j}$ may have different mean (location) since the intercepts are not being subtracted and hence θ is not always 1/2 even if the ϵ had the same distribution in the two groups. Of course, a null hypothesis of equality of the distribution of R^* leads to a null hypothesis of $\theta = 1/2$, which can be nonparametrically tested by the Mann–Whitney test. Assuming ϵ has the same distribution in the two groups, a parametric way of testing the equality of the distribution of R^* is to test $H_0 : \alpha_1 = \alpha_2$, using a standardized version of $\widehat{\alpha}_1 - \widehat{\alpha}_2$, using its asymptotic normality. If potential non-normality of the error distribution is a concern, a robust estimating equation should be in place for estimation of α in each group. It is fair to say, however that the Mann–Whitney (or equivalently, the Wilcoxon) test is the more common procedure than this later approach, if non-normal errors are suspected.

Coming back to the setup of the present paper, the β coefficients are unknown; furthermore, the $W^*_{i_j,j}$ are not always observed due to censoring. Thus we proceed in two steps to achieve of goal of inferring θ. First, we develop a robust estimating equation for the parameters in the AFT regression models that can be used in presence of right censoring. Second, using the resulting estimators of the β coefficients, we compute a modified version of the Mann–Whitney U-statistic that can be computed with the available data. As we will see, the common technique used in dealing with right censoring in both setups is "inverse probability of censoring weighting" (IPCW). However, the weighting principle for the U-statistic is subtle and attempts to take advantage of the form of the kernel so that maximum available information in the data are used including some pairs where the exit time of one may be unobserved due to censoring.

2.3 Robust Estimation in the AFT Model of Waiting Times in Presence of censoring

As the waiting times of the subjects in the two groups are right censored, the standard M-estimating equations for these two models need to be modified following the IPCW reweighting principle (see, e.g., [3, 4]). Thus the estimators $(\widehat{\alpha}_1, \widehat{\beta}_1)$ and $(\widehat{\alpha}_2, \widehat{\beta}_2)$ are obtained from the estimating equations given below:

$$\Delta_1 = \sum_{i_1=1}^{n_1} \psi\left(\log(W^*_{i_1,1}) - \alpha_1 - Z^T_{i_1,1}\beta_1\right) \tilde{Z}_{i_1,1} \frac{\delta_{i_1,1}}{\widehat{K}_1(V_{i_1,1}-)},$$

$$\Delta_2 = \sum_{i_2=1}^{n_2} \psi\left(\log(W^*_{i_2,1}) - \alpha_2 - Z^T_{i_2,2}\beta_2\right) \tilde{Z}_{i_2,2} \frac{\delta_{i_2,2}}{\widehat{K}_2(V_{i_2,2}-)}.$$

Here, ψ is a score function satisfying $E[\psi(\epsilon_j)] = 0$, $E[\{\psi'(\epsilon_j)\}^2] < \infty$, $\tilde{Z}_{i_j,j} = (1, Z^T_{i_j,j})^T$; for our simulation study and data analysis, we have used $\psi(x) = \tanh(x)$. Also, $\widehat{K}_1, \widehat{K}_2$ are the estimated survival functions for the censoring distributions in the two groups; they are calculated using the Kaplan–Meier formulas applied to the data from two groups separately where censoring events are taken as corresponding to the censoring indicators $\delta_{i_1,1}, \delta_{i_2,2}$ for the two groups. Note that $E(\delta_{i_j,j}|V^*_{i_j,j}) = P(C_{i_j,j} \geq V^*_{i_j,j}|V^*_{i_j,j}) = K_j(V^*_{i_j,j}-)$, since the censoring is independent, which in turns equals $K_j(V_{i_j,j}-)$, if $\delta_{i_j,j} = 1$. Thus, large sample consistency of the regression parameter estimators obtained from (2) follows from approximate unbiasedness of the corresponding estimating functions which in turn follows from the mean preserving property of IPCW averages [3].

2.4 IPCW Mann–Whitney

We denote the corresponding censored versions of the covariate adjusted log-waiting times by $R_{i_j,j}$ =log$(W_{i_j,j}) - Z^T_{i_j,j}\widehat{\beta}_j$; $j = 1, 2$. Let $C(z;\beta) = C(z_1, z_2; \beta_1, \beta_2) = e^{-(z_1^T\beta_1 - z_2^T\beta_2)}$. We define the following U-statistic based on $R_{i_1,1}$ and $R_{i_2,2}$ obtained after fitting the reweighted accelerated failure time (AFT) models on the waiting times of the individuals in the two groups:

$$\begin{aligned}\widehat{U} &= \frac{1}{n_1 n_2} \sum_{i_1=1}^{n_1}\sum_{i_2=1}^{n_2} \frac{I(R_{i_1,1} \leq R_{i_2,2})\delta_{i_1,1}\eta_{i_2,2}}{\widehat{K}_1(V_{i_1,1}-)\widehat{K}_2(X_{i_2,2} + C(Z_i;\widehat{\beta})W_{i_1,1}-)} \\ &= \frac{1}{n_1 n_2} \sum_{i_1=1}^{n_1}\sum_{i_2=1}^{n_2} \frac{I(C(Z_i;\widehat{\beta})W_{i_1,1} \leq W_{i_2,2})\delta_{i_1,1}\eta_{i_2,2}}{\widehat{K}_1(V_{i_1,1}-)\widehat{K}_2(X_{i_2,2} + C(Z_i;\widehat{\beta})W_{i_1,1}-)} \qquad (3)\end{aligned}$$

where $C(Z_i;\widehat{\beta}) = C(Z_{i_1,1}, Z_{i_2,2}; \widehat{\beta}_1, \widehat{\beta}_2)$, $\delta_{i_1,1} = I(C_{i_1,1} \geq V^*_{i_1,1})$ and $\eta_{i_2,2} = I(C_{i_2,2} \geq X^*_{i_2,2})$ are the two censoring indicators corresponding to the two groups as defined earlier.

This particular U-statistic enables us to compare all observed waiting time pairs, plus some additional pairs when the exit time of the larger waiting time may not be observed. The two weights in the denominator of the summands compensate for the presence of the two censoring indicators $\delta_{i_1,1}$ and $\eta_{i_2,2}$. With some conditioning arguments and algebra, one can see that these weights estimate the conditional expectations of these indicators given the true event times on the set $\{\delta_{i_1,1}\eta_{i_2,2} = 1\}$.

Next, we describe the asymptotic distribution of $\widehat{U}$. For this purpose we provide a brief description of the different technical quantities used in its formulation. Let $N^c_{i,j}(t) = I(V_{i,j} \le t, \delta_{i,j} = 0)$ be the counting process of censoring in group j and $M^c_{i,j}(t) = N^c_{i,j}(t) - \int_0^t Y_{i,j}(u)d\Lambda^c_j(u)$ be the associated martingale ; $j = 1, 2$, where $Y_{i,j}(t) = I(V_{i,j} \ge t)$ is the corresponding at-risk process. Let $Y_j(t) = \sum_{i=1}^{n_j} Y_{i,j}(t)$.

Now, define the following set of functions which will be used in describing the asymptotic distribution of $\widehat{U}$:

$$\begin{aligned} G_1(v, w, z) &= P\{V_1 \le v, W_1 \le w, Z_1 \le z, \delta_1 = 1\} \\ G_2(x, w, z) &= P\{X_2 \le x, W_2 \le w, Z_2 \le z, \eta_2 = 1\} \\ S_2(u) &= E\{I(e^{-Z_2^T\beta_2} W_2 \le u)\} \\ F_1(w, z) &= P\{W_1 \le w, Z_1 \le z\} \end{aligned}$$

Define the following additional functions over the domain $[0, \infty)$:

$$\omega_1(s) = \frac{1}{y_1(s)} \int I(v > s) \frac{S_2(e^{-z^T\beta_1} w)}{K_1(v-)} dG_1(v, w, z),$$

$$\omega_2(s) = \frac{1}{y_2(s)} \iint \frac{I(x_2 + C(z; \beta)w_1 > s) I(C(z; \beta)w_1 < w_2)}{K_2(x_2 + C(z; \beta)w_1-)} dF_1(w_1, z_1)\, G_2(x_2, w_2, z_2)$$

with $y_j(s) = P(V_j \ge s)$; $j = 1, 2$.

For the theory, we assume that the covariables $Z_{i,j}$ are i.i.d. for $1 \le i \le n_j$, for each $j = 1, 2$.

Theorem 1 *Under suitable regularity conditions (see the Appendix A.1) as n_j tends to ∞, such that $n_j/n \to c_j$, with $n = n_1 + n_2$, we have $\sqrt{n}(\widehat{U} - \theta) \to N(0, \sigma^2)$, where*

$$\begin{aligned} \sigma^2 = c_1^{-1} Var&\left[\frac{\delta_{1,1}\{S_2(e^{-Z_{1,1}^T\beta_1} W_{1,1}) + AC_{Z_1}^{-1}\psi(\epsilon_{1,1})Z_{1,1}\}}{K_1(V_{1,1}-)} + \int_0^\infty \omega_1(s)dM^c_{1,1}(s)\right] \\ + c_2^{-1} Var&\left[\eta_{1,2}\int \frac{I(C(z, Z_{1,2}; \beta_1, \beta_2)w < W_{1,2})}{K_2(X_{1,2} + C(z, Z_{1,2}; \beta_1, \beta_2)w-)} dF_1(w, z) \right. \\ &\left. + \frac{\delta_{1,2}BC_{Z_2}^{-1}\psi(\epsilon_{1,2})Z_{1,2}}{K_2(V_{1,2}-)} + \int_0^\infty \omega_2(s)dM^c_{1,2}(s)\right], \end{aligned}$$

where the vectors A, B, and the matrices C_{Z_1}, C_{Z_2}, are defined in the appendix.

From the structure of the asymptotic variance of $\widehat{U}$ we can derive its estimator using the corresponding sample counterparts of the expressions in the above formula (see Appendix A.2). However, the corresponding formulas are computationally tedious. Hence we recommend using bootstrap to estimate the variance. Note that a nonparametric bootstrap can be easily implemented by resampling the 5-tuples $(X_{i,j}, \delta_{i,j}, V_{i,j}, \eta_{i,j}, Z_{i,j})$ within each group and recomputing the U-statistic for each bootstrap resample.

3 Simulation Studies

We conduct a thorough simulation study to investigate the sampling properties of our regression parameter estimators and that of the covariate adjusted U-statistic $\widehat{U}$. We also investigate the power of detecting differences in the waiting times in the two groups after adjusting for other group specific covariates using a test statistic that is a symmetrized version of our adjusted U-statistic.

3.1 Bias and Variance Study

Group specific covariates Z_{ij} are generated from $N(1.5, 1)$ for group 1 individuals, and from $N(2.5, 1)$ for group 2 individuals. Next, we generate the entry and waiting times from a bivariate normal distribution as follows: $X_{ij} = exp(\sim X_{ij})$, $W_{ij} = exp(\alpha_j + \beta_j Z_{ij} + \epsilon_{ij})$, where $\sim X_{ij}, \epsilon_{ij})$ are generated from a bivariate normal with zero mean vector and dispersion matrix $\begin{pmatrix} 1 & \rho \\ \rho & 1 \end{pmatrix}$. We also add contamination by changing a certain percentage of the ϵ_{ij} to random generations of a discrete uniform on the set of four values $\{-6, -4, 4, 6\}$. The regression parameters were $\alpha_1 = \alpha_2 = 0.5$; $\beta_1 = 0.3$, $\beta_2 = 0.5$.

We have used three choices of the sample size per group $n_j = 25, 50$ and 200. Two values of ρ (= −0.5 and 0.5), two choices of the score function $\phi(x) = x$, and $\phi(x) = \tanh(x)$, and two values of the contamination percentage ($\gamma = 0$ and 10 %) were used. The censoring times in the two groups are generated from two lognormal distributions with unit log-scale, but with varying log-mean parameters depending on the desired censoring rates in the two groups. We report the performance results for light (25 %) and heavy (50 %) censoring, respectively.

Results for $\rho = 0.5$ are provided in Table 1a and b. The conclusions for $\rho = -0.5$ were similar and are not shown. Overall, the variance terms decrease with the sample size. The bias terms, though small, do not always show a monotonic pattern in the range we have attempted. The intercept terms are difficult to estimate when data have contamination. Use of a robust score function helps the situation greatly.

Table 1 (a) Empirical bias and standard deviation (in parentheses) of various estimators when $\phi(x) = x$ and $\rho = 0.5$. All entries are based on 5000 Monte Carlo iterations. (b) Empirical bias and standard deviation (in parentheses) of various estimators when $\phi(x) = \tanh(x)$ and $\rho = 0.5$. All entries are based on 5000 Monte Carlo iterations

(a)	Contamination % = 0					
	Censoring = 25 %			Censoring = 50 %		
Parameter	$n_j = 25$	$n_j = 50$	$n_j = 200$	$n_j = 25$	$n_j = 50$	$n_j = 200$
θ	0.015 (0.210)	0.023 (0.156)	0.015 (0.088)	0.022 (0.247)	0.027 (0.195)	0.029 (0.125)
α_1	−0.039 (0.461)	−0.034 (0.326)	−0.009 (0.181)	−0.192 (0.564)	−0.121 (0.421)	−0.053 (0.255)
β_1	−0.046 (0.266)	−0.031 (0.189)	−0.018 (0.107)	−0.073 (0.337)	−0.066 (0.249)	−0.048 (0.152)
α_2	0.048 (0.684)	0.062 (0.493)	0.043 (0.264)	0.035 (0.853)	0.052 (0.632)	0.073 (0.382)
β_2	−0.066 (0.266)	−0.057 (0.190)	−0.033 (0.101)	−0.136 (0.344)	−0.113 (0.251)	−0.079 (0.151)
	Contamination % = 10					
	Censoring = 25 %			Censoring = 50 %		
Parameter	$n_j = 25$	$n_j = 50$	$n_j = 200$	$n_j = 25$	$n_j = 50$	$n_j = 200$
θ	0.016 (0.220)	0.013 (0.172)	0.010 (0.102)	0.002 (0.258)	0.015 (0.208)	0.020 (0.133)
α_1	−0.324 (0.696)	−0.274 (0.482)	−0.251 (0.271)	−0.468 (0.883)	−0.396 (0.615)	−0.309 (0.352)
β_1	−0.034 (0.402)	−0.034 (0.279)	−0.020 (0.155)	−0.084 (0.529)	−0.085 (0.365)	−0.050 (0.209)
α_2	−0.175 (1.028)	−0.189 (0.715)	−0.199 (0.415)	−0.198 (1.351)	−0.162 (0.890)	−0.183 (0.516)
β_2	−0.075 (0.396)	−0.057 (0.276)	−0.034 (0.158)	−0.164 (0.551)	−0.137 (0.355)	−0.083 (0.202)
(b)	Contamination % = 0					
	Censoring = 25 %			Censoring = 50 %		
Parameter	$n_j = 25$	$n_j = 50$	$n_j = 200$	$n_j = 25$	$n_j = 50$	$n_j = 200$
θ	0.019 (0.218)	0.019 (0.163)	0.012 (0.086)	0.007 (0.250)	0.027 (0.205)	0.032 (0.128)
α_1	−0.030 (0.475)	−0.015 (0.333)	−0.002 (0.176)	−0.134 (0.597)	−0.082 (0.443)	−0.034 (0.260)
β_1	−0.040 (0.279)	−0.027 (0.195)	−0.015 (0.103)	−0.072 (0.353)	−0.066 (0.263)	−0.039 (0.161)
α_2	0.079 (0.733)	0.063 (0.506)	0.040 (0.262)	0.041 (0.915)	0.103 (0.682)	0.100 (0.401)
β_2	−0.066 (0.284)	−0.049 (0.194)	−0.027 (0.101)	−0.122 (0.370)	−0.113 (0.270)	−0.078 (0.160)

(continued)

Table 1 (continued)

	Contamination % = 10					
	Censoring = 25 %			Censoring = 50 %		
Parameter	$n_j = 25$	$n_j = 50$	$n_j = 200$	$n_j = 25$	$n_j = 50$	$n_j = 200$
θ	0.008 (0.207)	0.011 (0.151)	0.007 (0.086)	0.001 (0.249)	0.013 (0.200)	0.020 (0.124)
α_1	−0.148 (0.607)	−0.109 (0.397)	−0.085 (0.206)	−0.332 (0.845)	−0.206 (0.561)	−0.112 (0.305)
β_1	−0.034 (0.340)	−0.025 (0.229)	−0.012 (0.123)	−0.065 (0.487)	−0.060 (0.322)	−0.044 (0.184)
α_2	−0.107 (0.927)	−0.064 (0.628)	−0.048 (0.323)	−0.145 (1.290)	−0.080 (0.866)	−0.003 (0.463)
β_2	−0.050 (0.346)	−0.042 (0.236)	−0.023 (0.127)	−0.138 (0.517)	−0.111 (0.336)	−0.076 (0.186)

3.2 Power Study

We perform a size and power analysis using a test statistic based on $\widehat{U}$ in order to examine whether the waiting time distributions in the two groups exhibit any significant difference after adjusting for the individual specific covariates. For this purpose we consider the same data generation scheme as before except we let $\alpha_2 = \alpha_1 + \Delta$. We only consider the choice $\phi(x) = \tanh(x)$ and set the censoring rate to 25 %. The test statistic we consider is a symmetrized version of $\widehat{U}$:

$$T = 0.5[\widehat{U}(1,2) + 1 - \widehat{U}(2,1)],$$

where $\widehat{U}(1,2)$ is the value of $\widehat{U}$ computed with the observations on the 5-tuples, from the two groups 1 and 2: $(X_{i,j}, \delta_{i,j}, V_{i,j}, \eta_{i,j}, Z_{i,j})$, $(i = 1, 2 \ldots n_j;\ j = 1, 2;\ n_1 + n_2 = n)$, being in their natural order. $\widehat{U}(2,1)$ is the version of $\widehat{U}$ with this order being reversed.

Large sample theory of T can be carried out by linearizing each of $\widehat{U}(1,2)$ and $\widehat{U}(2,1)$. It will be asymptotically normally distributed under H_0, with mean 0.5 and variance σ_T^2, say. We estimate the asymptotic variance of T by implementing the bootstrap resampling technique. For this purpose we generate a resample of size 500 (with replacement) from the observations on the 5-tuple $(X_{i,j}, \delta_{i,j}, V_{i,j}, \eta_{i,j}, Z_{i,j})$; $i = 1, 2, \ldots, n_j; j = 1, 2$, simulated in each of 1000 Monte Carlo replications. We compute the values of T for each of the bootstrap samples and take their sample variance as the estimated asymptotic variance of T ($\widehat{\sigma}_T^2$). With these estimates we construct the 95 % bias corrected confidence interval for the actual population mean of $T(\mu_T)$: $[\widehat{\mu}_T - 1.96\widehat{\sigma}_T, \widehat{\mu}_T + 1.96\widehat{\sigma}_T]$. We calculate the proportion of times the mean of $T(\mu_T)$ under the null distribution (0.5) is not covered by this interval, in order to get the size and power values for the corresponding model settings controlled by the parameters α_1 and α_2. These values are reported in Table 2 for two choices of per group sample sizes ($n_i = 50$ and 200).

Table 2 Empirically estimated size and power of test of equality of adjusted waiting times

Δ	Size/Power	
	$n_1/n_2 = 50$	$n_1/n_2 = 200$
−2.3	0.890	0.998
−1.8	0.734	0.990
−1.3	0.526	0.942
−0.8	0.250	0.642
0.0	0.080	0.070
0.8	0.338	0.704
1.3	0.616	0.962
1.8	0.782	1.000
2.3	0.894	1.000

Each value is based on 1000 Monte Carlo iterations

For the case of $n_i = 50$, size of the test is slighted inflated compared to the the nominal size (5 %) and the power increases gradually as the waiting time distributions in the two groups differ more and more owing to the extent of variation in the intercept parameters α_1 (= 0.5) and α_2 corresponding to groups 1 and 2, respectively. Moreover, the size gets closer to the nominal size and the test exhibits great power as the number of subjects in the two groups increases to 200.

4 Application to Spinal Cord Injury Data

In this section, we illustrate an application of our covariate adjusted Mann–Whitney test on a data set of NeuroRecovery Network (NRN) [8, 9] patients. The sample consists of 269 patients who are enrolled in a locomotor training program after suffering a spinal cord injury (a grade of C or D on the International Standards for Neurological Classification of Spinal Cord Injury scale). Out of these, 148 were in Phase 1 and 121 were in Phase 2 at enrolment; the concept of initial phase is given in Datta et al. [6].

Continued evaluation of the walking speed constitutes a fundamental part of the entire monitoring process of the patients following their time of enrollment in the program. On the basis of the performances in a walking test these individuals are subsequently classified into different speed categories by virtue of their maximum walking speed. Following clinical benchmarks these categories are represented as specific speed limits which can be jointly visualized as a three-state progressive model (Fig. 1). Overall, 16 individuals made a transition from state 1 to 2, 33 individuals moved from state 2 to 3 and 8 individuals moved directly from state 1 to 3. We consider three individual specific covariates that may potentially control the movement of the individuals along the different states in the model. These covariates are: (1) time from the spinal cord injury to enrollment in the program, (2) lower motor score from the international standards for neurological classification of spinal cord injury (ISNCSCI) exam and (3) treatment intensity given by the ratio of the cumulative number of training sessions received by the individual and his/her duration of enrollment in the program.

Now, we create two different groups of injured patients depending on their initial phase at the time of enrollment (1 or 2). Our objective is to use the modified IPCW-based U-statistic to compare the sojourn time distributions at stage 2 between these two categories of patients after adjusting for their information on the three covariates described above.

Application of the Mann–Whitney test based on our U-statistic (as discussed in Sect. 3.2) gives the absolute value of the test statistic T as 0.639 (< 1.96). But, using the Fan-Datta U-statistic [7] we get $|T| = 4.950\,(> 1.96)$. This demonstrates that there is indeed a substantial difference between the overall sojourn time distributions for the injured patients enrolled in the initial Phases 1 and 2. However, this effect can either be due to the difference between the covariate distributions in the two groups of patients or a variation in its impact over the two groups (characterized

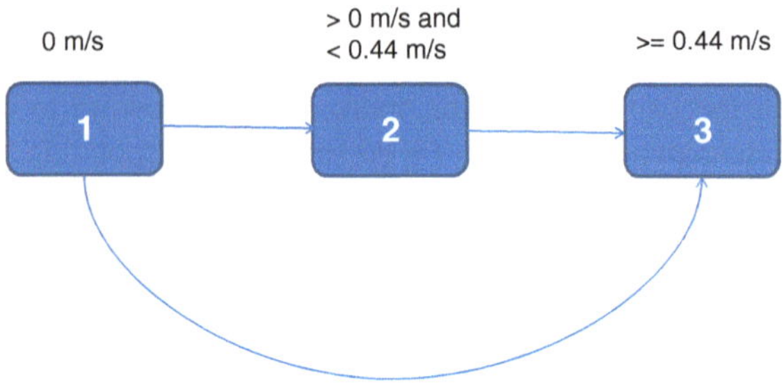

Fig. 1 Network showing the different states and their transition paths for the spinal cord injury data

by the regression coefficients) or a combined effect from both of them. There is no significant difference in the covariate adjusted log-waiting times in the two groups.

5 Discussion

U-statistics are fundamental objects in theoretical statistics and provide a broad generalization of different types of commonly used measures in the statistical analyses (sample mean, variance, etc.) [10, 15, 16]. Different types of statistics with complicated expressions (that are not readily amenable to algebraic treatments) can be expressed as U-statistics, or approximate U-statistics [16], thereby facilitating their asymptotic treatments (consistency, asymptotic normality, etc.) in an unified fashion.

Mann–Whitney U-statistics [12] are well known in this context and can be used to test the equality of two probability distributions by formulating an indicator kernel function in terms of the observed sample values on their corresponding random variables. Fan and Datta [7] initiated the development of a modified Mann–Whitney U-statistic from a right censored data on the sojourn times of individuals classified into two groups. Specifically, their work was focused on the use of this modified statistic to compare the stage waiting time distributions between two groups of subjects/individuals progressing through the different branches of a multistate network affected by right censoring. In the present context we have pursued an extension of this work to build a covariate adjusted version of the Mann–Whitney U-statistic to ensure a more detailed inference on the comparison of the waiting time distributions between the two groups of individuals.

We have demonstrated the performance of our modified U-statistic in terms of its empirical bias and standard deviation through extensive simulation studies involving different censoring patterns. We have illustrated the usefulness of the test based on our modified Mann–Whitney U-statistic by a power analysis. Moreover, application of our modified Mann–Whitney U-test on the spinal cord injury data shows that the apparent difference between the sojourn time distributions of the two categories of patients is actually explained by the three observed covariates.

Another byproduct of the current paper is the methodology of robust estimation of the regression parameters in an accelerated failure time model for the state waiting times under right censoring. Large sample theory of the proposed estimators can be developed following classical techniques of M-estimators combined with the martingale approaches of [14]. Bootstrap remains a viable alternative for inference on the regression parameters as well.

It will be interesting to extend this methodology to other form of censored data problems such as current status or interval censored entry/exit times. It is not immediately obvious how to do this. We hope to explore this direction in future research work.

In this development, we have made the implicit assumption that every individual in the uncensored experiment eventually enter and exit the stage whose sojourn times are being compared. However, in a general multistate system, this may not always be the case. In such situations, one may compare the sojourn times conditional on stage entry using our statistic.

Acknowledgments This research was supported in parts by grants from the US National Science Foundation (DMS-0706965) and the National Security Agency (H98230-11-1-0168). Chakraborty acknowledges fellowship support and a dissertation completion award from University of Louisville. The authors thank an anonymous reviewer for helpful comments; they also thank the Christopher and Dana Reeve Foundation and all current and past members of the NeuroRecovery Network.

Appendix

A.1 Proof of Theorem 1

Define the following auxiliary "U-statistics":

$$U^p = \frac{1}{n_1 n_2} \sum_{i_1=1}^{n_1} \sum_{i_2=1}^{n_2} \frac{I(e^{-Z_{i_1,1}^T \beta_1} W_{i_1,1} \le e^{-Z_{i_2,2}^T \beta_2} W_{i_2,2}) \delta_{i_1,1} \eta_{i_2,2}}{K_1(V_{i_1,1}-) K_2(X_{i_2,2} + C(Z_i; \beta) W_{i_1,1}-)},$$

$$U^h = \frac{1}{n_1 n_2} \sum_{i_1=1}^{n_1} \sum_{i_2=1}^{n_2} \frac{I(e^{-Z_{i_1,1}^T \widehat{\beta}_1} W_{i_1,1} \le e^{-Z_{i_2,2}^T \widehat{\beta}_2} W_{i_2,2}) \delta_{i_1,1} \eta_{i_2,2}}{K_1(V_{i_1,1}-) K_2(X_{i_2,2} + C(Z_i; \widehat{\beta}) W_{i_1,1}-)}.$$

Then our primary U-statistic $\widehat{U}$ can be written as

$$\sqrt{n}(\widehat{U} - \theta) = \sqrt{n}(U^p - \theta) + \sqrt{n}(U^h - U^p) + \sqrt{n}(\widehat{U} - U^h). \quad (4)$$

As in Fan and Datta [7], we will derive the large sample linear approximations of the three expressions on the right hand side of (4) and combine them to obtain the asymptotic distribution of $\widehat{U}$. By the projection theorem of U-statistics ([16], page 188) and the definitions of various functions and weights, we get after some expectation calculations

$$\sqrt{n}(U^p - \theta) = \frac{1}{n}\Bigg[\frac{1}{c_1}\sum_{i_1=1}^{n_1}\left\{\frac{\delta_{i_1,1}S_2(e^{-Z_{i_1,1}^T\beta_1}W_{i_1,1})}{K_1(V_{i_1,1}-)} - \theta\right\}$$

$$+ \frac{1}{c_2}\sum_{i_2=1}^{n_2}\left\{\eta_{i_2,2}\int\frac{I(C(z, Z_{i_2,2};\beta_1,\beta_2)w < W_{i_2,2})}{K_1(v-)K_2(X_{i_2,2}+C(z, Z_{i_2,2};\beta_1,\beta_2)w-)}dG_1(v,w,z) - \theta\right\}\Bigg] + o_p(1).$$

For the second part, we apply the delta method type approximation on U^h along with laws of large numbers for U-statistics to get

$$\sqrt{n}(U^h - U^p) = \left[\frac{1}{n_1 n_2}\sum_{i_1=1}^{n_1}\sum_{i_2=1}^{n_2}A_{i_1,i_2}\right]\sqrt{n}(\widehat{\beta_1} - \beta_1)$$

$$+ \left[\frac{1}{n_1 n_2}\sum_{i_1=1}^{n_1}\sum_{i_2=1}^{n_2}B_{i_1,i_2}\right]\sqrt{n}(\widehat{\beta_2} - \beta_2) + o_p(1),$$

where

$$A_{i_1,i_2} = \frac{K_2'(X_{i_2,2} + C(\beta;Z_i)W_{i_1,1}-)C(\beta;Z_i)W_{i_1,1}I(C(\beta;Z_i)W_{i_1,1} < W_{i_2,2})Z_{i_1,1}^T}{K_1(V_{i_1,1}-)K_2^2(X_{i_2,2} + C(\beta;Z_i)W_{i_1,1}-)}$$

and

$$B_{i_1,i_2} = \frac{-K_2'(X_{i_2,2} + C(\beta;Z_i)W_{i_1,1}-)C(\beta;Z_i)W_{i_1,1}I(C(\beta;Z_i)W_{i_1,1} < W_{i_2,2})Z_{i_2,2}^T}{K_1(V_{i_1,1}-)K_2^2(X_{i_2,2} + C(\beta;Z_i)W_{i_1,1}-)}.$$

By the Weak Law of Large Numbers and the theory of estimating equations the above is asymptotically equivalent (i.e., up to $o_p(1)$ terms) to

$$\frac{1}{n}\left[\frac{1}{c_1}AC_{Z_1}^{-1}\sum_{i_1=1}^{n_1}\psi(\epsilon_{i_1})\frac{\delta_{i_1,1}}{K_1(V_{i_1,1}-)}Z_{i_1,1} + \frac{1}{c_2}BC_{Z_2}^{-1}\sum_{i_2=1}^{n_2}\psi(\epsilon_{i_2})\frac{\delta_{i_2,2}}{K_2(V_{i_2,2}-)}Z_{i_2,2}\right],$$

where $A = E(A_{i_1,i_2})$, $B = E(B_{i_1,i_2})$, and $C_{Z_h} = E(\psi'(\epsilon_{i_h})Z_h\tilde{Z}_h^T)$, $h = 1, 2$.

Now for the third and final part, we may replace treat β by $\widehat{\beta}$ leading to an error that is $o_p(1)$, since the quantities are centered and the analysis will hold uniformly in a small neighborhood of the true β. Thus, we get

$$\sqrt{n}(\widehat{U} - U^h) = -\frac{\sqrt{n}}{n_1 n_2} \sum_{i_1=1}^{n_1} \sum_{i_2=1}^{n_2} \Bigg[\frac{I(C(\beta; Z_i) W_{i_1,1} < W_{i_2,2}) \delta_{i_1,1} \eta_{i_2,2}}{\widehat{K}_1(V_{i_1,1}-)}$$
$$\times \left\{ \frac{\widehat{K}_2(X_{i_2,2} + C(\beta; Z_i) W_{i_1,1}-) - K_2(X_{i_2,2} + C(\beta; Z_i) W_{i_1,1}-)}{\widehat{K}_2(X_{i_2,2} + C(\beta; Z_i) W_{i_1,1}-) K_2(X_{i_2,2} + C(\beta; Z_i) W_{i_1,1}-)} \right\}$$
$$+ \left\{ \frac{I(e^{-Z_{i_1,1}\beta_1} W_{i_1,1} < e^{-Z_{i_2,2}\beta_2} W_{i_2,2}) \delta_{i_1,1} \eta_{i_2,2}}{K_2(X_{i_2,2} + C(\beta; Z) W_{i_1,1}-)} \right\} \left\{ \frac{\widehat{K}_1(V_{i_1,1}-) - K_1(V_{i_1,1}-)}{\widehat{K}_1(V_{i_1,1}-) K_1(V_{i_1,1}-)} \right\} \Bigg]. \tag{5}$$

Now by an L_1 analysis of the difference (as in Fan and Datta [7]) we can replace $\widehat{K}_j$ by $K_j, j = 1, 2$ in the last two expressions on the R.H.S. of (5). Again, by delta method we have

$$\sqrt{n_j}(\widehat{K}_j - K_j) = -\sqrt{n_j} K_j (\widehat{\Lambda}_j^c - \Lambda_j^c) + o_p(1),$$

where Λ_j^c is the cumulative censoring hazard and $\widehat{\Lambda}_j^c$ is the corresponding Nelson–Aalen estimator.

Thus

$$\sqrt{n}(\widehat{U} - U^h) = \frac{\sqrt{n}}{n_1 n_2} \sum_{i_1=1}^{n_1} \sum_{i_1=1}^{n_2} \frac{I(C(\beta; Z_i) W_{i_1,1} < W_{i_2,2}) \delta_{i_1,1} \eta_{i_2,2}}{K_1(V_{i_1,1}-) K_2(X_{i_2,2} + C(Z; \beta) W_{i_1,1}-)}$$
$$\times \Bigg[\left\{ \widehat{\Lambda}_1^c(V_{i_1,1}-) - \Lambda_1^c(V_{i_1,1}-) \right\}$$
$$+ \left\{ \widehat{\Lambda}_2^c(X_{i_2,2} + C(Z_i; \beta) W_{i_1,1}-) - \Lambda_2^c(X_{i_2,2} + C(Z_i; \beta) W_{i_1,1}-) \right\} \Bigg] + o_p(1).$$

Now using projection calculations, the above expression boils down to

$$\frac{1}{\sqrt{n}} \Bigg[\frac{1}{c_1} \sum_{i_1=1}^{n_1} \frac{S_2(e^{-Z_{i_1,1}^T \beta_1} W_{i_1,1}) \delta_{i_1,1}}{K_1(V_{i_1,1}-)} \left\{ \widehat{\Lambda}_1^c(V_{i_1,1}-) - \Lambda_1^c(V_{i_1,1}-) \right\}$$
$$+ \frac{1}{c_2} \sum_{i_2=1}^{n_2} \eta_{i_2,2} \int \frac{I(C(z, Z_{i_2}; \beta_1, \beta_2) w < W_{i_2,2})}{K_2(X_{i_2,2} + C(z, Z_{i_2}; \beta_1, \beta_2) w-)} \times \Big\{ \widehat{\Lambda}_2^c(X_{i_2,2} + C(z, Z_{i_2}; \beta_1, \beta_2) w-)$$
$$- \Lambda_2^c(X_{i_2,2} + C(z, Z_{i_2}; \beta_1, \beta_2) w-) \Big\} dG_1(v, w, z) \Bigg]$$

which further equals by martingale representations of $\widehat{\Lambda}_j^c - \Lambda_j^c$ (Andersen et al. 1993),

$$\frac{1}{\sqrt{n}} \Bigg[\frac{1}{c_1 \sqrt{n_1}} \sum_{i_1=1}^{n_1} \frac{\delta_{i_1,1} S_2(e^{-Z_{i_1,1}^T \beta_1} W_{i_1,1})}{K_1(V_{i_1,1}-)} \left\{ \int_0^{V_{i_1,1}-} \frac{d\overline{M}_1^c(s)}{y_1(s)} \right\}$$
$$+ \frac{1}{c_2 \sqrt{n_2}} \sum_{i_2=1}^{n_2} \eta_{i_2,2} \Bigg\{ \int \frac{I(C(z, Z_{i_2}; \beta_1, \beta_2) w < W_{i_2,2})}{K_2(X_{i_2,2} + C(z, Z_{i_2}; \beta_1, \beta_2) w-)}$$

$$\times \left(\int_0^{X_{i_2,2}+C(z,Z_{i_2};\beta_1,\beta_2)w-} \frac{d\overline{M}_2^c(s)}{y_2(s)} \right) \Bigg\} dG_1(v,w,z) \Bigg] + o_p(1), \tag{6}$$

where $\overline{M}_j^c = n_j^{-1/2} \sum_{i_j=1}^{n_j} M_{i_j,j}^c$; $j = 1, 2$.

Now, by the asymptotically linear representation of an U-statistic, the first part of the above expression inside the square brackets equals:

$$\frac{\sqrt{n_1}}{c_1} \int \frac{S_2(e^{-z^T\beta_1}w)}{K_1(v-)} \left\{ \int_0^{v-} \frac{d\overline{M}_1^c(s)}{y_1(s)} \right\} dG_1(v,w,z) + o_p(\sqrt{n_1})$$

which by Fubini's theorem is equal to

$$\frac{1}{c_1} \sum_{i_1=1}^{n_1} \int_0^\infty \left\{ \frac{1}{y_1(s)} \int I(v > s) \frac{S_2(e^{-z^T\beta_1}w)}{K_1(v-)} dG_1(v,w,z) \right\} dM_{i_1,1}^c(s) + o_p(\sqrt{n_1})$$

$$= \frac{1}{c_1} \sum_{i_1=1}^{n_1} \int_0^\infty \omega_1(s) dM_{i_1,1}^c(s) + o_p(\sqrt{n_1}).$$

By a similar treatment as above the second part of the square brackets in RHS of (6) equals (up to $o_p(\sqrt{n_2})$)

$$\frac{1}{c_2} \iint \frac{I(C(z;\beta)w_1 < w_2)}{K_2(x_2 + C(z;\beta)w_1-)} \left\{ \int_0^{x_2+C(z;\beta)w_1-} \frac{d\overline{M}_2^c(s)}{y_2(s)} \right\} dG_1(v_1,w_1,z_1) dG_2(x_2,w_2,z_2)$$

which again by Fubini's theorem equals

$$\frac{\sqrt{n_2}}{c_2} \int_0^\infty \left\{ \frac{1}{y_2(s)} \iint \frac{I(s < x_2 + C(z;\beta)w_1) I(C(z;\beta)w_1 < w_2)}{K_2(x_2 + C(z;\beta)w_1-)} dG_1(v_1,w_1,z_1) \right.$$
$$\left. \times\, dG_2(x_2,w_2,z_2) \right\} d\overline{M}_2^c(s) + o_p(\sqrt{n_2})$$
$$= \frac{1}{c_2} \sum_{i_2=1}^{n_2} \int_0^\infty \omega_2(s) dM_{i_2,2}^c(s) + o_p(\sqrt{n_2}).$$

Now, combining the asymptotic expressions for all the three parts we finally get:

$$\sqrt{n}(\widehat{U} - \theta) = \frac{1}{\sqrt{n}} \left[\frac{1}{c_1} \sum_{i_1=1}^{n_1} \left\{ \frac{\delta_{i_1,1}\{S_2(e^{-Z_{i_1,1}^T\beta_1} W_{i_1,1}) + AC_{Z_1}^{-1}\psi(\epsilon_{i_1}) Z_{i_1,1}\}}{K_1(V_{i_1,1}-)} - \theta \right.\right.$$
$$\left. + \int_0^\infty \omega_1(s)\, dM_{i_1,1}^c(s) \right\}$$

$$+ \frac{1}{c_2} \sum_{i_2=1}^{n_2} \left\{ \eta_{i_2,2} \int \frac{I(C(z, Z_{i_2,2}; \beta_1, \beta_2)w < W_{i_2,2})}{K_1(v-)K_2(X_{i_2,2} + C(z, Z_{i_2,2}; \beta_1, \beta_2)w-)} dG_1(v, w, z) - \theta \right.$$

$$\left. + \frac{\delta_{i_2,2} BC_{Z_2}^{-1} \psi(\epsilon_{i_2}) Z_{i_2,2}}{K_2(V_{i_2,2}-)} + \int_0^\infty \omega_2(s) dM_{i_2,2}^c(s) \right\} \Bigg] + o_p(1), \tag{7}$$

which by the Central Limit Theorem converges asymptotically to a normal distribution with mean 0 and variance σ^2, with the expression for σ^2 being provided in the statement of Theorem 1.

A.2 Variance Estimation

We estimate each summand in the linear approximation of the U-statistic by their empirical versions and then use the sample variances of the estimated summands to compute the overall variance estimator. Thus we have

$$\widehat{\sigma}^2 = \frac{n}{n_1(n_1-1)} \sum_{i_1=1}^{n_1} (L_{i_1,1} - \bar{L}_1)^2 + \frac{n}{n_2(n_2-1)} \sum_{i_2=1}^{n_2} (L_{i_2,2} - \bar{L}_2)^2$$

where

$$L_{i_1,1} = \frac{\delta_{i_1,1}}{\widehat{K}_1(V_{i_1,1}-)} \left\{ \widehat{S}_2(e^{-Z_{i_1,1}^T \widehat{\beta}_1} W_{i_1,1}) + \left(\frac{1}{n_1 n_2} \sum_{i_1', i_2'} A_{i_1', i_2'} \right) \right.$$

$$\left. \times \left(\sum_{i_1'} \frac{\delta_{i_1',1}}{\widehat{K}_1(V_{i_1',1}-)} \psi'(R_{i_1',1} - \widehat{\alpha}_1) Z_{i_1',1} \tilde{Z}_{i_1',1}^T \right)^{-1} \psi(R_{i_1,1} - \widehat{\alpha}_1) Z_{i_1,1} \right\}$$

$$+ \widehat{\omega}_1(V_{i_1,1}) \bar{\delta}_{i_1,1} - \sum_{i_1'=1}^{n_1} \frac{\widehat{\omega}_1(V_{i_1',1}) I(V_{i_1,1} \geq V_{i_1',1}) \bar{\delta}_{i_1',1}}{Y_1(V_{i_1',1}-)},$$

and

$$L_{i_2,2} = \eta_{i_2,2} \frac{1}{n_1} \sum_{i_1=1}^{n_1} \frac{I(R_{i_1,1} \leq R_{i,2}) \delta_{i_1,1}}{\widehat{K}_1(V_{i_1,1}-) \widehat{K}_2(X_{i_2,2} + C(Z_i; \widehat{\beta}) W_{i_1,1}-)} + \frac{\delta_{i_2,2}}{\widehat{K}_2(V_{i_2,2}-)}$$

$$\left(\frac{1}{n_1 n_2} \sum_{i_1', i_2'} B_{i_1', i_2'} \right) \left(\sum_{i_2'} \frac{\delta_{i_2',2}}{\widehat{K}_2(V_{i_2',2}-)} \psi'(R_{i_2',2} - \widehat{\alpha}_2) Z_{i_2',2} \tilde{Z}_{i_2',2}^T \right)^{-1} \psi(R_{i_2,1} - \widehat{\alpha}_2) Z_{i_2,2}$$

$$+ \widehat{\omega}_2(V_{i_2,2}) \bar{\delta}_{i,2} - \sum_{i_2'=1}^{n_2} \frac{\widehat{\omega}_2(V_{i_2',2}) I(V_{i_2,2} \geq V_{i_2',2}) \bar{\delta}_{i_2',2}}{Y_2(V_{i_2',2}-)},$$

with

$$\widehat{\omega}_1(s) = \frac{1}{Y_1(s)} \sum_{i_1=1}^{n_1} I(V_{i_1,1} > s) \frac{\widehat{S}_2^h(e^{-Z_{i_1,1}^T \widehat{\beta}_1} W_{i_1,1})\delta_{i_1,1}}{\widehat{K}_1(V_{i_1,1}-)}$$

$$\widehat{\omega}_2(s) = \frac{1}{n_1 Y_2(s)} \sum_{i_1=1}^{n_1} \sum_{i_2=1}^{n_2} \frac{\delta_{i_1,1}\eta_{i_2,2} I(X_{i_2,2} + C(Z_i; \widehat{\beta}) W_{i_1,1} > s) I(R_{i_1,1} < R_{i_2,2})}{\widehat{K}_1(V_{i_1,1}-)\widehat{K}_2(X_{i_2,2} + C(Z_i; \widehat{\beta}) W_{i_1,1})},$$

$$\widehat{S}_2^h(e^{-Z_{i_1,1}^T \widehat{\beta}_1} W_{i_1,1}) = \frac{1}{n_2} \sum_{i_2=1}^{n_2} \frac{I(C(Z_i; \widehat{\beta}) W_{i_2,2} > W_{i_1,1})\delta_{i_2,2}}{\widehat{K}_2(V_{i_2,2}-)}.$$

References

1. Beran, R.: Nonparametric regression with randomly censored survival data. Technical report. University of California, Berkeley (1981)
2. Cox, D.R.: Regression models and life-tables. J. R. Stat. Soc. Ser. B **34**, 187–202 (1972)
3. Datta, S. Estimating the mean life time using right censored data. Stat. Methodol. **2**, 65–69 (2005)
4. Datta, S., Satten, G.A.: Validity of the Aalen-Johansen estimators of stage occupation probabilities and integrated transition hazards for non-Markov models. Stat. Probab. Lett. **55**, 403–411 (2001)
5. Datta, S., Bandyopadhyay, D., Satten, G.A.: Inverse probability of censoring weighted U-statistics for right censored data with applications. Scandinavian J. Stat. **37**, 680–700 (2010)
6. Datta, S., Lorenz, D.J., Harkema, S.J.: A dynamic longitudinal evaluation of the utility of the Berg Balance Scale in patients with motor incomplete spinal cord injury. Arch. Phys. Med. Rehabil. **93**, 1565–1573 (2012)
7. Fan, J., Datta, S.: Mann-Whitney tests for comparing sojourn time distributions when the transition times are right censored. Ann. Inst. Stat. Math. **65**, 149–166 (2013)
8. Harkema, S.J., Schmidt-Read, M., Behrman, A.L., Bratta, A., Sisto, S.A., Edgerton, V.R.: Establishing the NeuroRecovery Network: multisite rehabilitation centers that provide activity-based therapies and assessments for neurologic disorders. Arch. Phys. Med. Rehabil. **93**, 1498–1507 (2011)
9. Harkema, S.J., Schmidt-Read, M., Lorenz, D., Edgerton, V.R., Behrman, A.L.: Balance and ambulation improvements in individuals with chronic incomplete spinal cord injury using locomotor training based rehabilitation. Arch. Phys. Med. Rehabil. **93**, 1508–1517 (2011b)
10. Hoeffding, W.: A class of statistics with asymptotically normal distribution. Ann. Math. Stat. **19**, 293–325 (1948)
11. Huang, Y.: Censored regression with the multistate accelerated sojourn times model. J. R. Stat. Soc. Ser. B **64**, 17–29 (2002)
12. Mann, H.B., Whitney, D.R.: On a test of whether one of two random variables is stochastically larger than the other. Ann. Math. Stat. **18**, 50–60 (1947)
13. Satten, G.A., Datta, S.: Marginal estimation for multistage models: waiting time distributions and competing analyses. Stat. Med. **21**, 3–19 (2002)
14. Satten, G.A., Datta, S., Robins, J.M.: An estimator for the survival function when data are subject to dependent censoring. Stat. Probab. Lett. **54**, 397–403 (2001)
15. Schisterman, E., Rotnizky, A.: Estimation of the mean of a K-sample U-statistic with missing outcomes and auxiliaries. Biometrika **88**, 713–725 (2001)
16. Serfling, R.J.: Approximation Theorems of Mathematical Statistics. Wiley, New York (1980)
17. Strawderman, R.L.: The accelerated gap times model. Biometrika **92**, 647–666 (2005)

Claim Reserving Using Distance-Based Generalized Linear Models

Eva Boj and Teresa Costa

Abstract Generalized linear models (GLM) can be considered a stochastic version of the classical Chain-Ladder (CL) method of claim reserving in nonlife insurance. In particular, the deterministic CL model is reproduced when a GLM is fitted assuming over-dispersed Poisson error distribution and logarithmic link. Our aim is to propose the use of distance-based generalized linear models (DB-GLM) in the claim reserving problem. DB-GLM can be considered a generalization of the classical GLM to the distance-based analysis, because DB-GLM contains as a particular instance ordinary GLM when the Euclidean, l^2, metric is applied. Then, DB-GLM can be considered too a stochastic version of the CL claim reserving method. In DB-GLM, the only information required is a predictor distance matrix. DB-GLM can be fitted using the `dbstats` package for `R`. To estimate reserve distributions and standard errors, we propose a nonparametric bootstrap technique adequate to the distance-based regression models. We illustrate the method with a well-known actuarial dataset.

Keywords Reserving · Chain-Ladder · Generalized linear models · Distance-based prediction · dbstats.

1 Introduction

The objective of this work is to propose the DB-GLM [7] as an alternative methodology to solve the claim reserving problem. To complete the tool, we propose using the nonparametric technique of bootstrapping pairs [9] to estimate the predictive distribution of reserves. Bootstrapping pairs is an adequate bootstrap technique for DB-GLM as is proposed in [8].

E. Boj (✉) · T. Costa
Facultat d'Economia i Empresa, Universitat de Barcelona,
Avinguda Diagonal 690, 08034 Barcelona, Spain
e-mail: evaboj@ub.edu

T. Costa
e-mail: tcosta@ub.edu

R. Cao et al. (eds.), *Nonparametric Statistics*, Springer Proceedings
in Mathematics & Statistics 175, DOI 10.1007/978-3-319-41582-6_10

To contextualize the claim reserving problem we list some of the pioneers deterministic methods in the literature (see [1, 20]): Grossing-up, Link ratio, CL, variants of CL, de Vylder least squares, and arithmetic and geometric separation methods of Taylor. Later some stochastic methods were proposed, both probabilistic and Bayesian. A list of some of those methods is: the Mack model, the Bornhuetter-Ferguson model and the Munich Chain-Ladder method. The three methods have the common characteristic of generalizing (from a stochastic point of view) the CL deterministic method, because the estimation of the reserves coincides in all of them.

The same occurs when using GLM (see [15]) as a stochastic method of claim reserving, with the assumptions of over-dispersed Poisson distribution and logarithmic link. In this case, the estimation of the reserves coincides with that of the CL deterministic method (see [14, 18, 21, 22] for a detailed proof). Additionally, the GLM has as particular cases other deterministic methods as are: the least squares of de Vylder and the arithmetic and geometric separation methods of Taylor. Therefore, in the last decade, GLM has been studied in the actuarial literature in application to the claim reserving problem. We refer to [2, 10–12] or [5] for a detailed description of the calculus of prediction errors when using GLM and when the problem is estimating reserves.

When using the l^2 metric (the named Euclidean), DB-GLM reproduces ordinary GLM, and then we can consider that DB-GLM is a generalization of GLM in the context of distance-based analysis. If using the Euclidean metric and assuming the over-dispersed Poisson distribution and the logarithmic link with DB-GLM, the classical CL method is reproduced. In this study, we apply DB-GLM to the claim reserving problem assuming these hypothesis as an alternative method which generalizes the classical CL deterministic method to the stochastic framework. With DB-GLM we have the same particular cases than with GLM when using the Euclidean metric. But with DB-GLM we have more treatments for the analysis because we could use other metrics besides the Euclidean one.

The work is structured as follows: In Sect. 2 we describe the problem of claim reserving and the CL deterministic method. In Sect. 2.1 we show how GLM is as a stochastic generalization of the classical CL. In Sect. 3 we list the main characteristics of DB-GLM, we propose a nonparametric bootstrap technique for the DB-GLM and we describe function `dbglm` of the `dbstats` package for `R`. In Sect. 4 we make a numerical example to illustrate the proposed method of claim reserving. Finally, in Sect. 5 we summarize the main results of the study.

2 Claim Reserving: The Chain-Ladder Method

Consider a portfolio of risks and assume that each claim is settled either in the accident year or in the following k development years. Consider a family of random variables $\{c_{ij}\}_{i,j\in\{0,1,\dots,k\}}$, where c_{ij} is the amount of claim losses of accident year i which is paid with a delay of j years and hence in development year j and in calendar year $i+j$. We refer to c_{ij} as the incremental loss of accident year i and development year

Table 1 Run-off triangle with incremental losses

	0	1	...	j	...	$k-i$	...	$k-1$	k
0	$c_{0,0}$	$c_{0,1}$	...	$c_{0,j}$	...	$c_{0,k-i}$	...	$c_{0,k-1}$	$c_{0,k}$
1	$c_{1,0}$	$c_{1,1}$	...	$c_{1,j}$	...	$c_{1,k-i}$	...	$c_{1,k-1}$	
...	...	...	...	...	...	...	...		
i	$c_{i,0}$	$c_{i,1}$	...	$c_{i,j}$	...	$c_{i,k-i}$			
...	...	...	...	...	...				
$k-j$	$c_{k-j,0}$	$c_{k-j,1}$	...	$c_{k-j,j}$					
...	...	...	...						
$k-1$	$c_{k-1,0}$	$c_{k-1,1}$							
k	$c_{k,0}$								

j. Assume that the incremental losses c_{ij} are observable for calendar years $i+j \leq k$ and that they are collected in a run-off triangle as in Table 1.

In the run-off triangle, the numbers are grouped by year of origin i, by rows, and by development year j, by columns. The numbers on the against-diagonals with $i+j=t$ denote the payments that were made in the calendar year t. The incremental losses are unobservable for calendar years $i+j \geq k+1$ and the problem is to predict those nonobservable incremental losses.

There are various kinds of incurred but not reported (IBNR) claim reserves which are of interest. The reserves for the different accident years $i=1,\ldots,k$ which are obtained by adding the future incremental losses in the corresponding row of the square, the total reserve which is calculated by adding all the future incremental losses predicted in the bottom-right part of the run-off triangle, and the reserves for the different calendar years $t=k+1,\ldots,2k$ which are obtained by adding the incremental losses that were made in the future calendar years, i.e., the values of the same against-diagonal t. We refer to [2, 5] for a detailed description.

As explained in Sect. 1, one of the first methods proposed in the literature to estimate reserves was the CL method. The method works with cumulative losses

$$C_{ij} = \sum_{r=0}^{j} c_{ir}.$$

The idea behind the CL method is that, in any development year, about the same percentage of the claims from each year of origin will be settled. It is assumed that the development of the losses of every accident year follows a development pattern which is common to all accident years. A vector $m=(m_1,\ldots,m_k)$ of parameters is said to be a development pattern for factors if the identity

$$m_j = \frac{E\left[C_{ij}\right]}{E\left[C_{ij-1}\right]},$$

holds for all $j \in \{1, \ldots, k\}$ and for all $i \in \{0, 1, \ldots, k\}$. Thus, a development pattern for factors exists if, and only if, for every development year $j \in \{1, \ldots, k\}$ the individual factors $m_{ij} = \frac{E[C_{ij}]}{E[C_{ij-1}]}$ are identical for all accident years. Then, for every development year $j \in \{1, \ldots, k\}$, each of the empirical individual factors $\hat{m}_{ij} = \frac{C_{ij}}{C_{ij-1}}$ with $i \in \{0, 1, \ldots, k-j\}$ is a reasonable estimator of m_j and this is also true for every weighted mean $\hat{m}_j = \sum_{h=0}^{k-j} w_{hj}\hat{m}_{hj}$ with random variables (or constants) satisfying $\sum_{h=0}^{k-j} w_{hj} = 1$. The estimator of the CL factor is obtained by:

$$\hat{m}_j = \frac{\sum_{h=0}^{k-j} C_{hj}}{\sum_{h=0}^{k-j} C_{hj-1}}.$$

Then, CL predictors of the cumulative losses C_{ij} with $i + j \geq k + 1$ are defined as

$$\hat{C}_{ij} = C_{i\,k-i} \prod_{h=k-i+1}^{j} \hat{m}_h.$$

Finally, the incremental losses can be calculated by differences.

2.1 Generalized Linear Model: A Stochastic Chain-Ladder Method

Now, we describe how GLM is applied in the estimation of reserves. Assume a GLM to model the incremental losses of the run-off triangle. Assume an over-dispersed Poisson distribution

$$\mu_{ij} = E\left[c_{ij}\right] \text{and } Var\left[c_{ij}\right] = \left(\varphi/w_{ij}\right) V\left(\mu_{ij}\right) = \left(\varphi/w_{ij}\right) \mu_{ij}$$

where φ is the dispersion parameter and w_{ij} are a priori weights of the data, assumed equal to one, $w_{ij} = 1$, for the incremental claim losses of a run-off triangle. Assume for the GLM the logarithmic link function $\log \mu_{ij} = \eta_{ij}$.

Then, we can define $\log\left(\mu_{ij}\right) = c_0 + \alpha_i + \beta_j$, a GLM in which the responses c_{ij} are modeled as random variables with linear predictor $\eta_{ij} = c_0 + \alpha_i + \beta_j$, where α_i is the factor corresponding to the accident year $i = 1, \ldots, k$ and β_j is the factor corresponding to the development year $j = 1, \ldots, k$. The c_0 value is the term corresponding to the accident year 0 and development year 0. In the Poisson case, where $\varphi = 1$ is assumed, over-dispersion is taken into account by estimating the unknown scale parameter φ as a part of the fitting procedure. Then predicted values $\hat{c}_{ij}$ of the IBNR reserves are estimated from $\hat{c}_{ij} = \exp\left(\hat{c}_0 + \hat{\alpha}_i + \hat{\beta}_j\right)$.

The estimates of the cumulative losses $\hat{C}_{ik}$ with $i = 1, 2, \ldots, k$ in this GLM may be obtained from the sums

$$\hat{C}_{ik} = C_{i\ k-i} + \sum_{j=k-i+1}^{k} e^{\hat{c}_0 + \hat{\alpha}_i + \hat{\beta}_j}$$

where $\hat{c}_0, \hat{\alpha}_i$ and $\hat{\beta}_j$ are the maximum likelihood estimates of the parameters. In [18], maximizing a conditional likelihood that gives the same estimates than the Poisson model, authors obtained the following estimates of the cumulative losses $\hat{C}_{ik}$:

$$\hat{C}_{ik} = \frac{C_{i\ k-i}}{1 - \sum\limits_{j=k-i+1}^{k} \hat{p}_j} \qquad i = 1, 2, \ldots, k,$$

where $\hat{p}_j$ is the estimate of the (unconditional) probability that a claim is reported in development year j and $\sum\limits_{h=0}^{k} p_h = 1$. The estimate of the ultimate cumulative losses for accident year $k - j$ is:

$$\hat{C}_{k-j\ k} = \frac{C_{k-j\ j}}{1 - \sum\limits_{h=j+1}^{k} \hat{p}_h}.$$

In the CL, the estimates are

$$\hat{C}_{k-j\ k} = C_{k-j\ j} \prod_{h=j+1}^{k} \hat{m}_h.$$

Finally, it is shown in [18] that $\frac{1}{1 - \sum\limits_{h=j+1}^{k} \hat{p}_h} = \prod\limits_{h=j+1}^{k} \hat{m}_h$ and thus, the estimates in the CL method are equal that in the GLM model described above.

In this section, we have described the classical CL method and its generalization via GLM. In the next section, we propose the DB-GLM as a generalization of the CL method and as a generalization of the GLM for the solution of the claim reserving problem.

3 Distance-Based Generalized Linear Models

DB-GLM has been defined in [7] where we refer for a detailed description. In this section, we recall its main characteristics. DB-GLM could be fitted using the `dbglm` function of the `dbstats` package for `R` (see [6]). We refer to the help of the package for a detailed description of the function and its usage.

Let $\Omega = (\Omega_1, \ldots, \Omega_n)$ a population of n individuals; let $\mathbf{F} = \left(F_1, \ldots, F_p\right)$ the $n \times p$ matrix with a set of p mixed predictors; let w a priori weight of individuals of size $n \times 1$ with $w_i \in (0, 1)$; let y the response variable of size $n \times 1$; and let Δ the $n \times n$ matrix, whose entries are the squared distances $\delta^2\left(\Omega_i, \Omega_j\right)$. In the problem of claim reserving responses are the incremental losses and predictors are the origin and development years as is explained in Sect. 2 for ordinary GLM. The distances matrix contains predictor's information and it is the only object entered in the model for the predictor's space. In claim reserving, with the aim to reproduce the CL method we can use the Euclidean metric, but one of the advantages of this model is that we can choose a distance function more 'appropriate' than the Euclidean for a given set of predictors (e.g., [13]).

In the distance-based linear model (DB-LM) we calculate the $n \times n$ inner products matrix $G_w = -\frac{1}{2} J_w \cdot \Delta \cdot J_w$ where $J_w = I - 1 \cdot w'$ is the w-centering matrix. Let g_w the $n \times 1$ row vector containing the diagonal entries of G_w. Then the $n \times k$ latent Euclidean configuration matrix X_w w-centered is such that $G_w = X_w \cdot X_w'$. The DB-LM of response y with weights w and predictor matrix Δ is defined as the WLS regression of y on a w-centered Euclidean configuration X_w.

The hat matrix in a DB-LM is defined as $H_w = G_w \cdot \left(D_w^{1/2} \cdot F_w^+ \cdot D_w^{1/2}\right)$, where $D_w = \text{diag}(w)$, $F_w = D_w^{1/2} \cdot G_w \cdot D_w^{1/2}$ and F_w^+ is the Moore-Penrose pseudo-inverse of F_w. Then the predicted response is:

$$\hat{y} = \bar{y}_w \mathbf{1} + H_w \cdot (y - \bar{y}_w \mathbf{1}),$$

where $\bar{y}_w = w^T \cdot y$ is the w mean of y. The prediction of a new case Ω_{n+1} given δ_{n+1} the squared distances to the n previously known individuals is:

$$\hat{y}_{n+1} = \bar{y}_w + \frac{1}{2}(g_w - \delta_{n+1}) \cdot \left(D_w^{1/2} \cdot F_w^+ \cdot D_w^{1/2}\right) \cdot (y - \bar{y}_w \mathbf{1}).$$

DB-LM does not depend on a specific X_w, since the final quantities are obtained directly from distances. DB-LM contains WLS as a particular instance: if we start from a $n \times r$ w-centered matrix X_w of r continuous predictors corresponding to n individuals and we define Δ as the matrix of squared Euclidean distances between rows of X_w, then X_w is trivially a Euclidean configuration of Δ, hence the DB-LM hat matrix, response and predictions coincide with the corresponding WLS quantities of ordinary linear model.

Now, in DB-GLM we have the same elements as in DB-LM. Just as GLM with respect to LM, DB-GLM differs from DB-LM in two aspects: first, we assume the responses distribution is in an exponential dispersion family, as in any GLM; second, the relation between the linear predictor $\eta = X_w \cdot \beta$ obtained from the latent Euclidean configuration X_w, and the response y is given by a link function $g(\cdot)$: $y = g^{-1}(\eta)$.

To fit DB-GLM we use a standard IWLS algorithm, where DB-LM substitutes LM. This IWLS estimation process for DB-GLM does not depend on a specific X_w since the final quantities are obtained directly from distances.

DB-GLM contains GLM as a particular case: if we start from a $n \times r$ w-centered matrix X_w of r continuous predictors corresponding to n individuals and we define Δ as the matrix of squared Euclidean distances between rows of X_w, then X_w is trivially a Euclidean configuration of Δ, hence the DB-GLM hat matrix, response and predictions coincide with the corresponding IWLS quantities of ordinary GLM.

As a consequence, we can consider the DB-GLM as a generalization of the GLM to the distance-based analysis. In this line, we can consider the DB-GLM as a stochastic version of the CL deterministic method. We have shown in the last section, that when we assume in a GLM, an over-dispersed Poisson distribution and the logarithmic link we obtain the same estimations of reserves as those of the CL. Then, if we assume in a DB-GLM an over-dispersed Poisson distribution, the logarithmic link and the Euclidean metric, we will obtain the same estimations of reserves as those of the CL method.

To complete the methodology to estimate the predictive distribution of reserves, we propose to employ the resampling technique of bootstrapping pairs or resampling cases in which each bootstrap sample consists of n response-predictor pairs from the original data (see, e.g., [9]). This technique is adequate for the distance-based models as is shown in [4, 8]).

The mean squared error for the origin year reserves and for the total reserve, can be calculated in the Poisson case with logarithmic link with the following approximations (see, e.g., [2, 5]), which consist on the sum of two components: the process variance and the estimation variance. For the origin year reserves we have, for $i = 1, \ldots, k$:

$$E\left[\left(R_i - \hat{R}_i\right)^2\right] \approx \sum_{\substack{j=1,\ldots,k \\ i+j>k}} \varphi\mu_{ij} + \mu_i^T Var\left[\eta_i\right]\mu_i, \tag{1}$$

and for the total reserve we have:

$$E\left[\left(R - \hat{R}\right)^2\right] \approx \sum_{\substack{i,j=1,\ldots,k \\ i+j>k}} \varphi\mu_{ij} + \mu^T Var\left[\eta\right]\mu. \tag{2}$$

Then, the prediction error (PE) can be calculated by the square root of the mean squared errors (1) and (2). In the case in which we estimate by bootstrapping, the predictive distribution of the fitted values, we can approximate the estimation variance by the standard error (SE) of the bootstrapped predictive distribution. Then, PE for the origin year reserves for $i = 1, \ldots, k$ and for the total reserve can be calculated as follows:

$$PE^{boot}(R_i) \approx \sqrt{\sum_{\substack{j=1,\ldots,k \\ i+j>k}} \hat{\varphi}^P \hat{c}_{ij} + SE\left(\hat{R}_i^{boot}\right)^2}, \tag{3}$$

$$PE^{boot}(R) \approx \sqrt{\sum_{\substack{i,j=1,\ldots,k \\ i+j>k}} \hat{\varphi}^P \hat{c}_{ij} + SE\left(\hat{R}^{boot}\right)^2}. \tag{4}$$

4 Numerical Example

To illustrate the proposed methodology we use the triangle of [19] of Table 2 with incremental losses.

This dataset is used in many texts on IBNR problems as are [10, 11, 16, 17] to illustrate the use of the GLM and other claim reserving techniques. In Table 3 we show the estimation of reserves with the CL method. These estimations are equal for a GLM in which we assume an over-dispersed Poisson distribution and the logarithmic link, and are equal for a DB-GLM with the same assumptions of the GLM, and using the metric l^2 between factors. The instructions of the `dbstats` package to fit DB-GLM are in the Appendix.

In Table 4, we show the results for the GLM and for the DB-GLM when using analytic formulas of the Poisson distribution in the estimation variance. First, in the second column we show the estimation of the IBNR reserves (origin and total), in the third column we show the PE calculated using formulas (1) and (2), and in the fourth column, we show the named 'coefficient of variation', which is defined as the PE over the estimated IBNR (in per centage). In Table 5 we show the results for the GLM

Table 2 Run-off triangle of [19] with 55 incremental losses

	0	1	2	3	4	5	6	7	8	9
0	357848	766940	610542	482940	527326	574398	146342	139950	227229	67948
1	352118	884021	933894	1183289	445745	320996	527804	266172	425046	
2	290507	1001799	926219	1016654	750816	146923	495992	280405		
3	310608	1108250	776189	1562400	272482	352053	206286			
4	443160	693190	991983	769488	504851	470639				
5	396132	937085	847498	805037	705960					
6	440832	847361	1131398	1063269						
7	359480	1061648	1443370							
8	376686	986608								
9	344014									

Table 3 Fitted values using the CL method for the run-off triangle of Table 2

	0	1	2	3	4	5	6	7	8	9
0	357848	766940	610542	482940	527326	574398	146342	139950	227229	67948
1	352118	884021	933894	1183289	445745	320996	527804	266172	425046	**94633.8**
2	290507	1001799	926219	1016654	750816	146923	495992	280405	**375833.5**	**93677.8**
3	310608	1108250	776189	1562400	272482	352053	206286	**247190.0**	**370179.3**	**92268.5**
4	443160	693190	991983	769488	504851	470639	**334148.1**	**226674.1**	**339455.9**	**84610.6**
5	396132	937085	847498	805037	705960	**383286.6**	**351547.5**	**238477.3**	**357131.7**	**89016.3**
6	440832	847361	1131398	1063269	**605548.1**	**414501.0**	**389349.1**	**264120.5**	**395533.7**	**98588.2**
7	359480	1061648	1443370	**1310258.2**	**725788.5**	**508791.9**	**466660.0**	**316565.5**	**474072.7**	**118164.3**
8	376686	986608	**1018834.1**	**1089616.0**	**603568.6**	**423113.4**	**388076.4**	**263257.2**	**394240.8**	**98265.9**
9	344014	**856803.5**	**897410.1**	**959756.3**	**531635.7**	**372687.0**	**348125.7**	**231182.4**	**347255.4**	**86554.6**

Table 4 Origin year reserves and total reserve, prediction errors and coefficients of variation for the GLM and DB-GLM assuming an over-dispersed Poisson, the logarithmic link and the l^2 metric, using analytic formula

Origin year	Reserve	Prediction error	Coefficient of variation (%)
1	94634	110100	116.34
2	469511	216043	46.01
3	709638	260871	36.76
4	984889	303549	30.82
5	1419459	375013	26.42
6	2177641	495377	22.75
7	3920301	789960	20.15
8	4278972	1046512	24.46
9	4625811	1980101	42.81
Total	18680856	2945659	15.77

Table 5 Origin year mean reserves and total mean reserve, prediction errors and coefficients of variation for the GLM assuming an over-dispersed Poisson and the logarithmic link, using bootstrap with size 1000

Origin year	Mean reserve	Prediction error	Coefficient of variation (%)
1	100416	108422	114.57
2	477357	213629	45.50
3	727898	257700	36.31
4	978122	301693	30.63
5	1438384	369128	26.00
6	2194055	491174	22.55
7	3934897	787571	20.08
8	4236251	1032951	24.14
9	4711136	2081503	44.99
Total	18757856	2882413	15.43

using bootstrapping residuals (based on Pearson residuals) for the approximation of the estimation variance, and in Table 6 we show the results for the DB-GLM using bootstrapping pairs and calculating PE with formulas (3) and (4). In Tables 5 and 6 we include the mean reserves, the PE and the corresponding coefficients of variation, calculated over the IBNR estimated in Table 4.

If we compare the results shown in Tables 5 and 6 we observe differences. This is due to the different bootstrap methodologies. In both tables the fitted responses are the same as those of the CL classical method. But to estimate PE, in Table 5 we use bootstrapping residuals and in Table 6 bootstrapping pairs. The coefficients of variation of Table 5 are smaller for the initial origin years and greater for the latest origin years and for the total reserve than those coefficients of Table 6.

Table 6 Origin year mean reserves and total mean reserve, prediction errors and coefficients of variation for the DB-GLM assuming an over-dispersed Poisson, the logarithmic link and the l^2 metric, using bootstrap with size 1000

Origin year	Mean reserve	Prediction error	Coefficient of variation (%)
1	197097	155180	163.97
2	567832	229654	48.91
3	802434	292340	41.19
4	1096055	317125	32.19
5	1545744	391938	27.61
6	2310988	489300	22.46
7	3936212	835374	21.30
8	4316678	660744	15.44
9	4784830	677216	14.63
Total	19608104	2231054	11.94

Fig. 1 Predictive distribution of the total provision

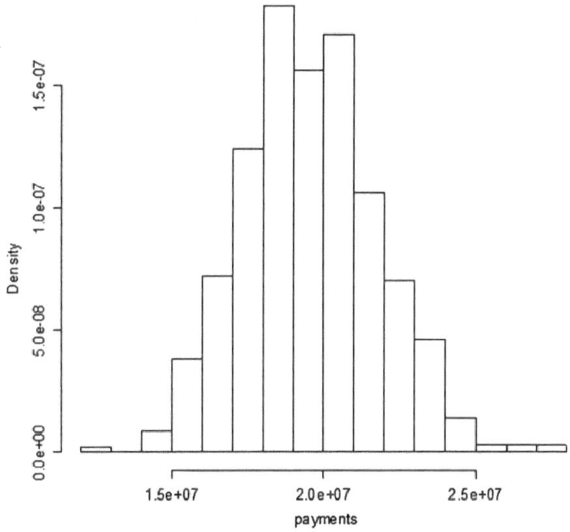

One deficiency of the bootstrapping pairs is that, compared with the bootstrapping residuals (when it is valid), generally it does not yield very accurate results. But bootstrapping pairs is less sensible to the hypotheses of the model, and the estimated standard error offers reasonable results when some hypotheses of the model are not satisfied. In the problem of claim reserving we always have a small dataset that probably does not follow the hypotheses of the GLM, then bootstrapping pairs is a reasonably methodology to estimate PE.

We show in Fig. 1 the histogram of the predictive distribution of the total reserve estimated with the DB-GLM and bootstrapping pairs. We include in the Appendix some descriptive statistics of this predictive distribution. We point out that the quantiles give the value at risk (VaR) of the losses of the portfolio. For example, the VaR with a confidence level of the 75 % is equal to 21093146.

5 Conclusions

We propose the use of the DB-GLM as an alternative methodology of claim reserving. Jointly with a bootstrapping pairs methodology we can estimate the predictive distribution of reserves and calculate prediction errors. The method is a tool for the actuary to take decisions about the best estimate of reserves and the solvency margins, and, therefore, about the financial inversions of the solvency capital required for the Company in the current context of Solvency II.

The method has the CL classical method as a particular case, when the over-dispersed Poisson distribution, the logarithmic link and the Euclidean distance between factors is assumed. The method has other particular cases (as has the GLM): the least squares method of de Vylder and the Taylor's separation (geometric and arithmetic) methods.

Additionally, our methodology generalizes the GLM to the distance-based analysis. Moreover, with the aim to obtain a best estimation of the reserves, it is possible to use another distance function instead of the Euclidean between factors (origin years and development years) of the run-of-triangle.

We illustrate the analysis with the triangle of [19]. We estimate origin year reserves and total reserve and its corresponding prediction errors (see Tables 4, 5 and 6). We show the histogram of the predictive distribution of the total reserve (see Fig. 1) and some statistics which describe the estimated distribution of the future losses of the Company. In particular it is of interest the study of the quantiles of the distribution, that provide to the actuary an estimation of the VaR of the portfolio, given a confidence level.

Acknowledgments Work supported by the Spanish Ministerio de Educación y Ciencia, grant MTM2014-56535-R.

Appendix

```
# Fitting DB-GLM
R> n<-length(cij)
R> k<-trunc(sqrt(2*n))
R> i<-rep(1:k,k:1);i<-as.factor(i)
R> j<-sequence(k:1);j<-as.factor(j)
R> orig.CL <- dbglm( cij ~ i + j, family = quasipoisson,
 metric = "euclidean", method = "rel.gvar", rel.gvar = 1)

# Descriptive statistics of the predictive distribution
```

```
# of the total reserve
R> quantile(payments, c(0.5,0.75,0.90,0.95,0.99))
      50%       75%       90%       95%       99%
 19541406  21093146  22518643  23512809  25248654
R> mean(payments)   # mean
[1] 19608104
R> sd(payments)   # standard deviation
[1] 2233737
R> cv<-(sd(payments)/mean(payments))*100  # cv in %
[1] 11.39191
R> pp<-(payments-mean(payments))/sd(payments)
R> sum(pp^3)/(nBoot-1)   # skewness
[1] 0.2290295
R> sum(pp^4)/(nBoot-1) -3  # kurtosis
[1] -0.1569525
```

References

1. Albarrán, I., Alonso, P.: Métodos estocásticos de estimación de las provisiones tenicas en el marco de Solvencia II. Cuadernos de la Fundación MAPFRE **158**. Fundación MAPFRE Estudios, Madrid (2010)
2. Boj, E., Costa, T.: Modelo lineal generalizado y cálculo de la provisión técnica. Depósito digital de la Universidad de Barcelona. Colección de objetos y materiales docentes (OMADO) (2014). http://hdl.handle.net/2445/49068
3. Boj, E., Costa, T.: Provisions for claims outstanding, incurred but not reported, with generalized linear models: prediction error formulation by calendar years. Cuad. Gestión (2015). (to appear)
4. Boj, E., Claramunt, M.M., Fortiana, J.: Selection of predictors in distance-based regression. Commun. Stat. A Theory Methods **36**, 87–98 (2007)
5. Boj, E., Costa, T., Espejo, J.: Provisiones técnicas por años de calendario mediante modelo lineal generalizado. Una aplicación con RExcel. An. Inst. Actuar. Esp. **20**, 83–116 (2014)
6. Boj, E., Caballé, A., Delicado, P., Fortiana, J.: dbstats: distance-based statistics (dbstats). R package version 1.4 (2014). http://CRAN.R-project.org/package=dbstats
7. Boj, E., Delicado, P., Fortiana, J., Esteve, A., Caballé, A.: Global and local distance-based generalized linear models. TEST (2015). doi:10.1007/s11749-015-0447-1
8. Boj, E., Costa, T., Fortiana, J., Esteve, A.: Assessing the importance of risk factors in distance-based generalized linear models. Methodol. Comput. Appl. **17**, 951–962 (2015)
9. Efron, B., Tibshirani, J.: An Introduction to the Bootstrap. Chapman and Hall, New York (1998)
10. England, P.D.: Addendum to 'Analytic and bootstrap estimates of prediction errors in claim reserving'. Insur. Math. Econ. **31**, 461–466 (2002)
11. England, P.D., Verrall, R.J.: Analytic and bootstrap estimates of prediction errors in claims reserving. Insur. Math. Econ. **25**, 281–293 (1999)
12. England, P.D., Verrall, R.J.: Predictive distributions of outstanding liabilities in general insurance. Ann. Actuar. Sci. **1:II**, 221–270 (2006)
13. Gower, J.C.: A general coeficient of similarity and some of its properties. Biometrics **27**, 857–874 (1971)
14. Kaas, R., Goovaerts, M., Dhaene, J., Denuit, M.: Modern Actuarial Risk Theory: Using R, 2nd edn. Springer, Heidelberg (2008)
15. McCullagh, P., Nelder, J.A.: Generalized Linear Models, 2nd edn. Chapman and Hall, London (1989)
16. Renshaw, A.E.: Chain ladder and interactive modelling (claims reserving and GLIM). J. Inst. Actuar. **116:III**, 559–587 (1989)

17. Renshaw, A. E.: On the second moment properties and the implementation of certain GLIM based stochastic claims reserving models. Actuarial Research Paper **65**. Department of Actuarial Science and Statistics, City University, London (1994)
18. Renshaw, A.E., Verrall, R.J.: A stochastic model underlying the Chain-Ladder technique. Br. Actuar. J. **4**, 903–923 (1998)
19. Taylor, G., Ashe, F.R.: Second moments of estimates of outstanding claims. J. Econom. **23**, 37–61 (1983)
20. van Eeghen, J., Greup, E.K., Nijssen, J.A.: Loss reserving methods. Surveys of Actuarial Studies 1, National Nederlanden (1981)
21. Verrall, R.J.: An investigation into stochastic claims reserving models and the chain-ladder technique. Insur. Math. Econ. **26**, 91–99 (2000)
22. Verrall, R.J., England, P.D.: Comments on: 'A comparison of stochastic models that reproduce chain ladder reserve estimates', by Mack and Venter. Insur. Math. Econ. **26**, 109–111 (2000)

Discrimination, Binomials and Glass Ceiling Effects

María Paz Espinosa, Eva Ferreira and Winfried Stute

Abstract We discuss dynamic models designed to describe the evolution of gender gaps deriving from the nature of the social decision processes. In particular, we study the committee choice function that maps a present committee composition to its future composition. The properties of this function and the decision mechanisms will determine the characteristics of the stochastic process that drives the dynamics over time and the long run equilibrium. We also discuss how to estimate the committee choice function parametrically and nonparametrically using conditional maximum likelihood.

Keywords Conditional nonparametric estimation · Gender gap dynamics

1 Introduction

The presence of gender gaps in the labour market has been well documented in the empirical literature. Female workers get lower wages and the differences seem to widen at upper levels (e.g., Arulampalam et al. [1]; De la Rica et al. [5]; Morgan [11]). Other authors have also identified a lower probability of females rising to the top positions on the corporate ladder (e.g., Bain and Cummings [2]). This paper seeks to shed some light on the dynamics of these gender gaps. First, we formalize decision processes that involve a gender bias and look at the implied dynamic models. The

M.P. Espinosa (✉)
Departamento de Fundamentos del Análisis Económico II, BRiDGE, BETS,
University of the Basque Country, Avenida Lehendakari Aguirre 83, 48015 Bilbao, Spain
e-mail: mariapaz.espinosa@ehu.es

E. Ferreira (✉)
Departamento de Economía Aplicada III & BETS, University of the Basque Country,
Avenida Lehendakari Aguirre 83, 48015 Bilbao, Spain
e-mail: eva.ferreira@ehu.es

W. Stute
Mathematical Institute, University of Giessen, Arndtstr. 2, 35392 Giessen, Germany
e-mail: Winfried.Stute@math.uni-giessen.de

R. Cao et al. (eds.), *Nonparametric Statistics*, Springer Proceedings
in Mathematics & Statistics 175, DOI 10.1007/978-3-319-41582-6_11

properties of the selection decision mechanisms will determine the characteristics of the stochastic process that drives the dynamics over time for the long run equilibrium. On the other hand, we also discuss the implied gender gaps in a hierarchical structure regarding the long run equilibrium, by formalizing statistical definitions of the glass ceiling effects.[1]

The model starts from a perception bias and analyzes how it affects the dynamics of promotion and the evolution of the gender gap. An important element of our model is that the decisions regarding the selection of candidates, for example for a promotion, are made by aggregating the preferences of the individual members of the committee. Thus, the composition of the committee is crucial for the decision made. Furthermore, the composition of the committee is endogenous since it comes from previous selection decisions; this feature of our model reflects the idea that when there is discrimination against a gender or an ethnic origin or a social group, this gender, ethnic origin or social group will be less present in the places where decisions are made.

The committee choice function maps the committee composition regarding a given decision and this decision will, in turn, determine the composition of future committees. Thus, we obtain a function that links the committee composition in one period with the committee composition in the previous one. We discuss the consequences of the characteristics of these functions in the long run. When abilities are not perfectly observable, meaning that candidates cannot be ranked, the selection process is conceived as a random draw from the population of participating individuals. This leads to a special autoregressive process that does not fit into the usual processes studied in time series analysis. We use Markov and martingale theory to derive the long run properties.

The paper is organized as follows. Section 2 sets up the framework and defines the committee choice function. Section 3 analyzes the time dynamics and the long run properties. Section 4 discusses estimation procedures and Sect. 5 concludes.

2 Decision Processes and Choice Function

This section formalizes the decision process that aggregates the preferences of the decision makers to yield a collective decision on the candidates to be selected. The properties of the preference aggregation procedure will determine the properties of the stochastic process characterizing the dynamics. The objective is to see how a group bias evolves over time depending on the selection process.

At each level n, there is a set of candidates $\{1, 2, 3, ..., F_n + M_n\}$ competing to become members of a committee of size z_n. There are two types of candidates, f and m; F_n candidates of type f and M_n of type m.

[1] A glass ceiling is described as a gender difference that increases along the corporate hierarchy and is not explained by other job-relevant characteristics of the employee [4].

At level n there is also a set of decision makers $\{1, 2, 3, ..., z_{n-1}\}$ in charge of the selection of candidates for the n-level committee. The set of decision makers has z_{n-1} members (the members of the committee at the previous level $n-1$) with defined preferences for the set of candidates. This modelling approach could correspond to a particular committee that renews over time and its members elect the new members, but, more generally, we try to represent how relevant social positions (in politics, business,...) are renewed over time. Obviously, the set of individuals with high decision-making power at a given period make selection decisions that influence who will be in power in the next period.

Definition 1 A decision maker preference relation (reflexive, complete and transitive) $\mathcal{P}_i$ is defined on the set of candidates $\{1, 2, 3, ..., F_n + M_n\}$.

Note that this formulation excludes the possibility of candidates being complementary. Candidates are independent in this model. We could also model preferences defined for the set of all possible teams of z candidates so that synergies among candidates could be considered, but this is beyond the scope of this paper.

Definition 2 $P^{z_{n-1}}_{(F_n+M_n)}$ is the set of all z_{n-1} decision makers' preference profiles defined for the $(F_n + M_n)$ candidates.

Definition 3 A Committee Choice Function (CCF) is a function

$$\mathrm{f} : P^{z_{n-1}}_{(F_n+M_n)} \quad \rightarrow \quad [\{z_n\}] \tag{1}$$

mapping the z_{n-1} members' preferences regarding a set of z_n individuals selected from the set of candidates; $[\{z_n\}]$ represents all possible sets of z_n candidates.

The committee preference profile, denoted $\left(\mathcal{P}_1, \mathcal{P}_2, \mathcal{P}_3, ..., \mathcal{P}_{z_{n-1}}\right)$, is a vector containing the preference profiles of the members of the committee.

Assumption 1 (*Anonymity*) If we permute the indexes of the members of the committee, the CCF assigns the same outcome, that is, f treats all decision makers identically, and no member has more power than another.

Assumption 2 (*Gender specific bias*) There are two types of members, f and m. All members of type f have the same preference profile $\mathcal{P}_f \in \mathcal{P}_{(F_n+M_n)}$, and all members of type m have the preference profile $\mathcal{P}_m \in \mathcal{P}_{(F_n+M_n)}$.

Under Assumption 2, there are only two types of preferences $\mathcal{P}_f$ and $\mathcal{P}_m$, where $\mathcal{P}_f(\mathcal{P}_m)$ denotes the preference profile of a female (male) member of the committee. Thus, taking into account Assumption 1, we may represent the committee preference profile as $(z^f_{n-1}\mathcal{P}_f, z^m_{n-1}\mathcal{P}_m)$, that is, z^f_{n-1} members with preferences $\mathcal{P}_f$ and z^m_{n-1} members with preferences $\mathcal{P}_m$.

We ignore any difference in abilities among the candidates. This allows us to isolate some consequences of gender bias.

Assumption 3 (*Unobservable abilities*) Decision makers cannot distinguish between any two candidates of the same type.

Given Assumption 3, the outcome can be represented as (z_n^f, z_n^m), that is, the number of females and males selected. The committee choice function f assigns an outcome (z_n^f, z_n^m) to each preference profile $(z_{n-1}^f \mathcal{P}_f, z_{n-1}^m \mathcal{P}_m)$

$$\mathrm{f}(z_{n-1}^f \mathcal{P}_f, z_{n-1}^m \mathcal{P}_m) = (z_n^f, z_n^m)$$

Assumption 4 (*Responsiveness*) The CCF is responsive to the number of men and women among the decision makers: Given two committee preference profiles $(z_{n-1}^f \mathcal{P}_f, z_{n-1}^m \mathcal{P}_m)$ and $(z_{n-1}^{f'} \mathcal{P}_f, z_{n-1}^{m'} \mathcal{P}_m)$ such that $z_{n-1}^{f'} > z_{n-1}^f$ and $z_{n-1}^f + z_{n-1}^m = z_{n-1}^{f'} + z_{n-1}^{m'} = z$, then the outcomes $(z_n^f, z_n^m) = \mathrm{f}(z_{n-1}^f \mathcal{P}_f, z_{n-1}^m \mathcal{P}_m)$ and $(z_n^{f'}, z_n^{m'}) = \mathrm{f}(z_{n-1}^{f'} \mathcal{P}_f, z_{n-1}^{m'} \mathcal{P}_m)$ are such that $z_n^f \leq z_n^{f'}$.

Assumption 5 (*Linearity*) $\mathrm{f}(z_{n-1}^f \mathcal{P}_f, z_{n-1}^m \mathcal{P}_m) = \frac{1}{\beta}\mathrm{f}(\beta z_{n-1}^f \mathcal{P}_f, \beta z_{n-1}^m \mathcal{P}_m)$ for any $\beta > 0$.

When Assumption 5 holds, only the proportion of men and women in the decision-making process matter regarding the outcome. From Assumption 5, $\mathrm{f}(z_{n-1}^f \mathcal{P}_f, z_{n-1}^m \mathcal{P}_m) = z\mathrm{f}(\frac{z_{n-1}^f}{z} \mathcal{P}_f, \frac{z_{n-1}^m}{z} \mathcal{P}_m) = z(\frac{z_n^f}{z}, \frac{z_n^m}{z})$. Denoting $\frac{z_{n-1}^f}{z_{n-1}} = \xi_{n-1}$ and $\frac{z_{n-1}^m}{z_{n-1}} = 1 - \xi_{n-1}$, and assuming $z_n = z$ for all n,

$$\mathrm{f}(\xi_{n-1} \mathcal{P}_f, (1 - \xi_{n-1}) \mathcal{P}_m) = (\xi_n, 1 - \xi_n)$$

Thus, the committee choice function selects for level n a proportion of members of type f which is a function of the proportion of members of type f at level $n-1$, ξ_{n-1}, and the preferences of each type $(\mathcal{P}_f, \mathcal{P}_m)$. The previous properties are similar to those required of social choice functions that aggregate the preferences of the voters or citizens (see May [8, 9]). We have adapted them to this context as the decision of a committee also aggregates the individual preferences of its members.

The rest of the paper considers that preferences can be summarized in a parameter vector θ and define $\mathrm{f}(\xi_{n-1} \mathcal{P}_f, (1 - \xi_{n-1}) \mathcal{P}_m) = f(\xi_{n-1}, \theta) = \xi_n$.

Using this notation, we summarize how the female proportion is affected by the preferences and the selection procedure through the parameter vector θ and the choice function f. If no parametric form is assumed, the notation will drop the parameter vector θ.

From Assumption 4 (Responsiveness) we know that the function $f(\xi_{n-1}, \theta)$ increases in ξ_{n-1}. We will now study which properties of this function would yield convergence.

2.1 Some Examples of Committee Choice Functions

(1) Majority rule and extreme preferences. Assume n is odd and decision makers of type m (f) strictly prefer candidates of type m (f). Then if $\xi_{n-1} > \frac{1}{2}$, $\xi_n = 1$; if $\xi_{n-1} < \frac{1}{2}$, $\xi_n = 0$.
With these extreme preferences, and depending on the initial gender composition, we will have only men or only women on the committee.
(2) When preferences are not extreme, decisions are made by consensus and each group influences the final decision according to its weight: $f(\xi_{n-1}, \theta) = \xi_n = \mu^f \xi_{n-1} + \mu^m (1 - \xi_{n-1})$, with $\theta = (\mu^f, \mu^m)$, where μ^f is the preferred female proportion chosen by females and μ^m is the preferred female proportion chosen by males.
(3) Each decision maker picks up a candidate and decision makers of type m(f) strictly prefer candidates of type m (f). Then, $f(\xi_{n-1}, \theta) = \xi_n = \xi_{n-1}$.
(4) Quota rules and majority. Assume ξ_n is restricted to $\xi_n \in [0.3, 0.7]$. Assume n is odd and decision makers of type m (f) strictly prefer candidates of type m (f). Then if $\xi_{n-1} > \frac{1}{2}$, $\xi_n = 0.7$; if $\xi_{n-1} < \frac{1}{2}$, $\xi_n = 0.3$.
(5) Borda rule. Assume there are 3 decision makers, 2 female and 1 male. 10 candidates (1–5 are female and 6–10 male) compete for 5 positions. Each member has to assign points (1 or 2) to candidates. Points are added up and the 5 candidates with the largest number of points get elected. Members 1 and 2 assign 1 point to all males and 2 points to all females; member 3 assigns 1 point to all females and 2 points to all males. Candidates 1–5 get 5 points; candidates 6–10 get 4 points. Candidates 1–5 get elected.

3 Gender Gap Dynamics and Conditional Binomials

According to the previous description, the choice function corresponds to a deterministic outcome. However, decision processes cannot be considered deterministic and are best described as being stochastic. Consider that, for a committee of a fixed size z, the number of females z_n at period n, follows a binomial distribution with probability of success equal to $f(\xi_{n-1})$, where f may be parametrically specified or not. In the parametric case, it can be written as $f(\xi_{n-1}, \theta)$ where θ is a parameter vector that accounts for the characteristics of the decision makers' preferences and other features such as the participation rate. That is, $z_n \sim B(z, f(\xi_{n-1}, \theta))$. Therefore, we will model the outcome of the decision process, the share of females, as an autoregressive process:

$$\xi_n = f(\xi_{n-1}, \theta) + \varepsilon_n \tag{2}$$

where $\mathbb{E}(\varepsilon_n | \xi_{n-1}) = 0$. Since $0 \leq \xi_n \leq 1$, model (2) is a restricted model.

In a time series context, we need to make some assumptions related to ergodicity or some type of stationarity of the process in order to undertake further statistical analysis. There is a vast literature on ergodicity of (2), according to the shape of f (linear and nonlinear) and the relationship between the main part and the error term. Recent papers such as Meitz and Salkonen [10] (and the references therein) discuss the geometric ergodicity for nonlinear AR-ARCH models. However, in this case ε_n crucially depends on the main part and the process described in (2) does not fit any of the nonlinear autoregressive models with conditionally heteroskedastic errors considered in those papers. That is, ε_n cannot be written in any GARCH form like $\sigma(\xi_{n-1}, \varepsilon_{n-1}, \ldots) u_n$, with an independent innovation u_n. Alternatively, to provide the limit structure we use the Markov structure of the process.

Since $z_n \sim B(z, f(\xi_{n-1}, \theta))$, $\{z_1, z_2, \ldots\}$ is a Markov sequence with initial value $Z_0 = z_0$ in $S = \{0, 1, \ldots, z\}$ and transition probabilities

$$\begin{aligned} p_{ij} &= \mathbb{P}(Z_n = j | Z_{n-1} = i) \\ &= \binom{z}{j} f^j\left(\frac{i}{z}\right)\left(1 - f\left(\frac{i}{z}\right)\right)^{z-j}, \end{aligned}$$

$i, j \in S$. Hence, conditionally on $Z_{n-1} = i$, Z_n is Binomial with parameters z and $f(i/z)$. Provided that

$$0 < f\left(\frac{i}{z}\right) < 1 \text{ for all } i \in S, \tag{3}$$

the chain is positive recurrent and aperiodic and admits a unique stationary distribution $\pi = (\pi_0, \pi_1, \ldots, \pi_z)$, i.e. there exists a unique probability vector π such that

$$\pi = \pi P, \tag{4}$$

where $P = (p_{ij})$ is the associated transition matrix from above. The vector π constitutes the unconditional distribution of Z_n in the long run.

Even though in theory π may be determined by solving Eq. (4), this is impossible in practice since a flexible modelling of f and hence P includes some unknown parameters. Therefore, a purely algebraic argument as to solving (4) is inappropriate to provide us with some insight into the equilibrium distribution of the Z-chain. Ferreira and Stute [6] provide some general results for the short and long run. Moreover, they provide a normal approximation that performs well even for small to moderate sizes.

For the normal approximation, we have that

$$\sqrt{z}(\xi_n - p_n) \to N(0, p_\infty(1 - p_\infty))$$

as $n, z \to \infty$, or,

$$\begin{aligned} \sqrt{z}\xi_n &\approx N\left(\sqrt{z}p_\infty, p_\infty(1 - p_\infty)\right) \\ z_n &\approx N(z\,p_\infty, z\,p_\infty(1 - p_\infty)) \end{aligned}$$

where p_∞ is the fixed point of the choice function f. Note that the only quantity that needs to be computed for the asymptotic distribution is the fixed point of f. Therefore, we do not require a parametric function to obtain the limit distribution, and nonparametric techniques can be applied to estimate that point. Once this is done, we can use the normal approximation to compute confidence intervals of the long run proportion of females on the committee. Note that this approach provides us with the unconditional distribution of ξ_n in the long run.

Using the long run distribution, we can then construct consistent estimators of the choice function f, also using nonparametric estimation.

However, estimating f consistently either parametrically or nonparametrically, for a finite sample where the stationarity of ξ_n cannot be assumed, becomes a very challenging problem.

4 Consistent Estimation

4.1 Parametric Case

In this section, we discuss how to estimate parameters using maximum likelihood. Denote by z_k the observed realization at step k, for $k = 0, 1, ..., n$. In general, denoting densities by $g_k(\cdot, \theta)$, we can write

$$g_k\left(z_1, z_2, ..., z_k, \theta\right) = g_{k-1}\left(z_1, z_2, ..., z_{k-1}, \theta\right) g_k\left(z_k | z_1, z_2, ..., z_{k-1}, \theta\right),$$

and the log-likelihood can be written as

$$L_n\left(\theta\right) = \sum_{k=1}^{n} \ln g_k\left(z_k, \theta | z_1, z_2, ..., z_{k-1}\right).$$

The maximum likelihood equation then becomes

$$\sum_{k=1}^{n} \frac{\partial}{\partial \theta} \ln g_k\left(z_k, \theta | z_1, z_2, ..., z_{k-1}\right) = 0. \tag{5}$$

In our case, the conditional density is:

$$g_k\left(z_k, \theta | z_{k-1}\right) \equiv g\left(z_k, \theta | z_{k-1}\right) = \binom{z}{z_k} f\left(\frac{z_{k-1}}{z}, \theta\right)^{z_k} \left(1 - f\left(\frac{z_{k-1}}{z}, \theta\right)\right)^{z - z_k},$$

where θ is an r-dimensional parameter.

Therefore, Eq. (5) leads in this case to:

$$\text{For } u = 1, \ldots, r \qquad \sum_{k=0}^{n} \frac{\partial}{\partial \theta_u} \ln g\left(z_k, \theta | z_{k-1}\right) = 0 \Leftrightarrow$$

$$\sum_{k=0}^{n} \left[z_k \frac{f_u\left(\frac{z_{k-1}}{z}, \theta\right)}{f\left(\frac{z_{k-1}}{z}, \theta\right)} - (z - z_k) \frac{f_u\left(\frac{z_{k-1}}{z}, \theta\right)}{1 - f\left(\frac{z_{k-1}}{z}, \theta\right)} \right] = 0 \Leftrightarrow$$

$$\sum_{k=0}^{n} \left[z_k \left(\frac{f_u\left(\frac{z_{k-1}}{z}, \theta\right)}{f\left(\frac{z_{k-1}}{z}, \theta\right)} + \frac{f_u\left(\frac{z_{k-1}}{z}, \theta\right)}{1 - f\left(\frac{z_{k-1}}{z}, \theta\right)} \right) - z \frac{f_u\left(\frac{z_{k-1}}{z}, \theta\right)}{1 - f\left(\frac{z_{k-1}}{z}, \theta\right)} \right] = 0 \Leftrightarrow$$

$$\sum_{k=0}^{n} \left[z_k - zf\left(\frac{z_{k-1}}{z}, \theta\right) \right] \frac{f_u\left(\frac{z_{k-1}}{z}, \theta\right)}{\left(1 - f\left(\frac{z_{k-1}}{z}, \theta\right)\right) f\left(\frac{z_{k-1}}{z}, \theta\right)} = 0 \tag{6}$$

where f_u denotes the partial derivative $\partial f / \partial \theta_u$. Since $0 < f < 1$, (6) can also be written as

$$\sum_{k=0}^{n} \left[\xi_k - f\left(\xi_{k-1}, \theta\right) \right] f_u\left(\xi_{k-1}, \theta\right) = 0$$

and the estimator becomes the usual OLS estimator in model (2).

Note that for any θ, the above sum is a martingale in differences, since $\mathbb{E}_\theta \left[z_k - zf\left(\frac{z_{k-1}}{z}, \theta\right) | \mathcal{F}_{k-1} \right] = 0$ and the weights

$$w_{u,\theta}\left(z_{k-1}\right) = \frac{f_u\left(\frac{z_{k-1}}{z}, \theta\right)}{\left(1 - f\left(\frac{z_{k-1}}{z}, \theta\right)\right) f\left(\frac{z_{k-1}}{z}, \theta\right)}$$

are $\mathcal{F}_{k-1}$-measurable. The consistency results use the SLLN for Markov chains (Billingsley [3]). Let $\{X_n\}_{n=1,0,\ldots}$ be a finite Markov chain with stationary probability π and transition matrix $\mathbb{P}$. Let $f(\cdot)$ be any function in L^1_π, then, given any initial probability measure π_0 we have that

$$\lim_{n \to \infty} \frac{1}{n} \sum_{k=1}^{n} f\left(X_k\right) = \mathbb{E}_\pi\left(X_1\right)$$

This result is very important since it means that it can begin under any initial probability measure, which is very useful for inference in our context.

Now, consider the following assumptions

(A1) For each θ, the stationary distribution π exists and it is unique,

(A2) For each m, $P(\cdot | z_{k-1} = m)$ is absolutely continuous with respect to π and $P(\cdot | z_{k-1}) << \pi(\cdot)$,

(A3) $\mathbb{E}_\theta \left[\left| \left[z_2 - zf\left(\frac{z_1}{z}, \theta\right)\right] w_{u,\theta}\left(z_1\right) \right|^{2+\epsilon} \right] < \infty$ for each θ and $u = 1, \ldots, r$.

Applying the results in Theorems 1.1, 1.2, 2.1 and 2.2 from Billingsley [3], the following results are straightforward:

Theorem 1 *Under assumptions* (A1) *and* (A2), *for any* θ,

$$\frac{1}{n}\sum_{k=1}^{n}\left[z_k - zf\left(\frac{z_{k-1}}{z},\theta\right)\right] w_{u,\theta}(z_{k-1}) \to \mathbb{E}_\theta\left[\left[z_2 - zf\left(\frac{z_1}{z},\theta\right)\right] w_{u,\theta}(z_1)\right] = 0$$

with probability one.

Theorem 2 *Under assumptions* (A1)–(A3),

$$\frac{1}{\sqrt{n}}\sum_{k=1}^{n}\left[z_k - zf\left(\frac{z_{k-1}}{z},\theta\right)\right] w_{u,\theta}(z_{k-1}) \to^{d} N(0,\sigma(\theta))$$

where $\sigma_{uv}(\theta) = \mathbb{E}_\theta\left[\left[z_2 - zf\left(\frac{z_1}{z},\theta\right)\right]^2 w_{u,\theta}(z_1)\, w_{v,\theta}(z_1)\right] < \infty.$

Theorem 3 *There exists a consistent maximum likelihood estimator of the true parameter* θ_0, θ_n, *which is a local maximum of the likelihood equation with probability going to one as* $n \to \infty$.

Theorem 4 *Let* y_n *and* l *be the vectors with components* $y_{n,u} = n^{-1}\sum_{k=1}^{n}\left[z_k - zf\left(\frac{z_{k-1}}{z},\theta_0\right)\right] w_{u,\theta_0}(z_{k-1})$ *and* $l_u = n^{1/2}\left(\theta_{n,u} - \theta_{0,u}\right)$. *Under* (A1) *to* (A3),

$$\begin{aligned}
y_n &\sim \sigma(\theta_0)\, l_n \\
y_n &\to {}^{d}N(0,\sigma(\theta_0)) \\
l_n &\to {}^{d}N\left(0,\sigma^{-1}(\theta_0)\right) \\
2\left[\max_\theta L_n - L_n^0\right] &\sim \langle \sigma(\theta_0)\, l_n, l_n\rangle \sim \left\langle y_n, \sigma^{-1}(\theta_0)\, y_n\right\rangle \\
2\left[\max_\theta L_n - L_n^0\right] &\to {}^{d}\chi_r^2
\end{aligned}$$

That is, the ML estimator for this model is the OLS estimator. Moreover, we have shown that this unusual autoregressive structure fulfils the conditions for consistency and asymptotic normality.

4.2 Nonparametric Case

To obtain a nonparametric estimation of f, we can use a nonparametric log-likelihood function as

$$L_n(f) = \sum_{k=1}^{n} \ln\left[\binom{z}{z_k} f(\xi)^{z_k}(1-f(\xi))^{z-z_k}\right] K\left(\frac{\xi - z_{k-1}/z}{nh}\right) \tag{7}$$

To maximize (7) is equivalent to maximizing

$$\sum_{k=1}^{n} \left[z_k \ln f(\xi) + (z - z_k)\ln(1 - f(\xi))\right] K\left(\frac{\xi - z_{k-1}/z}{nh}\right)$$

for each value $f(\xi) = a$. First note that $h(a) = z_k \ln a + (z - z_k)\ln(1-a)$ is a concave function, since

$$\begin{aligned} h'(a) &= z_k \frac{1}{a} - (z - z_k)\frac{1}{1-a} \\ h''(a) &= -z_k \frac{1}{a^2} - (z - z_k)\frac{1}{(1-a)^2} \\ &= -\frac{z_k(1-a)^2 + (z - z_k)a^2}{a^2(1-a)^2} < 0 \end{aligned}$$

By differentiating we obtain the maximum as

$$\sum_{k=1}^{n} \left[z_k \frac{1}{a} - (z - z_k)\frac{1}{1-a}\right] K\left(\frac{\xi - z_{k-1}/z}{nh}\right) = 0$$

Solving the equation for a we obtain

$$\sum_{k=1}^{n} \left[\frac{z_k(1-a) - (z - z_k)a}{a(1-a)}\right] K\left(\frac{\xi - z_{k-1}/z}{nh}\right) = 0$$

$$\sum_{k=1}^{n} [z_k - za] K\left(\frac{\xi - z_{k-1}/z}{nh}\right) = 0$$

and the nonparametric estimator of f becomes

$$\widehat{f}(\xi) = \frac{\sum_{k=1}^{n} K\left(\frac{\xi - z_{k-1}/z}{nh}\right)\frac{z_k}{z}}{\sum_{k=1}^{n} K\left(\frac{\xi - z_{k-1}/z}{nh}\right)}$$

Observe that the nonparametric estimator becomes a usual regression kernel, as expected from the parametric result. However, we have no guarantee of the sufficient conditions, such as a stationary density for the variables z_k/z or mixing conditions for the error term, to derive the usual asymptotic properties as in Härdle [7] or Wand and Jones [12]. Therefore, further study is required to obtain the consistency of the nonparametric estimator, a challenging topic for future research.

5 Concluding Remarks

This paper builds dynamic models designed to describe the evolution of gender gaps deriving from the nature of social decision processes. These processes may involve bias or discrimination against certain social groups in job promotion or other selection mechanisms. We study how the properties of the decision mechanisms and the preferences of the decision makers determine the characteristics of the stochastic process that drives the dynamics over time and the long run equilibrium. Since the autoregressive process that drives the dynamics does not have conditionally heterocedastic errors, the consistency and asymptotic normality must be specifically derived. We provide these results for the parametric estimation of the committee choice function using maximum likelihood, along with some insights on how to estimate it nonparametrically. The fact that there is no need to assume that the process starts from the stationary distribution, as in this model of gender dynamics, is remarkable.

For practitioners, given data on the observed percentage of females in a given position, or as members of a given committee, in n periods, the procedure to estimate the committee function becomes very simple. For the parametric case it is the OLS estimation, and it becomes the Nadaraya-Watson estimation for the nonparametric case, when kernels are considered.

Acknowledgments Financial support from ME (ECO2012-35820, ECO2011-29268 and ECO2014-51914-P), the Basque Government (DEUI, IT-313-07) and UPV/EHU (UFI 11/46 BETS) is gratefully acknowledged.

References

1. Arulampalam, W., Booth, A.L., Bryan, M.L.: Is there a glass ceiling over Europe? Exploring the gender pay gap across the wages distribution. Ind. Labor Relat. Rev. **60**(2), 163–186 (2007)
2. Bain, O., Cummings, W.: Academe's glass ceiling: societal, professional-organizational, and institutional barriers to the career advancement of academicwomen. Comp. Educ. Rev. **44**(4), 493–514 (2000)
3. Billingsley, P.: Statistical Inference for Markov Processes. University of Chicago Statistical Research Monographs, University of Chicago Press, Institute of Mathematical Statistics. MR 23, no. A746 (1961)
4. Cotter, D.A., Hermsen, J.M., Ovadia, S., Vanneman, R.: The glass ceiling effect. Soc. Forces **80**(2), 655–681 (2001)
5. De la Rica, S., Dolado, J.J., Llorens, V.: Ceilings or floors? Gender wage gaps by education in Spain. J. Popul. Econ. **21**(3), 751–776 (2008)
6. Ferreira, E., Stute, W.: Dynamic binomials with an application to gender bias analysis. J. Appl. Probab. **53**(1), 82–90 (2016)
7. Härdle, W.: Applied Nonparametric Regression. Cambridge University Press, Cambridge (1990)
8. May, K.O.: A set of independent, necessary and sufficient conditions for simple majority decision. Econometrica **20**, 680–684 (1952)
9. May, K.O.: Intransitivity, utility, and the aggregation of preference patterns. Econometrica **22**, 1–13 (1954)

10. Meitz, M., Saikkonen, P.: A note on the geometric ergodicity of a nonlinear AR-ARCH model. Stat. Probab. Lett. **80**, 631–638 (2010)
11. Morgan, L.A.: Glass-ceiling effect or cohort effect? A longitudinal study of the gender earnings gap for engineers, 1982 to 1989. Am. Sociol. Rev. **63**(4), 479–493 (1998)
12. Wand, M., Jones, M.: Kernel Smoothing. Chapman and Hall, London (1995)

Extrinsic Means and Antimeans

Vic Patrangenaru, K. David Yao and Ruite Guo

Abstract Often times object spaces are compact, thus allowing to introduce new location parameters, maximizers of the Fréchet function associated with a random object X on a compact object space. In this paper we focus on such location parameters, when the object space is embedded in a numeric space. In this case the maximizer, whenever the maximizer of this Fréchet function is unique, is called the *extrinsic mean* of X.

Keywords Random object · Fréchet mean set · Extrinsic mean · Hypothesis testing for VW antimeans

1 Introduction

Fréchet [10] noticed that for data analysis purposes, in case a list of numbers would not give a meaningful representation of the individual observation under investigation, it is helpful to measure not just vectors, but more complicated features, he used to call "elements", and are nowadays called *objects*. As examples he mentioned "the shape of an egg taken at random from a basket of eggs." A natural way of addressing this problem consists of regarding a *random object X* as a random point on a complete metric space $(\mathcal{M}, \rho)$ that often times has a manifold structure (see Patrangenaru and Ellingson [18]). Important examples of objects that arise from electronic image data are shapes of configurations extracted from digital images, or from medical imaging outputs. For such data, the associated object considered are points on Kendall shape spaces (see Kendall [14]), Dryden and Mardia [7]), or on affine shape spaces (see Patrangenaru and Mardia [22], Sughatadasa [23]), on projective shape spaces (see Mardia and Patrangenaru [17], Patrangenaru et al. [21]). Other examples of object spaces are spaces of axes (see Fisher et al. [9], Beran and Fisher [1]), spaces of

V. Patrangenaru (✉) · R. Guo
Department of Statistics, Florida State University, Tallahassee, USA
e-mail: vpatrang@math.ttu.edu; vic@stat.fsu.edu

K.D. Yao
Department of Mathematics, Florida State University, Tallahassee, USA

R. Cao et al. (eds.), *Nonparametric Statistics*, Springer Proceedings
in Mathematics & Statistics 175, DOI 10.1007/978-3-319-41582-6_12

directions (see Watson [25]), and spaces of trees (see Billera et al. [5], Wang and Marron [24], Hotz et al. [13]). The aforementioned object spaces have a structure of *compact symmetric spaces* (see Helgasson [12]); however, the use of a Riemannian distance on a symmetric space for the goal of mean data analysis, including for regression with a response on a symmetric space, is a statistician choice, as opposed to being imposed by the nature of the data.

Therefore for practical purposes, in this paper we consider object spaces provided with a "chord" distance associated with the embedding of an object space into a numerical space, and the statistical analysis performed relative to a chord distance is termed *extrinsic data analysis*. The expected square distance from the random object X to an arbitrary point p defines what we call the *Fréchet function* associated with X:

$$\mathcal{F}(p) = \mathbb{E}(\rho^2(p, X)), \tag{1.1}$$

and its minimizers form the *Fréchet mean set*. When ρ is the "chord" distance on $\mathcal{M}$ induced by the Euclidean distance in $\mathbb{R}^N$ via an embedding $j : \mathcal{M} \to \mathbb{R}^N$, the Fréchet function becomes

$$\mathcal{F}(p) = \int_{\mathcal{M}} \|j(x) - j(p)\|_0^2 Q(dx), \tag{1.2}$$

where $Q = P_X$ is the probability measure on $\mathcal{M}$, associated with X. In this case, the Fréchet mean set is called the *extrinsic mean set* (see Bhattacharya and Patrangenaru [4]), and if we have a unique point in the extrinsic mean set of X, this point is the *extrinsic mean* of X, and is labeled $\mu_E(X)$ or simply μ_E. Also, given $X_1, \ldots, X_n$ i.i.d random objects from Q, their *extrinsic sample mean (set)* is the extrinsic mean (set) of the empirical distribution $\hat{Q}_n = \frac{1}{n}\sum_{i=1}^n \delta_{X_i}$.

In this paper, we will assume that $(\mathcal{M}, \rho)$ is a compact metric space; therefore, the Fréchet function is bounded, and its extreme values are attained at points on $\mathcal{M}$. We are now introducing a **new location parameter** for X.

Definition 1.1 The set of maximizers of the Fréchet function, is called the *extrinsic antimean set*. In case the extrinsic antimean set has one point only, that point is called **extrinsic antimean** of X, and is labeled $\alpha\mu_{j,E}(Q)$, or simply $\alpha\mu_E$, when j is known.

The remainder of the paper is concerned with geometric descriptions, explicit formulas, and computations of extrinsic means and antimeans. Simple inference problems for extrinsic means and antimeans are also investigated. The paper ends with a discussion on future directions in extrinsic antimean analysis.

2 Geometric Description of the Extrinsic Antimean

Let $(\mathcal{M}, \rho_0)$ be a compact metric space, where ρ_0 is the chord distance via the embedding $j : \mathcal{M} \to \mathbb{R}^N$, that is

$$\rho_0(p_1, p_2) = \|j(p_1) - j(p_2)\| = d_0(j(p_1), j(p_2)), \forall (p_1, p_2) \in \mathcal{M}^2,$$

where d_0 is the Euclidean distance in $\mathbb{R}^N$.

Remark 2.1 Recall that a point $y \in \mathbb{R}^N$ for which there is a unique point $p \in \mathcal{M}$ satisfying the equality,

$$d_0(y, j(\mathcal{M})) = \inf_{x \in \mathcal{M}} \|y - j(x)\|_0 = d_0(y, j(p))$$

is called j-nonfocal. A point which is not j-nonfocal is said to be j-focal. And if y is a j-nonfocal point, its projection on $j(\mathcal{M})$ is the unique point $j(p) = P_j(y) \in j(\mathcal{M})$ with $d_0(y, j(\mathcal{M})) = d_0(y, j(p))$.

With this in mind we now have the following definition.

Definition 2.1 (a) A point $y \in \mathbb{R}^N$ for which there is a unique point $p \in \mathcal{M}$ satisfying the equality,

$$\sup_{x \in \mathcal{M}} \|y - j(x)\|_0 = d_0(y, j(p)) \tag{2.1}$$

is called αj-nonfocal. A point which is not αj-nonfocal is said to be αj-focal.

(b) If y is an αj-nonfocal point, its projection on $j(\mathcal{M})$ is the unique point $z = P_{F,j}(y) \in j(\mathcal{M})$ with $\sup_{x \in \mathcal{M}} \|y - j(x)\|_0 = d_0(y, j(p))$.

For example, if we consider the unit sphere S^m in $\mathbb{R}^{m+1}$, with the embedding given by the inclusion map $j : S^m \to \mathbb{R}^{m+1}$, then the only αj-focal point is 0_{m+1}, the center of this sphere; this point also happens to be the only j-focal point of S^m.

Definition 2.2 A probability distribution Q on $\mathcal{M}$ is said to be αj-nonfocal if the mean μ of $j(Q)$ is αj-nonfocal.

The figures below illustrate the extrinsic mean and antimean of distributions on a complete metric space $\mathcal{M}$ where the distributions are j-nonfocal and also αj-nonfocal (Fig. 1).

Theorem 2.1 *Let μ be the mean vector of $j(Q)$ in $\mathbb{R}^N$. Then the following hold true:*

(i) The extrinsic antimean set is the set of all points $x \in \mathcal{M}$ such that $\sup_{p \in \mathcal{M}} \|\mu - j(p)\|_0 = d_0(\mu, j(x))$.
(ii) If $\alpha\mu_{j,E}(Q)$ exists, then μ is αj-nonfocal and $\alpha\mu_{j,E}(Q) = j^{-1}(P_{F,j}(\mu))$.

Proof For part (i), we first rewrite the following expression:

$$\|j(p) - j(x)\|_0^2 = \|j(p) - \mu\|_0^2 - 2\,\langle j(p) - \mu, \mu - j(x)\rangle + \|\mu - j(x)\|_0^2 \tag{2.2}$$

Since the manifold is compact, μ exists, and from the definition of the mean vector we have

$$\int_{\mathcal{M}} j(x)\, Q(dx) = \int_{\mathbb{R}^N} y j(Q)(dy) = \mu. \tag{2.3}$$

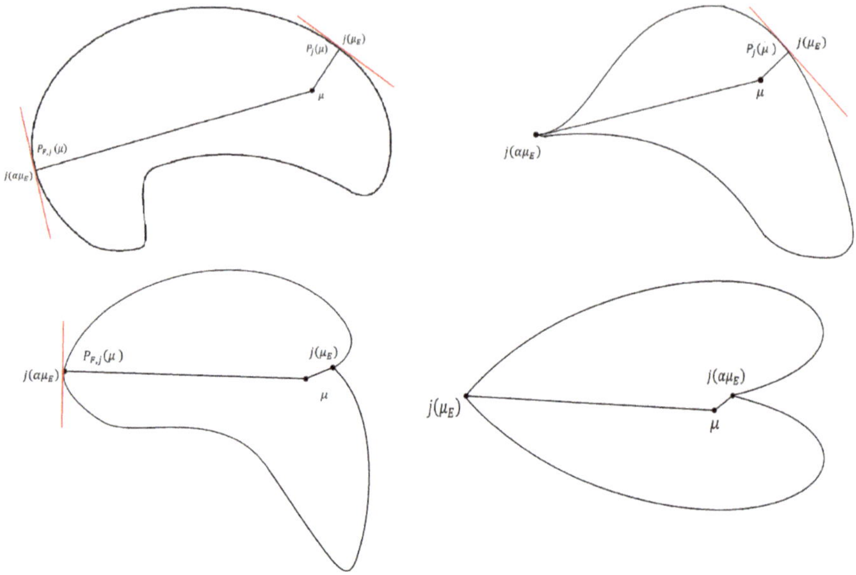

Fig. 1 Extrinsic mean and extrinsic antimean on a 1-dimensional topological manifold (*upper left* regular mean and antimean, *upper right* regular mean and sticky antimean, *lower left* sticky mean and regular antimean, *lower right* sticky mean and antimean

From Eqs. (2.2) and (2.3) it follows that

$$\mathcal{F}(p) = \|j(p) - \mu\|_0^2 + \int_{\mathbb{R}^N} \|\mu - y\|_0^2 j(Q)(dy) \tag{2.4}$$

Then, from (2.4),

$$\sup_{p\in\mathcal{M}} \mathcal{F}(p) = \sup_{p\in\mathcal{M}} \|j(p) - \mu\|_0^2 + \int_{\mathbb{R}^N} \|\mu - y\|_0^2 j(Q)(dy) \tag{2.5}$$

This then implies that the antimean set is the set of points $x \in \mathcal{M}$ with the following property;

$$\sup_{p\in\mathcal{M}} \|j(p) - \mu\|_0 = \|j(x) - \mu\|_0. \tag{2.6}$$

For Part (*ii*) if $\alpha\mu_{j,E}(Q)$ exists, then $\alpha\mu_{j,E}(Q)$ is the unique point $x \in \mathcal{M}$, for which Eq. (2.6) holds true, which implies that μ is αj-nonfocal and $j(\alpha\mu_{j,E}(Q)) = P_{F,j}(\mu)$.

Definition 2.3 Let $x_1, ..., x_n$ be random observations from a distribution Q on a compact metric space $(\mathcal{M}, \ \rho)$, then their extrinsic sample antimean set, is the set of maximizers of the Fréchet function $\hat{\mathcal{F}}_n$ associated with the empirical distribution $\hat{Q}_n = \frac{1}{n}\sum_{i=1}^n \delta_{x_i}$, which is given by

$$\hat{\mathcal{F}}_n(p) = \frac{1}{n}\sum_{i=1}^{n} \|j(x_i) - j(p)\|_0^2 \tag{2.7}$$

If $\hat{Q}_n$ has an extrinsic antimean, its extrinsic antimean is called extrinsic sample antimean, and it is denoted $a\bar{X}_{j,E}$.

Theorem 2.2 *Assume Q is an αj-nonfocal probability measure on the manifold $\mathcal{M}$ and $X = \{X_1, ..., X_n\}$ are i.i.d random objects from Q. Then,*

(a) If $\overline{j(X)}$ is αj-nonfocal, then the extrinsic sample antimean is given by $a\bar{X}_{j,E} = j^{-1}(P_{F,j}(\overline{j(X)}))$.
(b) The set $(\alpha F)^c$ of αj-nonfocal points is a generic subset of $\mathbb{R}^N$, and if $\alpha\mu_{j,E}(Q)$ exists, then the extrinsic sample antimean $a\bar{X}_{j,E}$ is a consistent estimator of $\alpha\mu_{j,E}(Q)$.

Proof (Sketch). (a) Since $\overline{j(X)}$ is αj-nonfocal the result follows from Theorem 2.1, applied to the empirical $\hat{Q}_n$, therefore $j(a\bar{X}_{j,E}) = P_{F,j}(\overline{j(X)})$.

(b) All the assumptions of the SLLN are satisfied, since $j(\mathcal{M})$ is also compact, therefore the sample mean estimator $\overline{j(X)}$ is a strong consistent estimator of μ, which implies that for any $\varepsilon > 0$, and for any $\delta > 0$, there is sample size $n(\delta, \varepsilon)$, such that $\mathbb{P}(\|\overline{j(X)} - \mu\| > \delta) \leq \varepsilon, \forall n > n(\delta, \varepsilon)$. By taking a small enough $\delta > 0$, and using a continuity argument for $P_{F,j}$, the result follows.

Remark 2.2 For asymptotic distributions of the extrinsic sample antimeans see Patrangenaru et al. [19].

3 VW Antimeans on $\mathbb{R}P^m$

In this section, we consider the case of a probability measure Q on the real projective space $\mathcal{M} = \mathbb{R}P^m$, which is the set of axes (1-dimensional linear subspaces) of $\mathbb{R}^{m+1}$. Here the points in $\mathbb{R}^{m+1}$ are regarded as $(m+1) \times 1$ vectors. $\mathbb{R}P^m$ can be identified with the quotient space $S^m/\{x, -x\}$; it is a compact homogeneous space, with the group $SO(m+1)$ acting transitively on $(\mathbb{R}P^m, \rho_0)$, where the distance ρ_0 on $\mathbb{R}P^m$ is induced by the chord distance on the sphere S^m. There are infinitely many embeddings of $\mathbb{R}P^m$ in a Euclidean space; however, for the purpose of two-sample mean or two-sample antimean testing, it is preferred to use an embedding j that is compatible with two transitive group actions of $SO(m+1)$ on $\mathbb{R}P^m$, respectively on $j(\mathbb{R}P^m)$, that is

$$j(T \cdot [x]) = T \otimes j([x]), \quad \forall\, T \in SO(m+1), \forall\, [x] \in \mathbb{R}P^m, \quad \textit{where } T \cdot [x] = [Tx]. \tag{3.1}$$

Such an embedding is said to be *equivariant* (see Kent [15]). For computational purposes, the equivariant embedding of $\mathbb{R}P^m$ that was used so far in the axial data analysis literature is the Veronese–Whitney (VW) embedding $j : \mathbb{R}P^m \to S_+(m+1, \mathbb{R})$, that associates to an axis the matrix of the orthogonal projection on this axis (see Patrangenaru and Ellingson [18] and references therein):

$$j([x]) = xx^T, \ \|x\| = 1, \tag{3.2}$$

Here $S_+(m+1, \mathbb{R})$ is the set of nonnegative definite symmetric $(m+1) \times (m+1)$ matrices, and in this case

$$T \otimes A = TAT^T, \quad \forall\, T \in SO(m+1), \forall\, A \in S_+(m+1, \mathbb{R}) \tag{3.3}$$

Remark 3.1 Let $N = \frac{1}{2}(m+1)(m+2)$. The space $\mathbb{E} = (S(m+1, \mathbb{R}),\ \langle\ ,\ \rangle_0)$ is an N-dimensional Euclidean space with the scalar product given by $\langle A,\ B\rangle_0 = Tr(AB),\ \ where\ \ A,\ B\ \in S(m+1, \mathbb{R})$. The associated norm $\|\cdot\|_0$ and Euclidean distance d_0 are given by respectively by $\|C\|_0^2 = \langle C,\ C\rangle_0$ and $d_0(A, B) = \|A - B\|_0, \forall C, A, B \in\ S(m+1, \mathbb{R})$.

With the notation in Remark 3.1, we have

$$\mathcal{F}([p]) = \|j([p]) - \mu\|_0^2 + \int_{\mathcal{M}} \|\mu - j([x])\|_0^2\, Q(d[x]), \tag{3.4}$$

and $\mathcal{F}([p])$ is maximized (minimized) if and only if $\|j([p]) - \mu\|_0^2$ is maximized (minimized) as a function of $[p] \in \mathbb{R}P^m$.

From Patrangenaru and Ellingson ([18], Chap. 4) and definitions therein, recall that the extrinsic mean $\mu_{j,E}(Q)$ of a j- nonfocal probability measure Q on $\mathcal{M}$ w.r.t. an embedding j, when it exists, is given by $\mu_{j,E}(Q) = j^{-1}(P_j(\mu))$ where μ is the mean of $j(Q)$. In the particular case, when $\mathcal{M} = \mathbb{R}P^m$, and j is the VW embedding, P_j is the projection on $j(\mathbb{R}P^m)$ and $P_j : S_+(m+1, \mathbb{R})\backslash\mathcal{F} \to j(\mathbb{R}P^m)$, where $\mathcal{F}$ is the set of j-focal points of $j(\mathbb{R}P^m)$ in $S_+(m+1, \mathbb{R})$. For the VW embedding, $\mathcal{F}$ is the set of matrices in $S_+(m+1, \mathbb{R})$ whose largest eigenvalues are of multiplicity at least 2. The projection P_j assigns to each nonnegative definite symmetric matrix A with highest eigenvalue of multiplicity 1, the matrix $\nu\nu^T$, where ν is a unit eigenvector of A corresponding to its largest eigenvalue.

Furthermore, the VW mean of a random object $[X] \in \mathbb{R}P^m$, $[X^TX] = 1$ is given by $\mu_{j,E}(Q) = [\gamma(m+1)]$ and $(\lambda(a), \gamma(a)),\ a = 1, .., m+1$ are eigenvalues and unit eigenvectors pairs (in increasing order of eigenvalues) of the mean $\mu = E(XX^T)$. Similarly, the VW sample mean is given by $\bar{x}_{j,E} = [g(m+1)]$ where $(d(a), g(a)),\ a = 1, \ldots, m+1$ are eigenvalues and unit eigenvectors pairs (in increasing order of eigenvalues) of the sample mean $J = \frac{1}{n}\sum_{i=1}^{n} x_i x_i^T$ associated with the sample $([x_i])_{i=\overline{1,n}}$, on $\mathbb{R}P^m$, where $x_i^T x_i = 1, \forall i = \overline{1,n}$.

Based on (3.4), we get similar results in the case of an αj-nonfocal probability measure Q:

Proposition 1 *(i) The set of αVW-nonfocal points in $S_+(m+1, \mathbb{R})$, is the set of matrices in $S_+(m+1, \mathbb{R})$ whose smallest eigenvalue has multiplicity 1.*

(ii) *The projection* $P_{F,j} : (\alpha F)^c \to j(\mathbb{R}P^m)$ *assigns to each nonnegative definite symmetric matrix A, of rank 1, with a smallest eigenvalue of multiplicity 1, the matrix* $j([\nu])$*, where* $\|\nu\| = 1$ *and* ν *is an eigenvector of A corresponding to that eigenvalue.*

We now have the following:

Proposition 2 *Let Q be a distribution on* $\mathbb{R}P^m$.

(a) *The VW-antimean set of a random object* $[X], X^TX = 1$ *on* $\mathbb{R}P^m$*, is the set of points* $p = [v] \in V_1$*, where* V_1 *is the eigenspace corresponding to the smallest eigenvalue* $\lambda(1)$ *of* $E(XX^T)$.

(b) *If in addition* $Q = P_{[X]}$ *is* αVW*-nonfocal, then*

$$\alpha\mu_{j,E}(Q) = j^{-1}(P_{F,j}(\mu)) = [\gamma(1)]$$

where $(\lambda(a), \gamma(a)), a = 1, .., m+1$ *are eigenvalues in increasing order and the corresponding unit eigenvectors of* $\mu = E(XX^T)$.

(c) *Let* $[x_1], \dots, [x_n]$ *be observations from a distribution Q on* $\mathbb{R}P^m$*, such that* $\overline{j(X)}$ *is* α*VW-nonfocal. Then the VW sample antimean of* $[x_1], \dots, [x_n]$ *is given by*

$$a\bar{x}_{j,E} = j^{-1}(P_{F,j}(\overline{j(x)})) = [g(1)]$$

where $(d(a), g(a))$ *are the eigenvalues in increasing order and the corresponding unit eigenvectors of* $J = \frac{1}{n}\sum_{i=1}^{n} x_i x_i^T$*, where* $x_i^T x_i = 1, \forall i = \overline{1, n}$.

4 Two-Sample Test for VW Means and Antimeans Projective Shapes in 3D

Recall that the space $P\Sigma_3^k$ of projective shapes of 3D k-ads in $\mathbb{R}P^3$, $([u_1], ..., [u_k])$, with $k > 5$, for which $\pi = ([u_1], \dots, [u_5])$ is a projective frame in $\mathbb{R}P^3$, is homeomorphic to the manifold $(\mathbb{R}P^3)^q$ with $q = k - 5$ (see Patrangenaru et al. [21]). Also recall that a Lie group, is a manifold $\mathcal{G}$, that has an additional group structure $\odot : \mathcal{G} \times \mathcal{G} \to \mathcal{G}$ with the inverse map $\iota : \mathcal{G} \to \mathcal{G}, \iota(g) = g^{-1}$, such that both operations $\odot$ and ι are differentiable functions between manifolds.

Note that S^3 regarded as set of quaternions of unit norm has a Lie group structure inherited from the quaternion multiplication, which yields a Lie group structure on $\mathbb{R}P^3$. This multiplicative structure turns the $(\mathbb{R}P^3)^q$ into a product Lie group $(\mathcal{G}, \odot_q)$ where $\mathcal{G} = (\mathbb{R}P^3)^q$ (see Crane and Patrangenaru [6], Patrangenaru et al. [20]). For the rest of this section $\mathcal{G}$ refers to the Lie group $(\mathbb{R}P^3)^q$. The VW embedding $j_q : (\mathbb{R}P^3)^q \to (S_+(4, \mathbb{R}))^q$ (see Patrangenaru et al. [20]), is given by

$$j_q([x_1], \dots, [x_q]) = (j([x_1]), \dots, j([x_q])), \tag{4.1}$$

with $j : \mathbb{R}P^3 \to S_+(4, \mathbb{R})$ the VW embedding given in (3.2), for $m = 3$ and j_q is also an equivariant embedding w.r.t. the group $(S_+(4, \mathbb{R}))^q$.

Given the product structure, it turns out that the VW mean μ_{j_q} of a random object $Y = (Y^1, \dots, Y^q)$ on $(\mathbb{R}P^3)^q$ is given by

$$\mu_{j_q} = (\mu_{1,j}, \cdots, \mu_{q,j}), \tag{4.2}$$

where, for $s = \overline{1, q},\ \mu_{s,j}$ is the VW mean of the marginal Y^s.
Assume $Y_a,\ a = 1, 2$ are random objects with the associated distributions $Q_a = P_{Y_a},\ a = 1, 2$ on $\mathcal{G} = (\mathbb{R}P^3)^q$. We now consider the two-sample problem for VW means and separately for VW-antimeans for these random objects.

4.1 Hypothesis Testing for VW Means

Assume the distributions $Q_a, a = 1, 2$ are in addition VW-nonfocal. We are interested in the hypothesis testing problem:

$$H_0 : \mu_{1,j_q} = \mu_{2,j_q} \text{ versus } H_a : \mu_{1,j_q} \neq \mu_{2,j_q}, \tag{4.3}$$

which is equivalent to testing the following

$$H_0 : \mu_{2,j_q}^{-1} \odot_q \mu_{1,j_q} = 1_{(\mathbb{R}P^3)^q} \text{ versus } H_a : \mu_{2,j_q}^{-1} \odot_q \mu_{1,j_q} \neq 1_{(\mathbb{R}P^3)^q} \tag{4.4}$$

1. Let $n_+ = n_1 + n_2$ be the total sample size, and assume $\lim_{n_+ \to \infty} \frac{n_1}{n_+} \to \lambda \in (0, 1)$. Let φ be the log chart defined in a neighborhood of $1_{(\mathbb{R}P^3)^q}$ (see Helgason [12]), with $\varphi(1_{(\mathbb{R}P^3)^q}) = 0$. Then, under H_0

$$n_+^{1/2}\, \varphi(\bar{Y}_{j_q,n_2}^{-1} \odot_q \bar{Y}_{j_q,n_1}) \to_d \mathcal{N}_{3q}(0_{3q}, \Sigma_{j_q}) \tag{4.5}$$

   where Σ_{j_q} depends linearly on the extrinsic covariance matrices Σ_{a,j_q} of Q_a.
2. Assume in addition that for $a = 1, 2$ the support of the distribution of $Y_{a,1}$ and the VW mean μ_{a,j_q} are included in the domain of the chart φ and $\varphi(Y_{a,1})$ has an absolutely continuous component and finite moment of sufficiently high order. Then the joint distribution

$$V = n_+^{\frac{1}{2}} \varphi(\bar{Y}_{j_q,n_2}^{-1} \odot_q \bar{Y}_{j_q,n_1}) \tag{4.6}$$

   can be approximated by the bootstrap joint distribution of

$$V^* = n_+^{1/2}\, \varphi(\bar{Y^*}_{j_q,n_2}^{-1} \odot_q \bar{Y}_{j_q,n_1}^*)$$

From Patrangenaru et al. [21], recall that given a random sample from a distribution Q on $\mathbb{R}P^m$, if J_s, $s = 1, \dots, q$ are the matrices $J_s = n^{-1}\sum_{r=1}^{n} X_r^s (X_r^s)^T$, and if for $a = 1, \dots, m+1$, $d_s(a)$ and $g_s(a)$ are the eigenvalues in increasing order and corresponding unit eigenvectors of J_s, then the VW sample mean $\bar{Y}_{j_q,n}$ is given by

$$\bar{Y}_{J_q,n} = ([g_1(m+1)], \dots, [g_q(m+1)]). \tag{4.7}$$

Remark 4.1 Given the high dimensionality, the VW sample covariance matrix is often singular. Therefore, for nonparametric hypothesis testing, Efron's nonpivotal bootstrap is preferred. For nonparametric bootstrap methods see e.g. Efron [8]. For details, on testing the existence of a mean change 3D projective shape, when sample sizes are not equal, using nonpivotal bootstrap, see Patrangenaru et al. [20].

4.2 Hypothesis Testing for VW Antimeans

Unlike in the previous subsection, we now assume that for $a = 1, 2$, Q_a are αVW-nonfocal. We are now interested in the hypothesis testing problem:

$$H_0 : \alpha\mu_{1,j_q} = \alpha\mu_{2,j_q} \text{ vs. } H_a : \alpha\mu_{1,j_q} \neq \alpha\mu_{2,j_q}, \tag{4.8}$$

which is equivalent to testing the following

$$H_0 : \alpha\mu_{2,j_q}^{-1} \odot_q \alpha\mu_{1,j_q} = 1_{(\mathbb{R}P^3)^q} \text{ vs. } H_a : \alpha\mu_{2,j_q}^{-1} \odot_q \alpha\mu_{1,j_q} \neq 1_{(\mathbb{R}P^3)^q} \tag{4.9}$$

1. Let $n_+ = n_1 + n_2$ be the total sample size, and assume $\lim_{n_+\to\infty} \frac{n_1}{n_+} \to \lambda \in (0, 1)$. Let φ be the log chart defined in a neighborhood of $1_{(\mathbb{R}P^3)^q}$ (see Helgason [12]), with $\varphi(1_{(\mathbb{R}P^3)^q}) = 0_{3q}$. Then, from Patrangenaru et al. [19], it follows that under H_0

$$n_+^{1/2}\, \varphi(a\bar{Y}_{j_q,n_2}^{-1} \odot_q a\bar{Y}_{j_q,n_1}) \to_d \mathcal{N}_{3q}(0_{3q}, \tilde{\Sigma}_{j_q}), \tag{4.10}$$

   for some covariance matrix $\tilde{\Sigma}_{j_q}$.
2. Assume in addition that for $a = 1, 2$ the support of the distribution of $Y_{a,1}$ and the VW antimean $\alpha\mu_{a,j_q}$ are included in the domain of the chart φ and $\varphi(Y_{a,1})$ has an absolutely continuous component and finite moment of sufficiently high order. Then the joint distribution

$$aV = n_+^{\frac{1}{2}} \varphi(a\bar{Y}_{j_q,n_2}^{-1} \odot_q a\bar{Y}_{j_q,n_1}) \tag{4.11}$$

   can be approximated by the bootstrap joint distribution of

$$aV^* = n_+^{1/2}\, \varphi(a\bar{Y}^{*\,-1}_{j_q,n_2} \odot_q a\bar{Y}^*_{j_q,n_1})$$

Now, from Proposition 2, we get the following result that is used for the computation of the VW sample antimeans.

Proposition 3 *Given a random sample from a distribution* Q *on* $\mathbb{R}P^m$, *if* $J_s, s = 1, \ldots, q$ *are the matrices* $J_s = n^{-1}\sum_{r=1}^{n} X_r^s (X_r^s)^T$, *and if for* $a = 1, \ldots, m+1$, $d_s(a)$ *and* $g_s(a)$ *are the eigenvalues in increasing order and corresponding unit eigenvectors of* J_s, *then the VW sample antimean* $a\bar{Y}_{j_q,n}$ *is given by*

$$a\bar{Y}_{j_q,n} = ([g_1(1)], \ldots, [g_q(1)]). \tag{4.12}$$

5 Two-Sample Test for Lily Flowers Data

In this section, we will test for the existence of 3D mean projective shape change to differentiate between two lily flowers. We will use pairs of pictures of two flowers for our study.

Our data sets consist of two samples of digital images. The first one consist of 11 pairs of pictures of a single lily flower. The second has 8 pairs of digital images of another lily flower (Figs. 2 and 3).

We will recover the 3D projective shape of a spatial k-ad (in our case $k = 13$) from the pairs of images, which will allow us to test for mean 3D projective shape change detection.

Flowers belonging to the genus Lilum have three petals and three petal-like sepals. It may be difficult to distinguish the lily petals from the sepals. Here all six are referred to as *tepals*. For our analysis, we selected 13 anatomic landmarks, 5 of which will be used to construct a projective frame. In order to conduct a proper analysis, we recorded the same labeling of landmarks and kept a constant configuration for both flowers.

Fig. 2 Flower sample 1

Fig. 3 Flower sample 2

Fig. 4 Landmarks for flower 1 and flower 2

The tepals where labeled 1 through 6 for both flowers. Also the six *stamens* (male part of the flower),were labeled 7 through 12 starting with the stamen that is closely related to tepal 1 and continuing in the same fashion. The landmarks were placed at the tip of the *anther* of each of the six stamens and in the center of the *stigma* for the *carpel* (the female part) (Fig. 4).

For 3D reconstructions of k-ads, we used the reconstruction algorithm in Ma et al. [16]. The first 5 of our 13 landmarks were selected to construct our projective frame π. To each projective point, we associated its projective coordinate with respect to π. The projective shape of the 3D k-ad, is then determined by the 8 projective coordinates of the remaining landmarks of the reconstructed configuration.

We tested for the VW mean change, since $(\mathbb{R}P^3)^8$ has a Lie group structure (Crane and Patrangenaru [6]). Two types of VW mean changes were considered: one for cross-validation, and the other for comparing the VW mean shapes of the two flowers.

Suppose Q_1 and Q_2 are independent r.o.'s, the hypothesis for their mean change is

$$H_0 : \mu_{1,j_8}^{-1} \odot_8 \mu_{2,j_8} = 1_{(\mathbb{R}P^3)^8}$$

Given φ, the affine chart on this Lie group, $\varphi(1_{(\mathbb{R}P^3)^8}) = 0_{24}$, we compute the bootstrap distribution

$$D_* = \varphi((\bar{Y}^*_{1,j_8,11})^{-1} \odot_8 \bar{Y}^*_{2,j_8,8})$$

We fail to reject H_0, if all simultaneous confidence intervals contain 0, and reject it otherwise. We construct 95 % simultaneous nonparametric bootstrap confidence intervals. We will then expect to fail to reject the null, if we have 0 in all of our simultaneous confidence intervals.

5.1 *Results for Comparing the Two Flowers*

We would fail to reject our null hypothesis

$$H_0 : \mu_{1,j_8}^{-1} \odot_8 \mu_{2,j_8} = 1_{(\mathbb{R}P^3)^8}$$

if all of our 24 confidence intervals would contain the value 0.

We notice that 0 is does not belong to 13 simultaneous confidence intervals in the table below. We then can conclude that there is significant mean VW projective shape

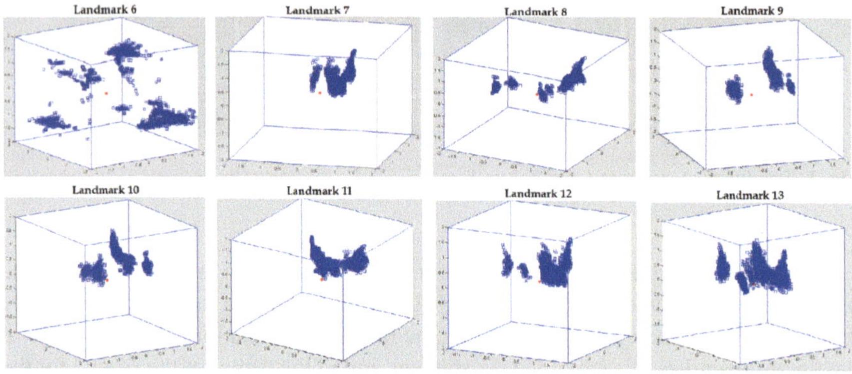

Fig. 5 Bootstrap projective shape marginals for lily data

Simultaneous Confidence Intervals for lily's landmarks 6 to 9				
	LM6	LM7	LM8	LM9
x	$(0.609514, 1.638759)$	$(0.320515, 0.561915)$	$(-0.427979, 0.821540)$	$(0.055007, 0.876664)$
y	$(-0.916254, 0.995679)$	$(-0.200514, 0.344619)$	$(-0.252281, 0.580393)$	$(-0.358060, 0.461555)$
z	$(-1.589983, 1.224176)$	$(0.177687, 0.640489)$	$(0.291530, 0.831977)$	$(0.213021, 0.883361)$

Simultaneous Confidence Intervals for lily's landmarks 10 to 13				
	LM10	LM11	LM12	LM13
x	$(0.060118, 0.822957)$	$(0.495050, 0.843121)$	$(0.419625, 0.648722)$	$(0.471093, 0.874260)$
y	$(-0.346121, 0.160780)$	$(-0.047271, 0.253993)$	$(-0.079662, 0.193945)$	$(-0.075751, 0.453817)$
z	$(0.198351, 0.795122)$	$(0.058659, 0.619450)$	$(0.075902, 0.569353)$	$(-0.146431, 0.497202)$

change between the two flowers. This difference is also visible within the figure of the boxes of the bootstrap projective shape marginals found in Fig. 5. The bootstrap projective shape marginals for landmarks 11 and 12, visually reinforce the rejection of the null hypothesis.

5.2 *Results for Cross-Validation of the Mean Projective Shape of the Lily Flower in Second Sample of Images*

One can show that, as expected, there is no mean VW projective shape change, based on the two samples with sample sizes respectively $n_1 = 5$ and $n_2 = 6$. In the tables below, 0 is contained in all of the simultaneous intervals. Hence, we fail to reject the null hypothesis at level $\alpha = 0.05$ (Fig. 6).

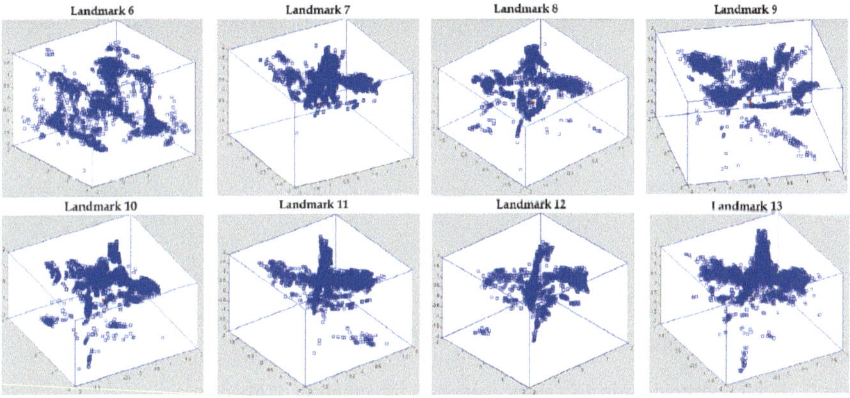

Fig. 6 Bootstrap projective shape marginals for cross-validation of lily flower

Simultaneous confidence intervals for lily's landmarks 6 to 9

	LM6	LM7	LM8	LM9
x	$(-1.150441, 0.940686)$	$(-1.014147, 1.019635)$	$(-0.960972, 1.142165)$	$(-1.104360, 1.162658)$
y	$(-1.245585, 2.965492)$	$(-1.418121, 1.145503)$	$(-1.250429, 1.300157)$	$(-1.078833, 1.282883)$
z	$(-0.971271, 1.232609)$	$(-1.654594, 1.400703)$	$(-1.464506, 1.318222)$	$(-1.649496, 1.396918)$

Simultaneous confidence intervals for lily's landmarks 10 to 13

	LM10	LM11	LM12	LM13
x	$(-1.078765, 1.039589)$	$(-0.995622, 1.381674)$	$(-0.739663, 1.269416)$	$(-1.015220, 1.132021)$
y	$(-1.126703, 1.140513)$	$(-1.210271, 1.184141)$	$(-1.324111, 1.026571)$	$(-1.650026, 1.593305)$
z	$(-1.092425, 1.795890)$	$(-1.222856, 1.963960)$	$(-1.128044, 1.762559)$	$(-1.035796, 2.227439)$

5.3 *Comparing the Sample Antimean for the Two Lily Flowers*

The Veronese–Whitney (VW) antimean is the extrinsic antimean associated with the VW embedding (see Patrangenaru et al. [20, 21] for details). The VW antimean changes were considered for comparing the VW antimean shapes of the two flowers. Suppose Q_1 and Q_2 are independent r.o.'s, the hypothesis for their mean change are

$$H_0 : \alpha\mu_{1,j_8}^{-1} \odot_8 \alpha\mu_{2,j_8} = 1_{(\mathbb{R}P^3)^8}$$

Let φ be the affine chart on this product of projective spaces, $\varphi(1_8) = 0_8$, we compute the bootstrap distribution,

$$\alpha D_* = \varphi(\overline{aY}^{*-1}{}_{1,j_8,11} \odot_8 \overline{aY}^{*}{}_{2,j_8,8})$$

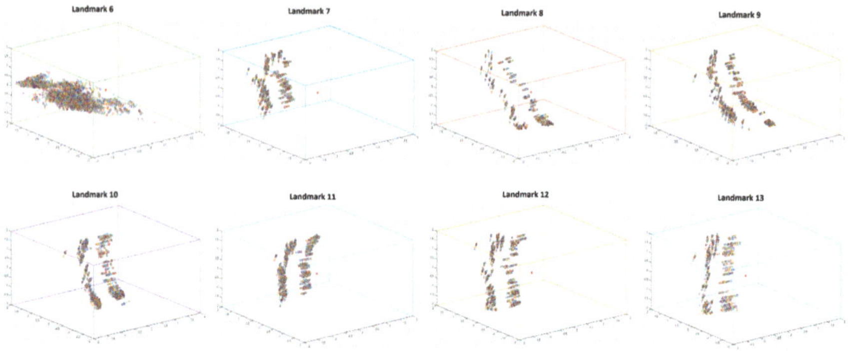

Fig. 7 Eight bootstrap projective shape marginals for antimean of lily data

and construct the 95 % simultaneous nonparametric bootstrap confidence intervals. We will then expect to fail to reject the null, if we have 0 in all of our simultaneous confidence intervals (Fig. 7).

Highlighted in blue are the intervals not containg $0 \in \mathbb{R}$.

simultaneous confidence intervals for lily's landmarks 6 to 9				
	LM6	LM7	LM8	LM9
x	$(-1.02, -0.51)$	$(-1.41, 0.69)$	$(-1.14, 0.40)$	$(-0.87, 0.35)$
y	$(0.82, 2.18)$	$(0.00, 0.96)$	$(-0.15, 0.92)$	$(-0.09, 0.69)$
z	$(-0.75, 0.36)$	$(-6.93, 2.83)$	$(-3.07, 3.23)$	$(-2.45, 2.38)$

Simultaneous confidence intervals for lily's landmarks 10 to 13				
	LM10	LM11	LM12	LM13
x	$(-0.61, 0.32)$	$(-0.87, 0.08)$	$(-0.99, 0.02)$	$(-0.84, -0.04)$
y	$(-0.07, 0.51)$	$(-0.04, 0.59)$	$(0.06, 0.75)$	$(0.18, 0.78)$
z	$(-3.03, 1.91)$	$(-5.42, 1.98)$	$(-7.22, 2.41)$	$(-4.91, 2.62)$

In conclusion, there is significant antimean VW projective shape change between the two flowers, showing that the extrinsic antimean is a sensitive parameter for extrinsic analysis.

6 Computational Example of VW Sample Mean and VW Sample Antimean on a Planar Kendall Shape Space

We use the VW embedding of the complex projective space (Kendall shape space) to compare VW means and VW antimeans for a configuration of landmarks on midface in a population of normal children, based on a study on growth measured from X-rays at 8 and 14 years of age (for data sets, see Patrangenaru and Ellingson [18], Chap. 1). The Fig. 8 is from lateral X-ray of a clinically normal skull (top, with landmarks).

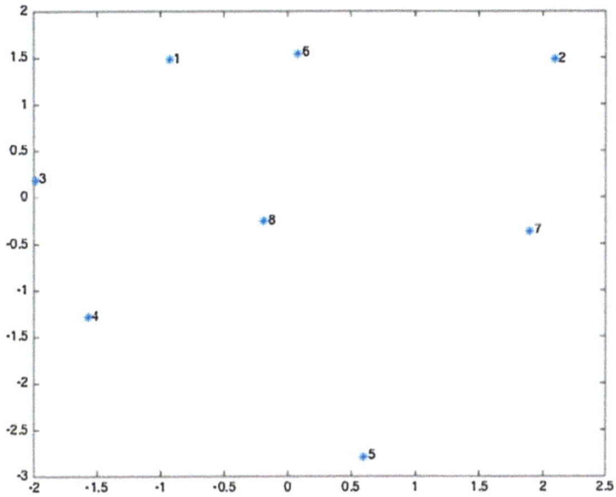

Fig. 8 The coordinates of first child's skull image

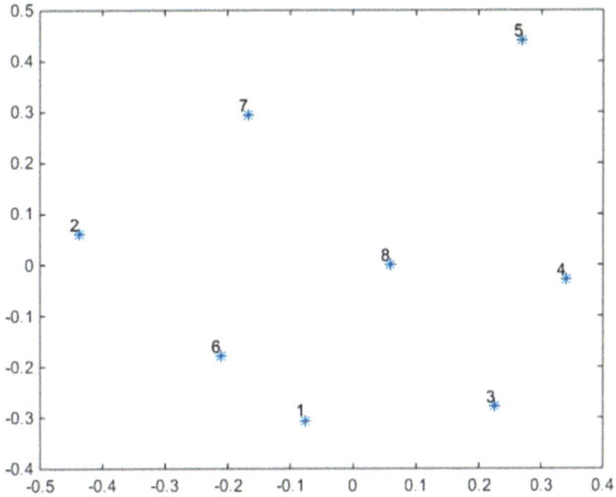

Fig. 9 Icon of extrinsic sample mean coordinates based on children midface skull data

In Fig. 9 one displays the coordinates of a representative of the extrinsic sample mean. You may find that with only a rotation, Figs. 8 and 9 looks very similar, as the extrinsic mean, is close to the sample observations. Here close is in the sense of small distance relative to the diameter of the object space.

On the other hand, we also have a sample VW-antimean, the representative of which is shown in Fig. 10. The VW-antimean statistic is far from the average, since according to the general results presented in this paper, the chord distance between the sample VW antimean and sample mean in the ambient spaces is maximized. The relative location of the landmarks is also different in antimean. The following result gives the coordinate of representatives (icons) of the VW mean and VW antimean Kendall shapes. Each coordinate of an icon is a complex number.

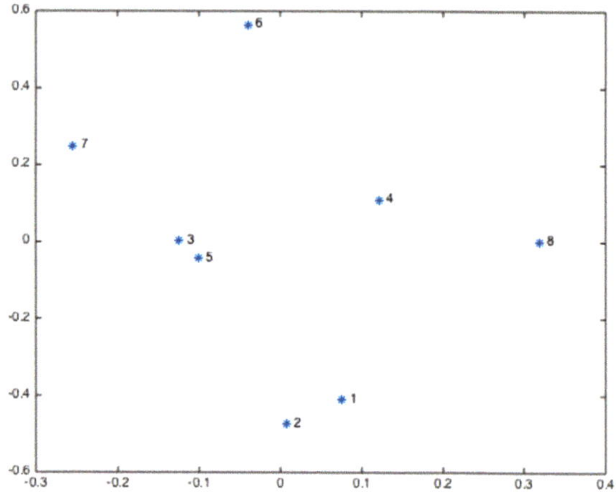

Fig. 10 Icon of extrinsic sample antimean coordinates based on children midface skull data

VW—sample mean $\bar{X}_E = (-0.0766 + 0.3066i, -0.4368 - 0.0593i, 0.2254 + 0.2786i, 0.3401 + 0.0298i, 0.2685 - 0.4409i, -0.2110 + 0.1791i, -0.1676 - 0.2939i, 0.0580 + 0.0000i)$.

VW—sample antimean $a\bar{X}_E = (0.0752 - 0.4103i, 0.0066 - 0.4731i, -0.1244 + 0.0031i, 0.1213 + 0.1102i, -0.1015 - 0.0422i, -0.0400 + 0.5639i, -0.2553 + 0.2485i, 0.3182 + 0.0000i)$.

7 Discussion and Thanks

In this paper, we introduce a new statistic, the sample extrinsic antimean. Just as with the extrinsic mean, the extrinsic antimean captures important features of a distribution on a compact object space. Certainly, the definitions and results extend to the general case of arbitrary Fréchet antimeans; however, based on the comparison between intrinsic and extrinsic sample means (see Bhattacharya et al. [2]), for the purpose of object data analysis (see Patrangenaru and Ellingson [18]), it is expected that intrinsic sample antimeans take way more time to compute than extrinsic sample means. Therefore, future research will parallel research on inference for extrinsic means. This includes results for stickiness of extrinsic means (see Hotz et al. [13]). The authors would like to thank Harry Hendricks and Mingfei Qiu for useful conversations on the subject of the stickiness phenomenon and antimeans and to the referee for useful comments that helped us improve the paper.

Acknowledgments Research supported by NSA-MSP-H98230-14-1-0135 and NSF-DMS-1106935.
Research supported by NSA-MSP- H98230-14-1-0135.
Research supported by NSF-DMS-1106935.

References

1. Beran, R., Fisher, N.I.: Nonparametric comparison of mean axes. Ann. Statist. **26**, 472–493 (1998)
2. Bhattacharya, R.N., Ellingson, L., Liu, X., Patrangenaru, V., Crane, M.: Extrinsic analysis on manifolds is computationally faster than intrinsic analysis, with applications to quality control by machine vision. Appl. Stoch. Models Bus. Ind. **28**, 222–235 (2012)
3. Bhattacharya, R.N., Patrangenaru, V.: Large sample theory of intrinsic and extrinsic sample means on manifolds-part II. Ann. Stat. **33**, 1211–1245 (2005)
4. Bhattacharya, R.N., Patrangenaru, V.: Large sample theory of intrinsic and extrinsic sample means on manifolds-part I. Ann. Stat. **31**(1), 1–29 (2003)
5. Billera, L.J., Holmes, S.P., Vogtmann, K.: Geometry of the space of phylogenetic trees. Adv. Appl. Math. **27**(4), 733–767 (2001)
6. Crane, M., Patrangenaru, V.: Random change on a Lie group and mean glaucomatous projective shape change detection from stereo pair images. J. Multivar. Anal. **102**, 225–237 (2011)
7. Dryden, I.L., Mardia, K.V.: Statistical Shape Analysis. Wiley, Chichester (1998)
8. Efron, B.: The Jackknife, the Bootstrap and Other Resampling Plans. CBMS-NSF Regional Conference Series in Applied Mathematics, vol. 38. SIAM, Philadelphia, Pa (1982)
9. Fisher, N.I., Hall, P., Jing, B.Y., Wood, A.T.A.: Properties of principal component methods for functional and longitudinal data analysis. J. Am. Stat. Assoc. **91**, 1062–1070 (1996)
10. Fréchet, M.: Les élements aléatoires de nature quelconque dans un espace distancié. Ann. Inst. H. Poincaré **10**, 215–310 (1948)
11. Guo, R., Patrangenaru, V., Lester, D.: Nonparametric Bootstrap test for Equality of Extrinsic Mean Reflection Shapes in Large vs Small Populations of Acrosterigma Magnum Shells, Poster, Geometric Topological and Graphical Model Methods in Statistics Fields Institute, Toronto, Canada, May 22–23, 2014
12. Helgason, S.: Differential Geometry and Symmetric Spaces. AMS Chelsea Publishing, AMS, Providence, Rhode Island (2001)
13. Hotz, T., Huckemann, S., Le, H., Marron, J.S., Mattingly, J.C., Miller, E., Nolen, J., Owen, M., Patrangenaru, V., Skwerer, S.: Sticky central limit theorems on open books. Ann. Appl. Probab. **23**, 2238–2258 (2013)
14. Kendall, D.G.: Shape manifolds, procrustean metrics, and complex projective spaces. Bull. Lond. Math. Soc. **16**, 81–121 (1984)
15. Kent, J.T.: New directions in shape analysis. In: The Art of Statistical Science, A Tribute to G. S. Watson, pp. 115–127 (1992)
16. Ma, Y., Soatto, A., Košecká, J., Sastry, S.: An Invitation to 3-D Vision: From Images to Geometric Models. Springer (2005)
17. Mardia, K.V., Patrangenaru, V.: Directions and projective shapes. Ann. Stat. **33**, 1666–1699 (2005)
18. Patrangenaru, V., Ellingson, L.L.: Nonparametric Statistics on Manifolds and Their Applications. Chapman & Hall/CRC Texts in Statistical Science (2015)
19. Patrangenaru, V., Guo, R., Yao, K.D.: Nonparametric inference for location parameters via Fréchet functions. In: Proceedings of *Second International Symposium on Stochastic Models in Reliability Engineering, Life Science and Operations Management, Beer Sheva*, Israel, pp. 254–262 (2016)

20. Patrangenaru, V., Qiu, M., Buibas, M.: Two sample tests for mean 3D projective shapes from digital camera images. Methodol. Comput. Appl. Probab. **16**, 485–506 (2014)
21. Patrangenaru, V., Liu, X., Sugathadasa, S.: Nonparametric 3D projective shape estimation from pairs of 2D images—I, in memory of W.P. Dayawansa. J. Multivar. Anal. **101**, 11–31 (2010)
22. Patrangenaru, V., Mardia, K.V.: Affine shape analysis and image analysis. In: Proceedings of the Leeds Annual Statistics Research Workshop, pp. 57–62. Leeds University Press (2003)
23. Sughatadasa, S.M.: Affine and Projective Shape Analysis with Applications. Ph.D. dissertation, Texas Tech University (2006)
24. Wang, H., Marron, J.S.: Object oriented data analysis: sets of trees. Ann. Stat. **35**, 1849–1873 (2007)
25. Watson, G.S.: Statistics on Spheres. Lecture Notes in the Mathematical Sciences. Wiley (1983)
26. Ziezold, H.: On expected figures and a strong law of large numbers for random elements in quasi-metric spaces. In: Transactions of Seventh Prague Conference on Information Theory, Statistical Decision Functions, Random Processes A, pp. 591–602 (1977)

Partial Distance Correlation

Gábor J. Székely and Maria L. Rizzo

Abstract Partial distance correlation measures association between two random vectors with respect to a third random vector, analogous to, but more general than (linear) partial correlation. Distance correlation characterizes independence of random vectors in arbitrary dimension. Motivation for the definition is discussed. We introduce a Hilbert space of U-centered distance matrices in which squared distance covariance is the inner product. Simple computation of the sample partial distance correlation and definitions of the population coefficients are presented. Power of the test for zero partial distance correlation is compared with power of the partial correlation test and the partial Mantel test.

Keywords Independence · Multivariate · Partial distance correlation · Dissimilarity · Energy statistics

1 Introduction

Distance correlation is a multivariate measure of dependence between random vectors in arbitrary, not necessarily equal dimension. Distance covariance (dCov) and the standardized coefficient, distance correlation (dCor), are nonnegative coefficients that characterize independence of random vectors; both are zero if and only if the random vectors are independent. The problem of defining a *partial distance correlation* coefficient analogous to the linear partial distance correlation coefficient had been an open problem since the distance correlation was introduced in 2007 [11]. For the definition of partial distance correlation, we introduce a new Hilbert space where

G.J. Székely
National Science Foundation, 4201 Wilson Blvd. #1025, Arlington, VA 22230, USA
e-mail: gszekely@nsf.gov

M.L. Rizzo (✉)
Department of Mathematics and Statistics, Bowling Green State University, Bowling Green, OH 43403, USA
e-mail: mrizzo@bgsu.edu
URL: http://personal.bgsu.edu/~mrizzo

R. Cao et al. (eds.), *Nonparametric Statistics*, Springer Proceedings in Mathematics & Statistics 175, DOI 10.1007/978-3-319-41582-6_13

the squared distance covariance is the inner product [15]. Our intermediate results include methods for applying distance correlation to dissimilarity matrices.

For background, we first review the definitions of population and sample dCov and dCor coefficients. In what follows, we suppose that X and Y take values in $\mathbb{R}^p$ and $\mathbb{R}^q$, respectively.

The distance covariance, $\mathcal{V}(X, Y)$, of two random vectors X and Y is a scalar coefficient defined by a weighted L_2 norm measuring the distance between the joint characteristic function $\phi_{X,Y}$ of X and Y, and the product $\phi_X \phi_Y$ of the marginal characteristic functions of X and Y. $\mathcal{V}(X, Y)$ is defined as the nonnegative square root of

$$\begin{aligned}\mathcal{V}^2(X, Y) &= \|\phi_{X,Y}(t, s) - \phi_X(t)\phi_Y(s)\|_w^2 \\ &:= \int_{R^{p+q}} |\phi_{X,Y}(t, s) - \phi_X(t)\phi_Y(s)|^2 w(t, s)\, dt\, ds,\end{aligned} \tag{1}$$

where $w(t, s) := (|t|_p^{1+p}|s|_q^{1+q})^{-1}$. The above integral exists if $|X|$ and $|Y|$ have finite first moments. The choice of weight function is not unique, but when we consider certain invariance properties that one would require for a measure of dependence it can be shown to be unique [13]. This particular weight function may have first appeared in this context in 1993 where Feuerverger [3] proposed a bivariate test of independence based on an L_2 norm (1).

The distance covariance coefficient can also be expressed in terms of expected distances, based on the following identity established in Székely and Rizzo [12, Theorem 8, p. 1250]. The notation X' indicates that X' is an independent and identically distributed (iid) copy of X. If (X, Y), (X', Y'), and (X'', Y'') are iid, each with joint distribution (X, Y), then

$$\begin{aligned}\mathcal{V}^2(X, Y) &= E|X - X'||Y - Y'| + E|X - X'| \cdot E|Y - Y'| \\ &\quad - E|X - X'||Y - Y''| - E|X - X''||Y - Y'|,\end{aligned} \tag{2}$$

provided that X and Y have finite first moments. Definition (2) can be extended to X and Y taking values in a separable Hilbert space. With that extension and our intermediate results, we can define and apply partial distance covariance (pdcov) and partial distance correlation (pdcor).

Distance correlation (dCor) $\mathcal{R}(X, Y)$ is a standardized coefficient,

$$\mathcal{R}(X, Y) = \begin{cases} \frac{\mathcal{V}(X,Y)}{\sqrt{\mathcal{V}(X,X)\mathcal{V}(Y,Y)}}, & \mathcal{V}(X, X)\mathcal{V}(Y, Y) > 0; \\ 0, & \mathcal{V}(X, X)\mathcal{V}(Y, Y) = 0. \end{cases}$$

(See [11] and [12])

The distance covariance and distance correlation statistics are functions of the double-centered distance matrices of the samples. For an observed random sample $\{(x_i, y_i) : i = 1, \ldots, n\}$ from the joint distribution of random vectors X and Y, Let

$(a_{ij}) = (|x_i - x_j|_p)$ and $(b_{ij}) = (|y_i - y_j|_q)$ denote the Euclidean distance matrices of the X and Y samples, respectively

Define the *double-centered distance matrix* of the X sample by

$$\widehat{A}_{ij} = a_{ij} - \bar{a}_{i.} - \bar{a}_{.j} + \bar{a}_{..}, \qquad i, j = 1, \ldots, n, \tag{3}$$

where

$$\bar{a}_{i.} = \frac{1}{n}\sum_{j=1}^{n} a_{ij}, \quad \bar{a}_{.j}, = \frac{1}{n}\sum_{i=1}^{n} a_{ij}, \quad \bar{a}_{..} = \frac{1}{n^2}\sum_{i,j=1}^{n} a_{ij}.$$

Similarly, define the double-centered distance matrix of the Y sample by $\widehat{B}_{ij} = b_{ij} - \bar{b}_{i.} - \bar{b}_{.j} + \bar{b}_{..}$, for $i, j = 1, \ldots, n$.

A double-centered distance matrix $\widehat{A}_{ij}$ has the property that all rows and columns sum to zero. Below we will introduce a modified definition $\mathcal{U}$-centering ("U" for unbiased) such that a $\mathcal{U}$-centered distance matrix $\widetilde{A}_{ij}$ has zero expected values of its elements $E[\widetilde{A}_{ij}] = 0$ for all i, j.

Sample distance covariance $\mathcal{V}_n(\mathbf{X}, \mathbf{Y})$ is the square root of

$$\mathcal{V}_n^2(\mathbf{X}, \mathbf{Y}) = \frac{1}{n^2}\sum_{i,j=1}^{n} \widehat{A}_{ij}\widehat{B}_{ij} \tag{4}$$

and sample distance correlation is the standardized sample coefficient

$$\mathcal{R}_n^2(\mathbf{X}, \mathbf{Y}) = \begin{cases} \frac{\mathcal{V}_n^2(\mathbf{X},\mathbf{Y})}{\sqrt{\mathcal{V}_n^2(\mathbf{X})\mathcal{V}_n^2(\mathbf{Y})}}, & \mathcal{V}_n^2(\mathbf{X})\mathcal{V}_n^2(\mathbf{Y}) > 0; \\ 0, & \mathcal{V}_n^2(\mathbf{X})\mathcal{V}_n^2(\mathbf{Y}) = 0. \end{cases} \tag{5}$$

The distance covariance test of multivariate independence is consistent against all dependent alternatives. Large values of the statistic $n\mathcal{V}_n^2(\mathbf{X}, \mathbf{Y})$ support the alternative hypothesis that X and Y are dependent (see [11, 12]). The test is implemented in the *energy* package [9] for R [8].

2 Partial Distance Correlation

To generalize distance correlation to partial distance correlation we require that essential properties of distance correlation are preserved, that pdcor has a meaningful interpretation as a population coefficient and as a sample coefficient, that inference is possible, and the methods are practical to apply. This generalization is not straightforward.

For example, one could try to follow the definition of partial correlation based on orthogonal projections in a Euclidean space, but this approach does not succeed. For partial distance covariance orthogonality means independence, but the orthogonal projection of a random variable onto the condition variable has a remainder that is typically not independent of the condition.

Alternately, because the product moment type of computing formula for the sample distance covariance (4) may suggest an inner product, one may consider defining the Hilbert space of double-centered distance matrices (3), where the inner product is (4). However, this approach also presents a problem, because in general it is not clear what the projection objects in this space actually represent. The difference of double-centered distance matrices is not a double-centered distance matrix of any sample except in some special cases. Although one could make the formal definitions, inference is not possible unless the sample coefficients have a meaningful interpretation as objects that arise from centering distance matrices of samples.

The sample coefficient $\mathcal{V}_n^2(\mathbf{X},\mathbf{Y})$ is a biased estimator of $\mathcal{V}^2(X,Y)$, so in a sense we could consider double centering to be a biased operation. We modify the inner product approach by first replacing double centering with $\mathcal{U}$-centering (1). The Hilbert space is the linear span of $n \times n$ "$\mathcal{U}$-centered" matrices. The inner product in this space is unbiased dCov; it is an unbiased estimator of $\mathcal{V}^2(X,Y)$. An important property of this space is that all linear combinations, and in particular all projections, are $\mathcal{U}$-centered matrices. (The corresponding property does not hold when we work with double-centered matrices.)

A representation theorem ([15]) connects the orthogonal projections to random samples in Euclidean space. With this representation result, methods for inference based on the inner product are defined and implemented. To obtain this representation we needed results for dissimilarity matrices. In many applications, such as community ecology or psychology, one has dissimilarity matrices rather than the sample points available, and the dissimilarities are often not Euclidean distances. Our intermediate results on dissimilarity matrices also extend the definitions, computing formulas, and inference to data represented by any symmetric, zero diagonal dissimilarity matrices.

2.1 The Hilbert Space of Centered Distance Matrices

Let $A = (a_{ij})$ be an $n \times n$ zero diagonal, symmetric matrix, $n > 2$ (a dissimilarity matrix). The *$\mathcal{U}$-centered* matrix $\widetilde{A}_{i,j}$ is defined by

$$\widetilde{A}_{i,j} = \begin{cases} a_{i,j} - \frac{1}{n-2}\sum_{\ell=1}^{n} a_{i,\ell} - \frac{1}{n-2}\sum_{k=1}^{n} a_{k,j} + \frac{1}{(n-1)(n-2)}\sum_{k,\ell=1}^{n} a_{k,\ell}, & i \neq j; \\ 0, & i = j. \end{cases} \tag{1}$$

Then

$$(\widetilde{A} \cdot \widetilde{B}) := \frac{1}{n(n-3)} \sum_{i \neq j} \widetilde{A}_{i,j} \widetilde{B}_{i,j} \tag{2}$$

is an unbiased estimator of squared population distance covariance $\mathcal{V}^2(X, Y)$.

The Hilbert space of $\mathcal{U}$-centered matrices is defined as follows. Let $\mathcal{H}_n$ denote the linear span of $\mathcal{U}$-centered $n \times n$ distance matrices, and for each pair (C, D) in $\mathcal{H}_n$, define their inner product by

$$(C \cdot D) = \frac{1}{n(n-3)} \sum_{i \neq j} C_{ij} D_{ij}. \tag{3}$$

It can be shown that every matrix $C \in \mathscr{H}_n$ is the $\mathcal{U}$-centered distance matrix of some n points in $\mathbb{R}^p$, where $p \leq n-2$.

The linear span of all $n \times n$ $\mathcal{U}$-centered matrices is a Hilbert space $\mathscr{H}_n$ with inner product defined by (3) [15].

2.2 *Sample PdCov and PdCor*

The projection operator (4) can now be defined in the Hilbert space $\mathscr{H}_n$, $n \geq 4$. Then partial distance covariance can be defined using projections in $\mathscr{H}_n$. Suppose that x, y, and z are samples of size n and $\widetilde{A}$, $\widetilde{B}$, $\widetilde{C}$ are their $\mathcal{U}$-centered distance matrices, respectively. Define the orthogonal projection

$$P_{z^\perp}(x) = \widetilde{A} - \frac{(\widetilde{A} \cdot \widetilde{C})}{(\widetilde{C} \cdot \widetilde{C})} \widetilde{C} \tag{4}$$

of $\widetilde{A}(x)$ onto $(\widetilde{C}(z))^\perp$, and

$$P_{z^\perp}(y) = \widetilde{B} - \frac{(\widetilde{B} \cdot \widetilde{C})}{(\widetilde{C} \cdot \widetilde{C})} \widetilde{C}, \tag{5}$$

the orthogonal projection of $\widetilde{B}(y)$ onto $(\widetilde{C}(z))^\perp$. If $(\widetilde{C} \cdot \widetilde{C}) = 0$, $P_{Z^\perp}(x) := \widetilde{A}$ and $P_{Z^\perp}(y) := \widetilde{B}$. Then $P_{z^\perp}(x)$ and $P_{z^\perp}(y)$ are elements of $\mathscr{H}_n$. Their inner product is as defined in (3).

Definition 1 (*Partial distance covariance*)

$$\text{pdCov}(x, y; z) = (P_{z^\perp}(x) \cdot P_{z^\perp}(y)), \tag{6}$$

where $P_{z^\perp}(x)$, and $P_{z^\perp}(y)$ are defined by (4) and (5), and

$$(P_{z^\perp}(x) \cdot P_{z^\perp}(y)) = \frac{1}{n(n-3)} \sum_{i \neq j} (P_{z^\perp}(x))_{i,j} (P_{z^\perp}(y))_{i,j}. \tag{7}$$

Definition 2 (*Partial distance correlation*). Sample partial distance correlation is defined as the cosine of the angle θ between the 'vectors' $P_{z^\perp}(x)$ and $P_{z^\perp}(y)$ in the Hilbert space $\mathscr{H}_n$:

$$R^*(x, y; z) := \cos\theta = \frac{(P_{z^\perp}(x) \cdot P_{z^\perp}(y))}{|P_{z^\perp}(x)||P_{z^\perp}(y)|}, \qquad |P_{z^\perp}(x)||P_{z^\perp}(y)| \neq 0, \tag{8}$$

and otherwise $R^*(x, y; z) := 0$.

2.3 Representation in Euclidean Space

If it is true that the projection matrices $P_{z^\perp}(x)$ and $P_{z^\perp}(y)$ are the $\mathcal{U}$-centered *Euclidean* distance matrices of samples of points in Euclidean spaces, then the sample partial distance covariance (7) is in fact the distance covariance (2) of those samples.

Our representation theorem [15] holds that given an arbitrary element H of $\mathscr{H}_n$, there exists a configuration of points $\mathbf{U} = [u_1, \dots, u_n]$ in some Euclidean space $\mathbb{R}^q$, for some $q \geq 1$, such that the $\mathcal{U}$-centered Euclidean distance matrix of sample $\mathbf{U}$ is exactly equal to the matrix H. In general, every element in $\mathscr{H}_n$, and in particular any orthogonal projection matrix, is the $\mathcal{U}$-centered distance matrix of some sample of n points in a Euclidean space.

The proof uses properties of $\mathcal{U}$-centered distance matrices and results from classical multidimensional scaling.

Lemma 1 *Let $\widetilde{A}$ be a $\mathcal{U}$-centered distance matrix. Then*

(i) Rows and columns of $\widetilde{A}$ sum to zero.
(ii) $\widetilde{(\widetilde{A})} = \widetilde{A}$. That is, if B is the matrix obtained by $\mathcal{U}$-centering an element $\widetilde{A} \in \mathscr{H}_n$, $B = \widetilde{A}$.
(iii) $\widetilde{A}$ is invariant to double centering. That is, if B is the matrix obtained by double centering the matrix $\widetilde{A}$, then $B = \widetilde{A}$.
(iv) If c is a constant and B denotes the matrix obtained by adding c to the off-diagonal elements of $\widetilde{A}$, then $\widetilde{B} = \widetilde{A}$.

In the proof, Lemma 1(iv) is essential for our results, which shows that we cannot apply double centering as in the original (biased) definition of distance covariance. Invariance with respect to the additive constant c in (iv) does not hold for double-centered matrices.

Our representation theorem applies certain results from classical MDS and Cailliez [1, Theorem 1].

Theorem 1 *Let H be an arbitrary element of the Hilbert space $\mathcal{H}_n$ of $\mathcal{U}$-centered distance matrices. Then there exists a sample $v_1, \ldots, v_n$ in a Euclidean space of dimension at most $n-2$, such that the $\mathcal{U}$-centered distance matrix of $v_1, \ldots, v_n$ is exactly equal to H.*

For details of the proof and an illustration see [15]. The details of the proof reveal why the simpler idea of a Hilbert space of double-centered matrices is not applicable here. The diagonals of $\widehat{A}$ are not zero, so we cannot get an exact solution by MDS. The inner product would depend on the additive constant c. Another problem is that $\mathcal{V}_n^2 \geq 0$, but the inner product of projections in that space can be negative.

An application of our representation theorem also provides methods for zero diagonal symmetric non-Euclidean dissimilarities. There exist samples in Euclidean space such that their $\mathcal{U}$-centered Euclidean distance matrices are equal to the dissimilarity matrices. There are existing software implementations of classical MDS that can obtain these sample points. The R function `cmdscale`, for example, includes options to apply the additive constant of Cailliez [1] and to specify the dimension.

Using the inner product (3), we can define a bias corrected distance correlation statistic

$$R^*_{x,y} := \begin{cases} \frac{(\widetilde{A}\cdot\widetilde{B})}{|\widetilde{A}||\widetilde{B}|}, & |\widetilde{A}||\widetilde{B}| \neq 0; \\ 0, & |\widetilde{A}||\widetilde{B}| = 0, \end{cases} \tag{9}$$

where $\widetilde{A} = \widetilde{A}(x)$, $\widetilde{B} = \widetilde{B}(y)$ are the $\mathcal{U}$-centered distance matrices of the samples x and y, and $|\widetilde{A}| = (\widetilde{A} \cdot \widetilde{A})^{1/2}$.

Here we should note R^* is a bias corrected statistic for the *squared* distance correlation (5) rather than the distance correlation.

An equivalent computing formula for pdCor(x, y, z) is

$$R^*_{x,y;z} = \frac{R^*_{x,y} - R^*_{x,z} R^*_{y,z}}{\sqrt{1-(R^*_{x,z})^2}\sqrt{1-(R^*_{y,z})^2}}, \tag{10}$$

$(1-(R^*_{x,z})^2)(1-(R^*_{y,z})^2) \neq 0$.

2.4 Algorithm to Compute Partial Distance Correlation $R^*_{x,y;z}$ from Euclidean Distance Matrices

Equation (10) provides a simple and familiar form of computing formula for the partial distance correlation. The following algorithm summarizes the calculations for distance matrices $A = (|x_i - x_j|)$, $B = (|y_i - y_j|)$, and $C = (|z_i - z_j|)$.

(i) Compute $\mathcal{U}$-centered distance matrices $\widetilde{A}$, $\widetilde{B}$, and $\widetilde{C}$ using

$$\widetilde{A}_{i,j} = a_{i,j} - \frac{a_{i.}}{n-2} - \frac{a_{.j}}{n-2} + \frac{a_{..}}{(n-1)(n-2)}, \qquad i \neq j,$$

and $\widetilde{A}_{i,i} = 0$.

(ii) Compute inner products and norms using

$$(\widetilde{A} \cdot \widetilde{B}) = \frac{1}{n(n-3)} \sum_{i \neq j} \widetilde{A}_{i,j} \widetilde{B}_{i,j}, \qquad |\widetilde{A}| = (\widetilde{A} \cdot \widetilde{A})^{1/2}$$

and $R^*_{x,y}$, $R^*_{x,z}$, and $R^*_{y,z}$ using $R^*_{x,y} = \frac{(\widetilde{A} \cdot \widetilde{B})}{|\widetilde{A}||\widetilde{B}|}$.

(iii) If $R^2_{x,z} \neq 1$ and $R^2_{y,z} \neq 1$

$$R^*_{x,y;z} = \frac{R^*_{x,y} - R_{x,z} R^*_{y,z}}{\sqrt{1 - (R^*_{x,z})^2} \sqrt{1 - (R^*_{y,z})^2}},$$

otherwise apply the definition (8).

In the above algorithm, it is typically not necessary to explicitly compute the projections, when (10) is applied. This algorithm has a straightforward translation into code. An implementation is provided in the *pdcor* package [10] (available upon request) or in the energy package for R.

3 Population Coefficients

Definition 3 (*Population partial distance covariance*) Introduce the scalar coefficients

$$\alpha := \frac{\mathcal{V}^2(X, Z)}{\mathcal{V}^2(Z, Z)}, \qquad \beta := \frac{\mathcal{V}^2(Y, Z)}{\mathcal{V}^2(Z, Z)}.$$

If $\mathcal{V}^2(Z, Z) = 0$ define $\alpha = \beta = 0$. The double-centered projections of A_X and B_Y onto the orthogonal complement of C_Z in Hilbert space $\mathscr{H}$ are defined

$$P_{Z^\perp}(X) := A_X(X, X') - \alpha C_Z(Z, Z'), \quad P_{Z^\perp}(Y) := B_Y(Y, Y') - \beta C_Z(Z, Z'),$$

or in short $P_{Z^\perp}(X) = A_X - \alpha C_Z$ and $P_{Z^\perp}(Y) = B_Y - \beta C_Z$, where C_Z denotes double centered with respect to the random variable Z.

The population partial distance covariance is defined by the inner product

$$(P_{Z^\perp}(X) \cdot P_{Z^\perp}(Y)) := E[(A_X - \alpha C_Z) \cdot (B_Y - \beta C_Z)].$$

Definition 4 (*Population pdCor*) Population partial distance correlation is defined

$$\mathcal{R}^*(X, Y; Z) := \frac{(P_{Z^\perp}(X) \cdot P_{Z^\perp}(Y))}{|P_{Z^\perp}(X)||P_{Z^\perp}(Y)|},$$

where $|P_{Z^\perp}(X)| = (P_{Z^\perp}(X) \cdot P_{Z^\perp}(X))^{1/2}$. If $|P_{Z^\perp}(X)||P_{Z^\perp}(Y)| = 0$ we define $\mathcal{R}^*(X, Y; Z) = 0$.

Theorem 2 (Population pdCor) *The following definition of population partial distance correlation is equivalent to Definition 4.*

$$\mathcal{R}^*(X, Y; Z) = \qquad (1)$$
$$\begin{cases} \frac{\mathcal{R}^2(X,Y) - \mathcal{R}^2(X,Z)\mathcal{R}^2(Y,Z)}{\sqrt{1-\mathcal{R}^4(X,Z)}\sqrt{1-\mathcal{R}^4(Y,Z)}}, & \mathcal{R}(X, Z) \neq 1 \, and \, \mathcal{R}(Y, Z) \neq 1; \\ 0, & \mathcal{R}(X, Z) = 1 \, or \, \mathcal{R}(Y, Z) = 1. \end{cases}$$

where $\mathcal{R}(X, Y)$ denotes the population distance correlation.

We have proved that projections can be represented as a $\mathcal{U}$-centered distance matrix of some configuration of n points $\mathbf{U}$ in a Euclidean space $\mathbb{R}^p$, $p \leq n - 2$. Hence a test for pdCov$(X, Y; Z) = 0$ (or similarly a test for pdCor$(X, Y; Z) = 0$) can be defined by applying the distance covariance test statistic $\mathcal{V}_n^2(\mathbf{U}, \mathbf{V})$, where U and V are a representation which exist by Theorem 1. This test can be applied to U, V using the `dcov.test` function of the energy package [9] or one can apply a test based on the inner product (6), which is implemented in the *pdcor* package [10].

4 Power Comparison

The tests for zero partial distance correlation are implemented as permutation (randomization) tests of the hypothesis of zero partial distance covariance. In these examples we used the `dcov.test` method described above, although in extensive simulations the two methods of testing this hypothesis are equivalent in their average power over 10,000 tests. The simulation design for the following examples used $R = 999$ replicates for the permutation tests and the estimated p-value is computed as

$$\widehat{p} = \frac{1 + \sum_{k=1}^{R} I(T^{(k)} \geq T_0)}{1 + R},$$

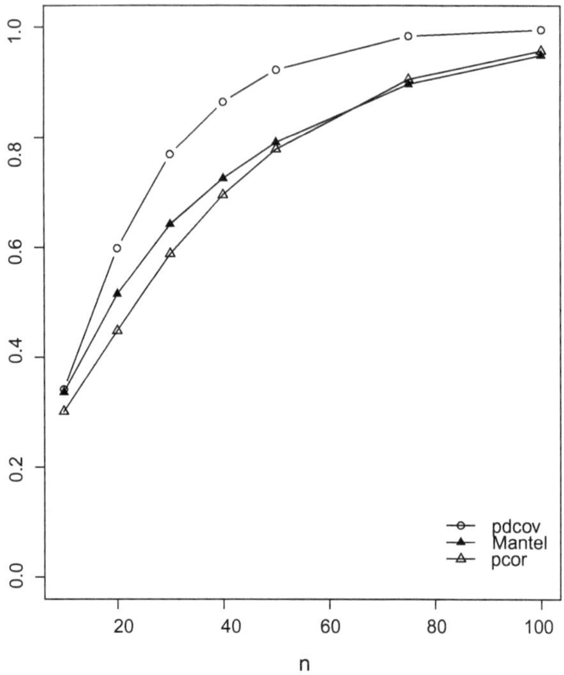

Fig. 1 Power comparisons for partial distance covariance, partial Mantel test, and partial correlation test at significance level $\alpha = 0.10$ (correlated standard normal data)

where $I(\cdot)$ is the indicator function, T_0 is the observed value of the test statistic, and $T^{(k)}$ is the statistic for the k-th sample. In each example 10,000 tests are summarized at each point in the plot, and the significance level is 10 %.

Example 1 In this example, power of tests is compared for correlated trivariate normal data with standard normal marginal distributions. The variables X, Y, and Z are each *correlated* standard normal. The pairwise correlations are $\rho(X, Y) = \rho(X, Z) = \rho(Y, Z) = 0.5$. The power comparison summarized in Fig. 1 shows that *pdcov* has higher power than *pcor* or *partial Mantel* tests.

Example 2 This example presents a power comparison for correlated non-normal data. The variables X, Y, and Z are each *correlated*, X is standard lognormal, while Y and Z are each standard normal. The pairwise correlations are $\rho(\log X, Y) = \rho(\log X, Z) = \rho(Y, Z) = 0.5$. The power comparison summarized in Fig. 2 shows that *pdcov* has higher power than *pcor* or *partial Mantel* tests.

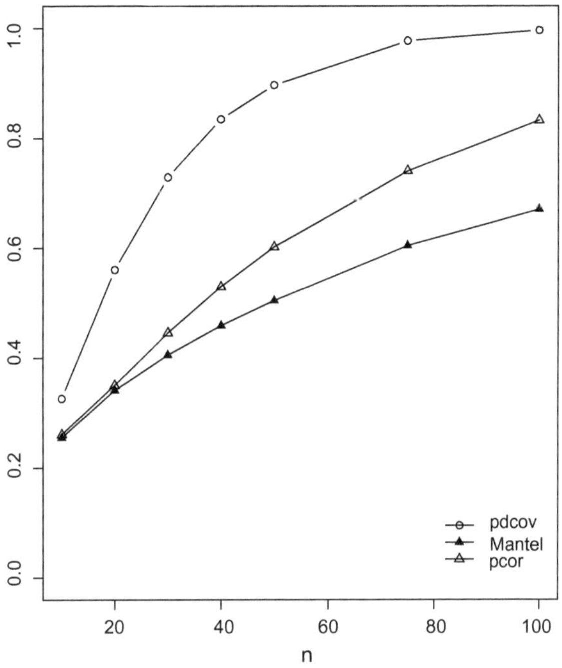

Fig. 2 Power comparisons for partial distance covariance, partial Mantel test, and partial correlation test at significance level $\alpha = 0.10$ (correlated non-normal data)

Acknowledgments Research of the first author was supported by the National Science Foundation, while working at the Foundation.

References

1. Cailliez, F.: The analytical solution of the additive constant problem. Psychometrika **48**, 343–349 (1983)
2. Cox, T.F., Cox, M.A.A.: Multidimensional Scaling, 2nd edn. Chapman and Hall (2001)
3. Feuerverger, A.: A consistent test for bivariate dependence. Int. Stat. Rev. **61**, 419–433 (1993)
4. Lyons, R.: Distance covariance in metric spaces. Ann. Probab. **41**(5), 3284–3305 (2013)
5. Mantel, N.: The detection of disease clustering and a generalized regression approach. Cancer Res. **27**, 209–220 (1967)
6. Mardia, K.V.: Some properties of classical multidimensional scaling. Commun. Stat. Theory and Methods **7**(13), 1233–1241 (1978)
7. Mardia, K.V., Kent, J.T., Bibby, J.M.: Multivar. Anal. Academic Press, London (1979)
8. Team, R.C.: A language and environment for statistical computing. R Foundation for Statistical Computing, Vienna, Austria (2013). http://www.R-project.org/
9. Rizzo, M.L., Székely, G.J.: Energy: E-statistics (energy statistics). R package version 1.6.1 (2014). http://CRAN.R-project.org/package=energy
10. Rizzo, M.L., Székely, G.J.: pdcor: Partial distance correlation. R package version 1, (2013)
11. Székely, G.J., Rizzo, M.L., Bakirov, N.K.: Measuring and testing independence by correlation of distances. Ann. Stat. **35**(6), 2769–2794 (2007). doi:10.1214/009053607000000505
12. Székely, G.J., Rizzo, M.L.: Brownian distance covariance. Ann. Appl. Stat. **3**(4), 1236–1265 (2009). doi:10.1214/09-AOAS312

13. Székely, G.J., Rizzo, M.L.: On the uniqueness of distance covariance. Stat. Probab. Lett. **82**(12), 2278–2282 (2012). doi:10.1016/j.spl.2012.08.007
14. Székely, G.J., Rizzo, M.L.: Energy statistics: statistics based on distances. J. Stat. Plan. Inference **143**(8), 1249–1272 (2013). doi:10.1016/j.jspi.2013.03.018
15. Székely, G.J., Rizzo, M.L.: Partial distance correlation with methods for dissimilarities. Ann. Stat. **32**(6), 2382–2412 (2014). doi:10.1214/14-AOS1255

Automatic Component Selection in Additive Modeling of French National Electricity Load Forecasting

Anestis Antoniadis, Xavier Brossat, Yannig Goude, Jean-Michel Poggi and Vincent Thouvenot

Abstract We consider estimation and model selection in sparse high-dimensional linear additive models when multiple covariates need to be modeled nonparametrically, and propose some multi-step estimators based on B-splines approximations of the additive components. In such models, the overall number of regressors d can be large, possibly much larger than the sample size n. However, we assume that there is a smaller than n number of regressors that capture most of the impact of all covariates on the response variable. Our estimation and model selection results are valid without assuming the conventional "separation condition"—namely, without assuming that the norm of each of the true nonzero components is bounded away from zero. Instead, we relax this assumption by allowing the norms of nonzero components to converge to zero at a certain rate. The approaches investigated in this paper consist of two steps. The first step implements the variable selection, typically by the Group Lasso, and the second step applies a penalized P-splines estimation to the selected additive components. Regarding the model selection task we discuss, the application of several criteria such as Akaike information criterion (AIC), Bayesian information criterion (BIC), and generalized cross validation (GCV) and study the consistency of BIC, i.e. its ability to select the true model with probability converging to 1. We then study post-model estimation consistency of the selected components. We end

A. Antoniadis
University Cape Town, Cape Town, South Africa
e-mail: Anestis.Antoniadis@imag.fr

A. Antoniadis
University Joseph Fourier, Grenoble, France

X. Brossat · Y. Goude
EDF R&D, Clamart, France

V. Thouvenot (✉)
Thales Communication & Security, Gennevilliers, France
e-mail: vincent.thouvenot@thalesgroup.com

Y. Goude · J.-M. Poggi · V. Thouvenot
University Paris-Sud, Orsay, France

J.-M. Poggi
University Paris Descartes, Paris, France

R. Cao et al. (eds.), *Nonparametric Statistics*, Springer Proceedings
in Mathematics & Statistics 175, DOI 10.1007/978-3-319-41582-6_14

the paper by applying the proposed procedure on some real data related to electricity load consumption forecasting: the EDF (Électricité de France) portfolio.

Keywords B-splines approximation · BIC · Consistency · Group LASSO · Multi-step estimator · Load forecasting · P-splines · Sparse additive model · Variable selection

1 Introduction

For electricity providers, forecasting electricity demand is a key activity as it is a crucial input of the production planning and energy trading. The literature on load forecasting is huge and a lot of statistical models have been applied to this problem. We focus here on additive models based on a spline basis decomposition of their additive components (see [15, 32]). These models combine the flexibility of fully nonparametric models and the simplicity of multiple regression models. They demonstrate their strong ability to cope with electricity data in previous work: Pierrot and Goude [24] applied it to national french consumption and Fan and Hyndman [13] show their interest for regional Australia's load forecasting. Additive models were used by three teams among the 10 bests (see [6, 23]) in the GEFCom2012 (see [16]). At least two of the first teams of the recent GEFCom2014 use these models too. One of the issue is the nonzero additive component selection, which is often realized by combining expert knowledge and stepwise selection procedures. This method is time consuming and hardly extendable. To avoid that, we establish automatic variable selection and estimation approach for sparse high-dimensional nonparametric additive models.

High-dimensional sparse additive model have been studied by many authors (see e.g., [7, 11, 19–21, 25, 26, 28]). The main issues are to select the nonzero components and to estimate well the identified additive model. The problem of selection is obviously more complex in this nonparametric context than in the parametric context. In particular, an approximation error needs to be controlled since one usually choses to approach the additive components by a B-splines basis decomposition. For selection, Fan and Jiang [12] use some testing methods. The probability distribution of the test statistic is estimated with some bootstrap methods which is computationally intensive and the significance of the covariates is tested one after one in a stepwise fashion which could converge to a suboptimal solution, particularly when high correlation exists among covariates. Penalized regression methods (see e.g., [1, 30]) are well known to have low computational cost and suffer less from correlated design. This is the approach we would like to use hereafter. Under suitable smoothness assumption, each covariate effect is modeled using a B-splines approximation (see [27]), and therefore each nonparametric component is represented by a linear combination of spline basis functions. Consequently, the problem of component selection becomes that of selecting the groups of coefficients in the linear combinations. A natural regularization method to deal with such a situation is the

Group LASSO estimator introduced by Yuan and Lin [33]. Group LASSO achieves group selection which is consistent under some conditions as explained in [4] but as for LASSO, Group LASSO introduces a bias. This fact motivates us to combine Group LASSO with another penalized regression method, namely the penalized splines (P-splines), introduced by Eilers and Marx in [10], or estimators like OLS (ordinary least squares). Although P-splines fitting is consistent in estimation (see [17]), it can not achieve covariate selection. In [22], the authors propose a small change in the penalty to achieve selection but we noticed on simulations that if the zero effects are estimated by values close to zero, the resulting penalisation does not shrink these effects to 0. What is desirable is to select the right model and then fit appropriately the resulting model. Despite its intuitive nature, the theoretical properties of such two-step procedures have to be carefully analyzed since the effect of the first step variable selection is random, and generally contains redundant additive components or misses significant additive components. Attempts implementing such two-step estimators have already been proposed in the recent literature (see, e.g., the NonNegative Garrote [2, 9]). Huang et al. [18] have used the adaptive Group LASSO. In the linear model context, Belloni et al. [5] combine LASSO and OLS and show the improvement of the correction, thanks to OLS, of the bias introduced by LASSO selection. Our two-step procedure developed in this paper is inspired by their work. One of the issue when penalized regression is used is selecting the regularization parameter. In the linear context, Wang et al. [31] demonstrate the consistency of BIC. We follow similar type of proofs than them.

Our paper is organized as follows. In Sect. 2, we present the statistical framework and a two-step estimator. In Sect. 3, we study some asymptotic properties of the resulting estimators. Their proofs are sketched in an appendix. Detailed theoretical results concerning the methodology can be found in detail in [3]. In Sect. 4, we apply the procedure on electricity load forecasting data. The interested reader is referred to [29] for some more detailed applications.

2 Additive Model and Two-Step Estimators

2.1 Additive Model

We consider additive models of the following form:

$$Y_i = \beta_0^* + \sum_{j=1}^{d} f_j^*(X_{i,j}) + \epsilon_i, \quad i = 1, \dots, n, \tag{1}$$

where $E(\epsilon_i|\underline{X}_i) = 0$, with $\underline{X}_i := (X_{i,1}, \dots, X_{i,d})^T$, ϵ_i i.i.d of variance σ^2. The nonlinear additive components $\{f_1^*, \dots, f_d^*\}$ are assumed to be smooth. More precisely, we assume that each function f_j^* lies in a Sobolev space:

$$H_2^2([0,1]) = \{f : [0,1] \to \mathbb{R} | f^{(1)} \text{abs. continuous and } \int_0^1 \left(f^{(2)}(x)\right)^2 dx < +\infty\}.$$

Moreover, for identifiability issues, we assume $E(f_j^*(X_j)) = 0$ for each j. We note $S^* = \{j \in \{1, \ldots, d\} | E(f_j^*(X_j)^2) \neq 0\}$ the index set of nonzero components and $s^* = card(S^*)$ the cardinal of S^*. We assume $s^* \ll n$ and $s^* \leq D < d = d(n)$.

We use B-splines to approximate the additive components. Using such spline truncated expansions for the additive components, the additive model (1) can be approached by the following model:

$$E\left(Y | (X_1, \ldots, X_d) = (x_1, \ldots, x_d)\right) = \beta_0 + \sum_{j=1}^{d} \sum_{k=1}^{m_j} \beta_{j,k} B_{j,k}^{q_j}(x_j), \tag{2}$$

where $\underline{B}_j(\cdot) = \{B_{j,k}^{q_j}(\cdot) | k = 1, \ldots, K_j + q_j = m_j\}$ is the basis of B-splines of degree q_j and fixed sequences of K_j knots, spanning the vector space onto which the jth additive component $f_j^*(\cdot)$ is projected. We assume $m_j = m_j(n) \to +\infty$ such as $m_j = o(n)$. We note $\mathbf{B_j} = (\underline{B}_j(x_{1j})^T, \ldots, \underline{B}_j(x_{nj})^T)^T \in \mathbb{R}^{n \times m_j}$, where $x_{i,j}$ is the ith observation of X_j. The model parameters are β_0 and $\boldsymbol{\beta} = (\beta_{1,1}, \ldots, \beta_{d,m_d})^T$. The covariates are $(X_i)_{i \in [\![1,d]\!]}$.

The main issues are to select the nonzero functions f_j^* and to estimate well the resulting model (1). We may use a penalized regression method, which tries to minimize a criterion like:

$$Q^{OLS}(\boldsymbol{\beta}) + \sum_{j=1}^{d} p_{\lambda_j}(\boldsymbol{\beta}_j),$$

where

$$Q^{OLS}(\boldsymbol{\beta}) = \sum_{i=1}^{n} \left(Y_i - \beta_0 - \sum_{j=1}^{d} C_{ij}(\boldsymbol{\beta}_j) \right)^2,$$

with $C_{ij}(\boldsymbol{\beta}_j) = \sum_{k=1}^{m_j} \beta_{j,k} B_{j,k}^{q_j}(X_{i,j})$ and p_λ some appropriate penalty function on the coefficients.

2.2 Variable Selection Algorithm: The Post1 Procedure

As LASSO, the Group LASSO is a powerful selection procedure similar to soft thresholding in the orthogonal design case but incurs poor performance in prediction because of the bias induced by such soft thresholding. In contrast, under good conditions, OLS estimators are known for being unbiased estimators that minimize the variance. In the high-dimensional linear model case, Belloni et al. [5] combine the LASSO with OLS estimator and obtain good performances on synthetic data. Inspired from their work, we propose in [3] several two-step procedures. Here, we will restrict our study to one of them, namely the one named Post1 in [3] and apply it on synthetic data. We give below a detailed description of this algorithm. We denote Λ_{GrpL} a uniform grid of λ values between $\lambda_{\min} = 0$ and $\lambda_{\max}$ which is the value of λ for which all the effects are shrank to 0. The set $S = \{1, \ldots, d\}$ denotes the index set made of the labels of the covariates.

Algorithm

1. **First step: subset selection (Group LASSO)**
   For each $\lambda_i \in \Lambda_{GrpL}$
   - Solve
   $$\hat{\boldsymbol{\beta}}^{\lambda_i} = \arg\min\{Q^{OLS}(\boldsymbol{\beta}) + \lambda_i \sum_{j=1}^{d} \sqrt{m_j}||\boldsymbol{\beta}_j||_2\}$$
   - Denote $S^{\lambda_i} = \{j | \hat{\boldsymbol{\beta}}_j^{\lambda_i} \neq 0\}$
2. **Second step: Estimation of the additive model (by OLS)**
   For each support set $S^{\lambda_s} \in \{S^{\lambda_{min}}, \ldots, S^{\lambda_{max}}\}$
   - Compute
   $$Q_{S_{\lambda_s}}^{OLS}(\boldsymbol{\beta}) = \sum_{i=1}^{n} \left(Y_i - \beta_0 - \sum_{j \in S_{\lambda_s}} C_{ij}(\boldsymbol{\beta}_j) \right)^2$$
   - Solve
   $$\tilde{\boldsymbol{\beta}}^{S_{\lambda_s}} = \arg\min\{Q_{S_{\lambda_s}}^{OLS}(\boldsymbol{\beta})\},$$
   - Compute the BIC (see Eq. (5)) for each $\tilde{\boldsymbol{\beta}}^{S_{\lambda_s}}$
3. **Third step: Selection of the final model** Select $\tilde{\boldsymbol{\beta}}^{S_{\lambda_b}}$ which minimizes the BIC

---

3 Asymptotic Study

3.1 Notation

We first fix the notation. For simplification, we assume, without loss of generality, that, for any $j = 1, \ldots, d$, $m_j = m = m_n$ and $\underline{B}_j(\cdot)$ are identical. We note, for any $1 \leq j \leq d$ and $i = 1, \ldots, n$:

$$\tilde{\underline{\mathbf{x}}}_{\mathbf{i},\mathbf{B_j}} = \underline{B}_j^T(x_{ij}) - \frac{1}{n} \sum_{i=1}^{n} \underline{B}_j^T(x_{ij}) \in \mathbf{R}^{m_n},$$

and

$$\tilde{\mathbf{X}}_{\mathbf{i}} = [\underline{\tilde{\mathbf{x}}}_{\mathbf{i},\mathbf{B_1}}^T, \ldots, \underline{\tilde{\mathbf{x}}}_{\mathbf{i},\mathbf{B_d}}^T]^T \in \mathbf{R}^{d\times m_n},$$

and for any $1 \leq j \leq d$:

$$\hat{\Sigma}_j = \frac{1}{n}\sum_{i=1}^{n} \underline{\tilde{\mathbf{x}}}_{\mathbf{i},\mathbf{B_j}} \underline{\tilde{\mathbf{x}}}_{\mathbf{i},\mathbf{B_j}}^T,$$

which is an $m_n \times m_n$ real matrix. We note the empirical variance covariance matrix, as:

$$\hat{\Sigma} = \frac{1}{n}\sum_{i=1}^{n} \tilde{\mathbf{X}}_{\mathbf{i}} \tilde{\mathbf{X}}_{\mathbf{i}}^{\mathbf{T}},$$

which is an $m_n d \times m_n d$ matrix. Finally, the unknown coefficients vector is denoted by:

$$\underline{\boldsymbol{\beta}} = (\underline{\boldsymbol{\beta_1}}^T, \ldots, \underline{\boldsymbol{\beta_d}}^T)^T \in \mathbb{R}^{dm_n},$$

and

$$\underline{\boldsymbol{\beta}}_{\boldsymbol{B}_j} = \underline{\boldsymbol{\beta}}_j = (\beta_{j1}, \ldots, \beta_{jm_n})^T.$$

The standardized Group LASSO solution is:

$$\hat{\beta}_0 = \frac{1}{n}\sum_{i=1}^{n} Y_i \text{ and } \hat{f}_j(X_j) = \sum_{k=1}^{m_n} \hat{\beta}_{jk}(B_{j,k}^{q_j}(X_j) - \bar{B}_{j,k}^{q_j}),$$

where $\bar{B}_{j,k}^{q_j} = 1/n \sum_{k=1}^{m_n} B_{j,k}^{q_j}(X_{i,j})$ and

$$\underline{\hat{\boldsymbol{\beta}}} = \arg \min_{\underline{\boldsymbol{\beta}} \in \mathbf{R}^{dm_n}} + \frac{1}{2}\sum_{i=1}^{n}(Y_i - \tilde{X}_i^T \underline{\boldsymbol{\beta}})^2 + \sqrt{m_n}\lambda \sum_{j=1}^{d} ||\hat{\Sigma}^{1/2}\underline{\boldsymbol{\beta}}_j||_2, \tag{3}$$

where $||\cdot||_2$ is the Euclidean norm of $\mathbb{R}^{m_n}$.

We note

$$||\mathbf{f}^*(\mathbf{x})|| = ||\mathbf{f}^*(\mathbf{x})||_2 = \sqrt{\sum_{i=1}^{n}\left(\sum_{j=1}^{d} f_j^*(x_{i,j})\right)^2} \text{ and } ||\mathbf{f}^*||_\infty = \max_{\mathbf{x_i}, i\in 1,\ldots n} |\mathbf{f}^*(\mathbf{x_i})|$$

$$= \max_{\mathbf{x_i}, i\in 1,\ldots n} |\sum_{j=1}^{d} f_j^*(x_{i,j})|.$$

Here $||\cdot||$ and $||\cdot||_2$ stand for the Euclidean norm.

Let, for $j = 1, \ldots, d$,

$$\mathbf{U_j} = \begin{pmatrix} B_{j1}(x_{1j}) - \bar{B}_{j1} & \ldots & B_{jm_n}(x_{1j}) - \bar{B}_{jm_n} \\ \vdots & & \vdots \\ B_{j1}(x_{nj}) - \bar{B}_{j1} & \ldots & B_{jm_n}(x_{nj}) - \bar{B}_{jm_n} \end{pmatrix},$$

which can be rewritten:

$$\mathbf{U_j} = \left(\tilde{\mathbf{x}}_{\mathbf{1},\mathbf{B_j}} \ldots \tilde{\mathbf{x}}_{\mathbf{n},\mathbf{B_j}}\right)^T.$$

Finally, we note $\underline{\mathbf{Y}} = (Y_1, \ldots Y_n)^T$ and $\mathbf{U} = [\mathbf{U_1} \ldots \mathbf{U_d}]_{n \times dm_n}$. For each design $S \subseteq \{1, \ldots, d\}$, $\mathbf{U_S}$ is the matrix extracted from $\mathbf{U}$ obtained by retaining the columns indexed by S, that is

$$\mathbf{U_S} = [\mathbf{U_j} | j \in S].$$

We define $\boldsymbol{\beta}_S = (\boldsymbol{\beta}_j)_{j \in S}$, $\hat{\boldsymbol{\beta}}_S = (\hat{\boldsymbol{\beta}}_j)_{j \in S}$ (estimated on S), and more generally, $\boldsymbol{\psi}_S = (\boldsymbol{\psi}_j)_{j \in S}$.

The vector $\mathbf{1_n}$ is the column vector of size n whose components are 1. To simplify the notation, we add $\mathbf{1_n}/\sqrt{m_n}$ as first column of $\mathbf{U}$ and of $\mathbf{U_S}$.

We note $\underline{\mathbf{b}} = \left(\sqrt{m_n}\beta_0, \boldsymbol{\beta}^T\right)^T$ and $\underline{\mathbf{b}}_\mathbf{S} = \left(\sqrt{m_n}\beta_0, \boldsymbol{\beta}_S{}^T\right)^T$.

With the above notation, for the design S, the OLS objective function can now be rewritten as:

$$Q_S^{OLS}(\mathbf{b_S}) = ||\underline{\mathbf{Y}} - \mathbf{U_S}\mathbf{b_S}||^2. \tag{4}$$

Let $\tilde{\underline{\mathbf{b}}}_\mathbf{S}$ be the vector which minimizes (4), that is $\hat{\underline{\mathbf{b}}}_\mathbf{S} = \arg\min_{\mathbf{b_S}} Q_S^{OLS}(\mathbf{b_S})$. We can now define the BIC criterion (see Wang et al. [31]) by:

$$BIC(S) = \log(||\underline{\mathbf{Y}} - \mathbf{U_S}\hat{\underline{\mathbf{b}}}_S||^2) + \text{card}(S)\, m_n \frac{\log(nd)}{n}. \tag{5}$$

The submodel $\hat{S}$ that achieves the minimum value of the above (over all submodels with $\text{card}(S) \leq D$) is chosen as the final model.

3.2 Assumptions

In this section, we introduce basic conditions commonly used in the analysis of the first and second step estimators. They do not need to hold simultaneously for getting a desired result, but it is clearer to collect all the assumptions here.

1. **Assumption on the random vector** $(Y_i, \underline{\mathbf{X_i}})_{i=1,\ldots,n}$: $(Y_i, \underline{\mathbf{X_i}})_{i=1,\ldots,n}$ are i.i.d. random vectors where $(Y_1, \underline{\mathbf{X_1}})$ satisfies the additive model stated in (1).

2. **Assumption on the distribution of errors**: the conditional distribution of ϵ_1 given $\underline{\mathbf{X_1}}$ is a $N(0, \sigma^2(\underline{\mathbf{X_1}}))$ and $\sigma_1^2 \leq \sigma^2(\underline{\mathbf{X_1}}) \leq \sigma^2$ a.s. with σ^2 and σ_1^2 fixed positive constants.
3. **Assumption on the distribution of $\underline{\mathbf{X_1}}$**: the support of $\underline{\mathbf{X_1}}$ is compact. For the sake of simplicity, we assume the support is bounded in $[0, 1]^d$. Moreover, the marginal density $\underline{\mathbf{X_1}}$ is bounded away from 0 and from infinity by two positives constants. The density of $\underline{\mathbf{X_1}}$ is absolutely continuous with respect to the Lebesgue measure.
4. **Assumption on smoothness of the additive components**: for all $j \in S^*$, $f_j^* \in H_2^\nu[0, 1]$ for some positive integer $\nu \geq 2$. This in particular implies that f_j^* is κ-Holder continuous with $\kappa > 1/2$.
5. **Assumptions on** $d, m_n,$ and s^*:

   (a) $m_n = An^{\frac{1}{2\nu+1}}$, with A constant.
   (b) $\frac{\log(d)}{n^{\frac{2\nu}{2\nu+1}}} \to 0$
   (c) $s^* \leq D \leq \min(d, \frac{n}{m_n})$
   (d) $\frac{m_n \log(nd)}{n \min_{j\in S^*} ||f_j^*||^2} \to 0$, that means $\frac{m_n \log(nd)}{n} = o(\min_{j\in S^*} ||f_j^*||^2)$.

The first three assumptions are standard. Assumption 4 is a classical hypothesis on the smoothness of the additive components. Assumption (5.a) is a consequence of Stone's choice about the B-splines approximation of the additive components. Assumption (5.b) lets the number of candidate covariates to increase not too quickly with the sample size. Assumption (5.c) ensures the uniqueness of the OLS estimator. Assumption (5.d) is sufficient to obtain the consistency results. Note that we do not assume that $\min_{j\in S^*} ||f_j^*||^2$ is bounded away from zero as in assumption (A1) of Huang et al. [18]. Instead, (5.d) makes it clear that this quantity is allowed to converge to zero at a certain rate.

3.3 BIC Consistency with OLS Estimator

We are now in position to state the main results of this paper. Let S a subset of $\{1, \ldots, d\}$ leading to the submodel of the additive model (1) obtained by retaining only the covariates whose index pertains to S. The proof is split into two parts, considering the underfitted models (some true nonzero components are not in S (false negative)) and overfitted models (some true zero components, as well as all true nonzero components, are included in S (false positive)), respectively.

3.3.1 $S^* \not\subseteq S$ Case (False Negative)

Assume $S^* \not\subseteq S$. Let $\underline{\tilde{\mathbf{b}}_S}$ and $\underline{\hat{\mathbf{b}}_{S^*}}$ be the OLS estimators when using for fitting the submodels S and S^* respectively. This means that we have:

$$\underline{\hat{\mathbf{b}}}_{\mathbf{S}} = \arg\min_{\underline{\mathbf{b}}} ||\underline{\mathbf{Y}} - \mathbf{U}_{\mathbf{S}}\underline{\mathbf{b}}||^2 \text{ and } \underline{\hat{\mathbf{b}}}_{\mathbf{S}^*} = \arg\min_{\underline{\mathbf{b}}} ||\underline{\mathbf{Y}} - \mathbf{U}_{\mathbf{S}^*}\underline{\mathbf{b}}||^2.$$

Remember that card$(S) \leq D$. Assumption (5.c) implies that $m_n D < n$. Denote $\bar{S} = S^* \cup S$, and let $\underline{\tilde{\mathbf{b}}}_{\mathbf{S}}$ be the vector of size card$(\bar{S})m_n + 1$ coinciding with $\underline{\hat{\mathbf{b}}}_{\mathbf{S}}$ for the coefficients in S and equal to zero otherwise. We can write this vector as $\underline{\tilde{\mathbf{b}}}_{\mathbf{S}} = \left(\bigcup_{j\in\bar{S}} \underline{\tilde{\mathbf{b}}}_{\mathbf{S,j}}^T\right)$ with $\underline{\tilde{\mathbf{b}}}_{\mathbf{S,j}} = \underline{\hat{\mathbf{b}}}_{\mathbf{S,j}}$ if $j \in S$, $\mathbf{0}$ otherwise. Likewise, let $\underline{\tilde{\mathbf{b}}}_{\mathbf{S}^*} = \left(\bigcup_{j\in\bar{S}} \underline{\tilde{\mathbf{b}}}_{\mathbf{S}^*,\mathbf{j}}^T\right)$ with $\underline{\tilde{\mathbf{b}}}_{\mathbf{S}^*,\mathbf{j}} = \underline{\hat{\mathbf{b}}}_{\mathbf{S}^*,\mathbf{j}}$ if $j \in S^*$, $\mathbf{0}$ otherwise and let $\underline{\mathbf{b}}_{\bar{\mathbf{S}}^*} = \left(\bigcup_{j\in\bar{S}} \underline{\mathbf{b}}_{\bar{\mathbf{S}},\mathbf{j}}^{*,T}\right)$ with $\underline{\mathbf{b}}_{\bar{\mathbf{S}},\mathbf{j}}^* = \underline{\mathbf{b}}_{\mathbf{S}^*,\mathbf{j}}{}^*$ if $j \in S^*$, $\mathbf{0}$ otherwise. That is, zero values are filled in the various sub-vectors to match the dimension whenever it is necessary to allow correct matrix products. The gap between the $BIC(S)$ and $BIC(S^*)$ is then given by:

$$\begin{aligned} BIC(S) - BIC(S^*) &= \log\left(1 + \frac{||\underline{\mathbf{Y}} - \mathbf{U}_{\bar{\mathbf{S}}}\underline{\tilde{\mathbf{b}}}_{\mathbf{S}}||^2/n - ||\underline{\mathbf{Y}} - \mathbf{U}_{\bar{\mathbf{S}}}\underline{\tilde{\mathbf{b}}}_{\mathbf{S}^*}||^2/n}{||\underline{\mathbf{Y}} - \mathbf{U}_{\bar{\mathbf{S}}}\underline{\tilde{\mathbf{b}}}_{\mathbf{S}^*}||^2/n}\right) \\ &\quad + \big(card(S) - card(S^*)\big)\, m_n \frac{\log(nd)}{n}. \end{aligned} \tag{6}$$

Using some elementary properties of the Euclidean norm, it is easy to see that:

$$\begin{aligned} &||\underline{\mathbf{Y}} - \mathbf{U}_{\bar{\mathbf{S}}}\underline{\tilde{\mathbf{b}}}_{\mathbf{S}}||^2 - ||\underline{\mathbf{Y}} - \mathbf{U}_{\bar{\mathbf{S}}}\underline{\tilde{\mathbf{b}}}_{\mathbf{S}^*}||^2 \\ &= -2\underline{\boldsymbol{\epsilon}}^T \mathbf{U}_{\bar{\mathbf{S}}}(\underline{\tilde{\mathbf{b}}}_{\mathbf{S}} - \underline{\tilde{\mathbf{b}}}_{\mathbf{S}^*}) + 2\left(\mathbf{U}_{\bar{\mathbf{S}}}\underline{\tilde{\mathbf{b}}}_{\mathbf{S}^*} - \underline{\mathbf{f}}^*(\underline{\mathbf{X}})\right)^T \mathbf{U}_{\bar{\mathbf{S}}}(\underline{\tilde{\mathbf{b}}}_{\mathbf{S}} - \underline{\tilde{\mathbf{b}}}_{\mathbf{S}^*}) \\ &\quad + ||\mathbf{U}_{\bar{\mathbf{S}}}(\underline{\tilde{\mathbf{b}}}_{\mathbf{S}} - \underline{\tilde{\mathbf{b}}}_{\mathbf{S}^*})||^2. \end{aligned} \tag{7}$$

Concerning the three terms of right hand side of Eq. (7) we have (see Lemmas 1, 2 and 3 of [3]):

- $||\mathbf{U}_{\bar{\mathbf{S}}}(\underline{\tilde{\mathbf{b}}}_{\mathbf{S}} - \underline{\tilde{\mathbf{b}}}_{\mathbf{S}^*})||^2 \geq A_2 \frac{n}{m_n} v_n^2$, and $v_n = A_3\left(\sqrt{m_n}\min_{j\in S^*}||f_j^*|| - \frac{m_n}{\sqrt{n}} - m_n^{-\nu+1/2}\right)$, where A_2 and A_3 are generic constants;
- $|2\left(\mathbf{U}_{\bar{\mathbf{S}}}\underline{\tilde{\mathbf{b}}}_{\mathbf{S}^*} - \underline{\mathbf{f}}^*(\underline{\mathbf{X}})\right)^T \mathbf{U}_{\bar{\mathbf{S}}}(\underline{\tilde{\mathbf{b}}}_{\mathbf{S}} - \underline{\tilde{\mathbf{b}}}_{\mathbf{S}^*})| \leq C_4\sqrt{n m_n^{-2\nu}}\sqrt{\frac{n}{m_n}}|v_n|$, with $C_4 > 0$ constant. This term is negligible compared to the previous term.
- $-2\underline{\boldsymbol{\epsilon}}^T \mathbf{U}_{\bar{\mathbf{S}}}(\underline{\tilde{\mathbf{b}}}_{\mathbf{S}} - \underline{\tilde{\mathbf{b}}}_{\mathbf{S}^*}) + ||\mathbf{U}_{\bar{\mathbf{S}}}(\underline{\tilde{\mathbf{b}}}_{\mathbf{S}} - \underline{\tilde{\mathbf{b}}}_{\mathbf{S}^*})||^2 \geq \frac{3}{4}||\mathbf{U}_{\bar{\mathbf{S}}}(\underline{\tilde{\mathbf{b}}}_{\mathbf{S}} - \underline{\tilde{\mathbf{b}}}_{\mathbf{S}^*})||^2 + o_P(m_n \log(nd)) \geq R_1 \frac{n}{m_n} v_n^2$, with R_1 constant.

Finally, we obtain

$$||\underline{\mathbf{Y}} - \mathbf{U}_{\bar{\mathbf{S}}}\underline{\tilde{\mathbf{b}}}_{\mathbf{S}}||^2 - ||\underline{\mathbf{Y}} - \mathbf{U}_{\bar{\mathbf{S}}}\underline{\tilde{\mathbf{b}}}_{\mathbf{S}^*}||^2 \geq N_1 \frac{n}{m_n} v_n^2 + o_p\left(\frac{n}{m_n} v_n^2\right), \tag{8}$$

with N_1 constant.

It follows (see Lemma 4 of [3]) that the sequences $||\underline{\mathbf{Y}} - \mathbf{U}_{\mathbf{S}^*}\mathbf{b}_{\mathbf{S}^*}||^2/n$ and $\inf_{S^*\subseteq S|card(S)\leq D} ||\underline{\mathbf{Y}} - \mathbf{U}_{\mathbf{S}}\hat{\mathbf{b}}_{\mathbf{S}}||^2/n$ are lower bounded by a strictly positive constant with a probability converging to one.

We therefore have, with a probability converging to one,

$$\frac{||\underline{\mathbf{Y}} - \mathbf{U}_{\bar{\mathbf{S}}}\tilde{\underline{\mathbf{b}}}_{\mathbf{S}}||^2/n - ||\underline{\mathbf{Y}} - \mathbf{U}_{\bar{\mathbf{S}}}\tilde{\underline{\mathbf{b}}}_{\mathbf{S}^*}||^2/n}{||\underline{\mathbf{Y}} - \mathbf{U}_{\bar{\mathbf{S}}}\tilde{\underline{\mathbf{b}}}_{\mathbf{S}^*}||^2/n} \geq N_2 \frac{v_n^2}{m_n} > \min ||f_j^*||^2 + o(\min ||f_j^*||^2) > 0.$$

It follows that

$$BIC(S) - BIC(S^*) \geq \min(\frac{L_1}{m_n} v_n^2, \log(2)) + (card(S) - card(S^*)) \frac{m_n \log(nd)}{n},$$

where $L_1 > 0$ is a generic constant. Thus $|\text{card}(S) - \text{card}(S^*)| \frac{m_n \log(nd)}{n} \; n \overrightarrow{\rightarrow +\infty} \; 0$, since by assumption card(S) and card(S^*) are both bounded by a constant, and, thanks to assumptions (5), we have $|card(S) - card(S^*)| \frac{m_n \log(nd)}{n} = o(\min ||f_j^*||^2)$. Thence, Theorem 1, stated below, follows.

Theorem 1 *Let S such as $S^* \nsubseteq S$. Then,*

$$P(\min_{S^* \nsubseteq S | card(S) \leq M} BIC(S) - BIC(S^*) > 0) \; n \overrightarrow{\rightarrow +\infty} \; 1,$$

when M constant.

In particular, Theorem 1 implies that there are not false negatives with probability converging to one when the true design is candidate.

3.3.2 $S^* \subseteq S$ case (False Positive)

Here, we assume that $S^* \subseteq S$. As before, we note $\underline{\mathbf{b}}_{\mathbf{S}}^*$ the vector $\underline{\mathbf{b}}_{\mathbf{S}^*}^*$ extended by some zeros for each j such as $j \in S$ and $j \notin S^*$. We study the expression:

$$BIC(S^*) - BIC(S) = \log\left(\frac{||\underline{\mathbf{Y}} - \mathbf{U}_{\mathbf{S}^*}\hat{\underline{\mathbf{b}}}_{\mathbf{S}^*}||^2}{||\underline{\mathbf{Y}} - \mathbf{U}_{\mathbf{S}}\hat{\underline{\mathbf{b}}}_{\mathbf{S}}||^2}\right) + \big(card(S^*) - card(S)\big) m_n \frac{\log(nd)}{n}.$$

According to the definition of the OLS estimator, $||\underline{\mathbf{Y}} - \mathbf{U}_{\mathbf{S}^*}\hat{\underline{\mathbf{b}}}_{\mathbf{S}^*}||^2 \leq ||\underline{\mathbf{Y}} - \mathbf{U}_{\mathbf{S}^*}\underline{\mathbf{b}}_{\mathbf{S}^*}^*||^2$.
It follows that

$$\begin{aligned} BIC(S^*) - BIC(S) \leq & \log\left(1 + \frac{||\underline{\mathbf{Y}} - \mathbf{U}_{\mathbf{S}^*}\underline{\mathbf{b}}_{\mathbf{S}^*}^*||^2 - ||\underline{\mathbf{Y}} - \mathbf{U}_{\mathbf{S}}\hat{\underline{\mathbf{b}}}_{\mathbf{S}}||^2}{||\underline{\mathbf{Y}} - \mathbf{U}_{\mathbf{S}}\hat{\underline{\mathbf{b}}}_{\mathbf{S}}||^2}\right) \\ & + \big(card(S^*) - card(S)\big) m_n \frac{\log(nd)}{n}. \end{aligned}$$

We can state (see Lemma 5 of [3]) the following result:

$$||\underline{\mathbf{Y}} - \mathbf{U}_{\mathbf{S}^*}\underline{\mathbf{b}}^*_{\mathbf{S}^*}||^2/n - ||\underline{\mathbf{Y}} - \mathbf{U}_{\mathbf{S}}\hat{\underline{\mathbf{b}}}_{\mathbf{S}}||^2/n = ||\underline{\mathbf{Y}} - \mathbf{U}_{\mathbf{S}}\underline{\mathbf{b}}^*_{\mathbf{S}}||^2/n - ||\underline{\mathbf{Y}} - \mathbf{U}_{\mathbf{S}}\hat{\underline{\mathbf{b}}}_{\mathbf{S}}||^2/n$$
$$= O_P(m_n^{-2\nu}) + o_P(\frac{m_n}{n}\log(nd)) \xrightarrow{P} 0.$$

The result is derived by controlling the projection of the squared bias due to the approximation of the components of (1) by a truncated B-splines series decomposition and by bounding the variance in the space spanned by the columns of $\mathbf{U}^T_{\bar{\mathbf{S}}}$.

As $\inf_{S^* \subseteq S | card(S) \leq D} ||\underline{\mathbf{Y}} - \mathbf{U}_{\mathbf{S}}\hat{\mathbf{b}}_{\mathbf{S}}||^2/n$ is lower bounded by a strictly positive constant with a probability converging to one, we obtain that:

$$\frac{||\underline{\mathbf{Y}} - \mathbf{U}_{\mathbf{S}^*}\underline{\mathbf{b}}^*_{\mathbf{S}^*}||^2 - ||\underline{\mathbf{Y}} - \mathbf{U}_{\mathbf{S}}\hat{\underline{\mathbf{b}}}_{\mathbf{S}}||^2}{||\underline{\mathbf{Y}} - \mathbf{U}_{\mathbf{S}}\hat{\underline{\mathbf{b}}}_{\mathbf{S}}||^2} \in [0, 1], \tag{9}$$

with a probability converging to 1. It follows that

$$BIC(S^*) - BIC(S) \leq O_P(m_n^{-2\nu}) + o_P(\frac{m_n}{n}\log(nd)) + \left(card(S^*) - card(S)\right) m_n \frac{\log(nd)}{n}.$$

As $(card(S^*) - card(S)) < 0$ and $O_P(m_n^{-2\nu}) = o_P(m_n \frac{\log(nd)}{n})$, we obtain Theorem 2.

Theorem 2

$$P(\min_{S^* \subset S | card(S) \leq M} BIC(S^*) - BIC(S) < 0) \; n \overrightarrow{\rightarrow +\infty} \; 1.$$

This result, combined with the statement of Theorem 1, allows us to see that there is no false positive.

Consistency for the BIC Theorems 1 and 2 guarantee that there is no false positive neither false negative. We can therefore derive Theorem 3 which states the consistency of the BIC procedure.

Theorem 3 *Under Assumptions 1–5, the BIC defined by (5) is variable selection consistent, i.e.,*

$$P(\hat{S} = S^*) \; n \overrightarrow{\rightarrow \infty} \; 1.$$

The BIC defined by (5) is therefore variable selection consistent. However, it is not sufficient for proving the consistency of the two-step estimator. We have just shown, without using any penalization, that when we estimate an additive model whose components are approximated in B-splines bases with OLS estimators, then the true model is selected with a probability converging to one if it is a candidate. It now remains to prove that the BIC criterion is consistent for our Group LASSO penalized criterion.

3.4 Penalized BIC Consistency

Under our assumptions, we can rewrite (3) as (see Kato [19]):

$$\tilde{\underline{\mathbf{b}}}_\lambda = \arg \min_{\underline{b} \in \mathbf{R}^{dm_n+1}} ||\underline{\mathbf{Y}} - \mathbf{U}^{\mathbf{T}}\underline{\mathbf{b}}||^2 + \sqrt{m_n}\lambda \sum_{j=1}^{d} ||\underline{\mathbf{b}}_{\mathbf{j}}||_2. \tag{10}$$

The parameter λ is chosen along a given grid by minimizing the BIC criterion. Let $S_\lambda = \{j; ||\tilde{\underline{\mathbf{b}}}_{\lambda,j}|| \neq 0\}$. We have

$$\hat{\underline{\mathbf{b}}}_\lambda = (\mathbf{U}_{\mathbf{S}_\lambda}{}^T \mathbf{U}_{\mathbf{S}_\lambda})^{-1}(\mathbf{U}_{\mathbf{S}_\lambda}{}^T \underline{\mathbf{Y}} + \underline{\mathbf{r}}) = \hat{\underline{\mathbf{b}}}_{\mathbf{S}_\lambda} + (\mathbf{U}_{\mathbf{S}_\lambda}{}^T \mathbf{U}_{\mathbf{S}_\lambda})^{-1}\underline{\mathbf{r}},$$

where $\underline{\mathbf{r}} = -\lambda\sqrt{m_n}\frac{\partial}{\partial \underline{\mathbf{b}}}\sum_{j\in S_\lambda} ||\underline{\mathbf{b}}_{\mathbf{j}}|| \big|_{\underline{\mathbf{b}}=\hat{\underline{\mathbf{b}}}_\lambda} = -\lambda\sqrt{m_n}\left(0 \dots (\frac{\tilde{\underline{\mathbf{b}}}_{\lambda,j}^T}{||\hat{\underline{\mathbf{b}}}_{\lambda,j}||})_{j\in S_\lambda}\right)^T$.

Variables selection consistency We consider $BIC(\lambda) = \log(||\underline{\mathbf{Y}} - \mathbf{U}\tilde{\underline{\mathbf{b}}}_\lambda||^2) + card(S_\lambda)m_n\frac{\log(nd)}{n}$. Moreover, we note that $\hat{\underline{\mathbf{b}}}_\lambda = (\tilde{\underline{\mathbf{b}}}_{\lambda,j})_{j\in S_\lambda}$.

It is therefore equivalent to solve the optimization problem with $\mathbf{U}$ or $\mathbf{U}_{\mathbf{S}_\lambda}$. Thus we have

$$\begin{aligned} BIC(\lambda) &= \log(||\underline{\mathbf{Y}} - \mathbf{U}_{\mathbf{S}_\lambda}\hat{\underline{\mathbf{b}}}_\lambda||^2) + card(S_\lambda)m_n\frac{\log(nd)}{n} \\ &= \log(||\underline{\mathbf{Y}} - \mathbf{U}_{\mathbf{S}}\tilde{\underline{\mathbf{b}}}_\lambda||^2) + card(S_\lambda)m_n\frac{\log(nd)}{n}. \end{aligned}$$

We will show first that if $\lambda_n \sim \sqrt{n\log(m_n^2 d)/m_n}$, then $P(S_{\lambda_n} = S^*) \to 1$. Note (see [18] for example) that $S_{\lambda_n} = S^*$ if

$$\begin{cases} ||\underline{\mathbf{b}}_{\mathbf{j}}^*|| - ||\tilde{\underline{\mathbf{b}}}_{\lambda_n,j}|| < ||\underline{\mathbf{b}}_{\mathbf{j}}^*|| \text{ si} j \in S^* \\ ||\mathbf{U}_{\mathbf{j}}{}^T(\underline{\mathbf{Y}} - \mathbf{U}_{\mathbf{S}^*}\tilde{\underline{\mathbf{b}}}_{\lambda_n,S^*})|| \leq \lambda_n\sqrt{m_n} \text{ si} j \notin S_{\lambda_n} \end{cases}$$

Hence, the conclusion that $P(S_{\lambda_n} = S^*) \to 1$ will follow by proving that $\lambda_n \sim \sqrt{n\log(m_n^2 d)/m_n}$ implies the next two assertions which are proved in the Appendix, namely:

- $P(\exists j \in S^*, ||\underline{\mathbf{b}}_{\mathbf{j}}^* - \tilde{\underline{\mathbf{b}}}_{\lambda_n,j}|| \geq ||\underline{\mathbf{b}}_{\mathbf{j}}^*||) \to 0.$
- $P(\exists j \notin S^*, ||\mathbf{U}_{\mathbf{j}}{}^T(\underline{\mathbf{Y}} - \mathbf{U}_{\mathbf{S}^*}\tilde{\underline{\mathbf{b}}}_{\lambda_n,S^*})|| > \lambda_n\sqrt{m_n}) \to 0.$

The first assertion states there is no false negative with a probability converging to one for Group LASSO using λ_n. We use Stone's results to prove it. To prove the second assertion, which deals with false positives, we need to add the assumption that if $j \neq j'$, then $||\mathbf{U}_{\mathbf{j}}{}^T\mathbf{U}_{\mathbf{j}'}|| = o_P(\frac{n}{m_n})$. This type of assumption is classical in variable selection (see for example [8]). It means that the design obeys a kind of "restricted isometry hypothesis." It essentially means that each couple $\mathbf{U}_{\mathbf{j}}{}^T\mathbf{U}_{\mathbf{j}'}$ behaves like an orthonormal system. In what follows we sketch the various steps that allow us to derive the proof of the second assertion.

A first step is to control the gap between the quadratic errors associated to the Group LASSO and the OLS estimators. In fact, we have

$$||\underline{\mathbf{Y}} - \mathbf{U}\hat{\underline{\mathbf{b}}}_{\lambda_n}||^2 - ||\underline{\mathbf{Y}} - \mathbf{U}_{\mathbf{S}_{\lambda_n}}\hat{\underline{\mathbf{b}}}_{\mathbf{S}_{\lambda_n}}||^2 = o_P(m_n\sqrt{\log(m_n^2 d)\log(nd)}),$$

which follows by expressing the Group LASSO estimator as a function of the OLS estimator.

As $\text{card}(S_{\lambda_n}) = card(S^*)$, we have $BIC(\lambda_n) - BIC(S^*) = \log(||\underline{\mathbf{Y}} - \mathbf{U}\hat{\underline{\mathbf{b}}}_{\lambda_n}||^2) - \log(||\underline{\mathbf{Y}} - \mathbf{U}_{\mathbf{S}_{\lambda_n}}\hat{\underline{\mathbf{b}}}_{\mathbf{S}_{\lambda_n}}||^2)$.

Following similar lines as in the previous subsection allows us to obtain:

$$BIC(\lambda_n) - BIC(S^*) = o_P(m_n\frac{\sqrt{\log(m_n^2 d)\log(nd)}}{n}) + o(o_P(m_n\frac{\sqrt{\log(m_n^2 d)\log(nd)}}{n})).$$

Note that $m_n\frac{\sqrt{\log(m_n^2 d)\log(nd)}}{n}$ is negligible compared to $\min(\log(2), v_n^2/m_n)$.

By the OLS definition, we necessary have $BIC(\lambda) \geq BIC(S_\lambda)$. Therefore,

$$BIC(\lambda) - BIC(\lambda_n) \geq BIC(S_\lambda) - BIC(S_{\lambda_n}) + o_P(m_n\frac{\sqrt{\log(m_n^2 d)\log(nd)}}{n}).$$

Now, $o_P(m_n\frac{\sqrt{\log(m_n^2 d)\log(nd)}}{n})$ is negligible compared to $\min(1, \frac{v_n^2}{m_n})$ and to $\frac{m_n\log(nd)}{n}$. It follows that, if $S^* \not\subseteq S_\lambda$, thanks to Theorem 1,

$$P(\min_{\lambda|card(S_\lambda)\leq M \text{ et } S^*\not\subseteq S_\lambda} BIC(\lambda) - BIC(\lambda_n) > 0) \; n \overrightarrow{\rightarrow +\infty} \; 1,$$

and if $S^* \subseteq S_\lambda$, thanks to Theorem 2,

$$P(\min_{\lambda|S^*\subset S_\lambda \text{ et } card(S_\lambda)\leq M} BIC(\lambda_n) - BIC(\lambda) < 0) \; n \overrightarrow{\rightarrow +\infty} \; 1.$$

The above facts allow now to state the main Theorem of this section.

Theorem 4 *Under our assumptions, take* $\lambda_n \sim \sqrt{n\log(m_n^2 d)/m_n}$, *then*

$$P(S_{\lambda_n} = S^*) \rightarrow 1,$$

and for all λ *such as* $S_\lambda \neq S^*$ *and* $card(S_\lambda) \leq M$

$$P(BIC(\lambda) - BIC(\lambda_n) > 0) \rightarrow 1.$$

Estimation consistency We adopt here the conditions of Theorem 4.2 of Kato [19]. Let A_f be a constant. The mentioned Theorem 4.2 shows that the penalized B-splines post-model allows a consistent estimation of the regression function with an empirical mean square error of the order bounded above by:

$$\begin{aligned}\frac{s^* m_n}{n^2} A_f n(1+\sqrt{\log(d)/m_n})^2 &= O(1)\frac{m_n}{n}(1+\sqrt{\log(d)/m_n})^2 \\ &= O(1)\frac{(1+\sqrt{\log(d)/m_n})^2}{n^{2\nu/(2\nu+1)}} \\ &\rightarrow 0 \text{ (thanks to assumption 5.b)}\end{aligned}$$

The estimation consistency of our two-step estimator thus follows from Theorem 4.2 of Kato [19].

4 Numerical Experiment

We shortly present here a numerical experiment on EDF data. For conciseness, since this paper is mostly devoted to theoretical material, we summarize below the main results of this application that is further detailed in the forthcoming publication [29]. We apply here the Post1 procedure. We use BIC, AIC, and GCV for the model selection criterion and we name Post1Bic, Post1Aic and Post1Gcv the resulting models. We focus on middle term (from one day to one year ahead) load forecasting on EDF portfolio data (the overall sum of EDF costumers' consumption). Forecasting load at this horizon is a fundamental task for maintenance planning and more generally risk management of electricity providers. At this horizon, forecasting models are based on deterministic information (calendar events) and meteorological data that could be simulated to derive a probabilistic view of load consumption. We consider in this study real meteorological data to overcome weather forecast errors when comparing our models. As consumption habits depend strongly on the hour of the day we fit one model per instant (see [24]). The data are measured each 30 min so we fit 48 models corresponding to the 48 instants per day. The training set stands from 2007 to August 2012 and the testing set from September 2012 to July 2013. We compare our procedures with a benchmark additive model, called EDF model, which is inspired by a model obtained by an EDF expert who use business expertise to select among a lot of possible combinations of models. Of course, this model performs very well and our main goal is to automatically obtain some models with similar performances to avoid this costly and time consuming expert selection. The study is motivated by the fact that the French load consumption changes and will change because of new habits (for example electrical car) and because of opening of market, which will change EDF portfolio. Here, we focus on meteorological covariates selection because of their large influence on electricity demand and build a dictionary of 30 covariates from the temperature, the cloud cover and the wind speed. We also work on detrended load to lower the large nonstationnarity level and use the EDF model to correct the trend.

Covariates selection There is a physical explanation for the selected covariates. For example, Post1Bic selects cloud cover during the day and not during the night, which can be easily explained because cloud cover influences mainly the lighting. As the cloud cover, the wind is selected almost for all instants of the day, which can be explained by its impact on the air feels. Daily maximum are selected around mid-day and the evening, which is consistent with known properties of temperature. Use smoothed temperature is very classical in load modeling and allows to have more regularized effects. The question is what smoothing coefficient to use. Many smoothed temperatures are candidates. Post1Bic selects two smoothed temperatures: one with a small smoothing coefficients and an other with smoothing coefficient near the one used in EDF model for almost all the instants. Real temperature is not selected for each instant because many variables explain the effect of temperature. Post1Aic and Post1Gcv are close in terms of selection which is coherent with the known properties of GCV and AIC, which are asymptotically similar under good conditions. On the contrary, BIC is known to be more severe than AIC and GCV and this is the case here for Post1Bic compared to Post1Aic and Post1Gcv.

Middle term performance To quantify the forecasting efficiency, we use the MAPE (Mean Absolute Percentage Error) and the RMSE (Root Mean Square Error). Table 1 summarizes the forecasting performances. As expected, the EDF model performs very well and it is a hard task to perform better. With the automatic selection, which can be different between two instants, Post1Bic procedure achieves a little better forecasting performance than the EDF model. We explain worst performance of Post1Aic and Post1Gcv by the fact their effects in models are not well estimated, as there are more variables. To avoid this issue, other estimators such as OLS estimators can be used. In [29], we apply the procedure we named Post2, which substitutes the OLS estimator by P-splines for the second step of Algorithm given in Sect. 2.2, to avoid the risk of overfitting and to regularize the effect. To forecast load demand with additive models, it is very common to use some regularized methods like P-Splines (see e.g., [13, 14, 24]) and seems necessary to have good forecasting performance.

Table 1 Forecasting MAPE and RMSE

Criterion	MAPE	RMSE
Post1Bic	1.43	830
EDF model	1.50	888
Post1Gcv	1.63	966
Post1Aic	1.75	1119

5 Conclusion

In this paper, we propose a two-step estimator for additive models and prove some variable selection properties of our procedures. The assumption, the most notable is that we assume that the Euclidean norm of nonzero components is bounded away from a sequence which can converge to zero when the number of observations grows. First, we show that the BIC defined by (5) is variable selection consistent, and then show that the penalized BIC is variable consistent. Finally, we show the estimation consistency of the resulting two-step estimator. The reader can find detailed proofs in [3]. Note that these theoretical results are valid only for the i.i.d. case. An extension to the dependent case is not an easy task. However, a paper by Zhou [34] could provide the appropriate tools to get an analogous of Propositions 1, 2, and 3 of [3], to the case of correlated observations using Gaussian comparison inequalities and some concentration results for the case of Gaussian distributed data. This is an interesting topic for future research.

We end the paper by providing a numerical experiment on French load consumption forecasting. We show that a two-step estimator using Group LASSO and OLS estimators allows to get automatically as good forecasting performances as a model selected by an EDF expert. We also notice some selection and forecasting performances differences when using BIC, AIC, or GCV, and point out the physical coherence in the resulting selected variables. The interested reader can find in [29] some more detailed applications.

Acknowledgments The authors thank the Editor and a reviewer for their comments on the original manuscript.

Appendix: Sketch of Proofs

Here, we sketch for the reader some elements of proof of Theorem 4. Detailed proofs are given in [3].

We first state the following lemma:

Lemma 1 *If* $\lambda_n = \sqrt{n \log(m_n^2 d)/m_n}$, *then*

- $P(\exists j \in S^*, ||\underline{\mathbf{b}}_{\mathbf{j}}^* - \tilde{\mathbf{b}}_{\lambda_{\mathbf{n}},\mathbf{j}}|| \geq ||\underline{\mathbf{b}}_{\mathbf{j}}^*||) \to 0.$
- $P(\exists j \notin S^*, ||\mathbf{U_j}^T(\underline{\mathbf{Y}} - \mathbf{U}_{\mathbf{S}^*}\tilde{\mathbf{b}}_{\lambda_{\mathbf{n}},\mathbf{S}^*})|| > \lambda_n\sqrt{m_n}) \to 0.$

Proof of Lemma 1 We begin with the term $P(\exists j \in S^*, ||\underline{\mathbf{b}}_{\mathbf{j}}^* - \tilde{\mathbf{b}}_{\lambda_{\mathbf{n}},\mathbf{j}}|| \geq ||\underline{\mathbf{b}}_{\mathbf{j}}^*||)$. Take $j \in S^*$. Note $T_j = (\mathbf{0}_{\mathbf{m_n}\times\mathbf{m_n}}, \ldots, \mathbf{0}_{\mathbf{m_n}\times\mathbf{m_n}}, \mathbf{I}_{\mathbf{m_n}\times\mathbf{m_n}}, \mathbf{0}_{\mathbf{m_n}\times\mathbf{m_n}}, \ldots, \mathbf{0}_{\mathbf{m_n}\times\mathbf{m_n}})$, where $\mathbf{0}_{\mathbf{m_n}\times\mathbf{m_n}}$ is the null matrix of size $m_n \times m_n$ and $\mathbf{I}_{\mathbf{m_n}\times\mathbf{m_n}}$ the identity matrix of size $m_n \times m_n$. The jth block of T_j is $\mathbf{I}_{\mathbf{m_n}\times\mathbf{m_n}}$.

We have $\tilde{\mathbf{b}}_{\lambda_{\mathbf{n}},\mathbf{j}} = T_j\tilde{\mathbf{b}}_{\lambda_n} = T_j(\hat{\underline{\mathbf{b}}}_{\mathbf{S}^*} - (\mathbf{U}_{\mathbf{S}^*}^T\mathbf{U}_{\mathbf{S}^*})^{-1}\underline{\mathbf{r}})$. Noting $C_{S^*} = \frac{1}{n}\mathbf{U}_{\mathbf{S}^*}^T\mathbf{U}_{\mathbf{S}^*}$, we have $||\tilde{\mathbf{b}}_{\lambda_{\mathbf{n}},\mathbf{j}} - \underline{\mathbf{b}}_{\mathbf{S}^*,\mathbf{j}}^*|| = ||T_j(\hat{\underline{\mathbf{b}}}_{\mathbf{S}^*} - \frac{1}{n}C_{S^*}^{-1}\underline{\mathbf{r}} - \underline{\mathbf{b}}_{\mathbf{S}}^*)|| \leq ||T_j(\hat{\underline{\mathbf{b}}}_{\mathbf{S}^*} - \underline{\mathbf{b}}_{\mathbf{S}}^*)|| + \frac{1}{n}||T_jC_{S^*}^{-1}\underline{\mathbf{r}}||$.

As $||\hat{\underline{\mathbf{b}}}_{\mathbf{S}^*} - \underline{\mathbf{b}}^*_{\mathbf{S}}|| = O(\frac{m_n}{\sqrt{n}})$, (see [27]), we obtain $\max_j ||T_j(\hat{\underline{\mathbf{b}}}_{\mathbf{S}^*} - \underline{\mathbf{b}}^*_{\mathbf{S}})|| \leq \frac{m_n}{\sqrt{n}}$.

Moreover, thanks to Lemma 3 of [18], we have $\frac{1}{n}||T_j C_{S^*}^{-1}\underline{\mathbf{r}}|| \leq \frac{1}{n}||T_j||||C_{S^*}^{-1}||||\underline{\mathbf{r}}|| \leq o(\sqrt{m_n}\min||f_j^*||)$. and therefore

$$||\tilde{\mathbf{b}}_{\lambda_{\mathbf{n}},\mathbf{j}} - \underline{\mathbf{b}}^*_{\mathbf{S}^*,\mathbf{j}}|| \leq \frac{m_n}{\sqrt{n}} + o(\sqrt{m_n}\min||f_j^*||).$$

We have $\min_{j\in S^*}||\underline{\mathbf{b}}^*_{\mathbf{j}}|| \geq O(1)\left(\sqrt{m_n}\min_{j\in S^*}||f_j^*|| - O(1)\frac{m_n}{\sqrt{n}}\right) \geq O(1)(\sqrt{m_n}\sqrt{\frac{m_n}{n}\log(nd)} + O(\frac{m_n}{\sqrt{n}})) \geq O(1)\left(\frac{m_n}{\sqrt{n}}\sqrt{\log(nd)} + O(\frac{m_n}{\sqrt{n}})\right)$. Given the assumptions it is clear that $\frac{m_n}{\sqrt{n}} = o(\frac{m_n}{\sqrt{n}}\sqrt{\log(nd)})$, which allows us to conclude.

We now examine the other assertion: $P(\exists j \notin S^*, ||\mathbf{U_j}^T(\underline{\mathbf{Y}} - \mathbf{U}_{\mathbf{S}^*}\tilde{\mathbf{b}}_{\lambda_{\mathbf{n}},\mathbf{S}^*})|| > \lambda_n\sqrt{m_n})$.

Take $j \notin S^*$. We have

$$\begin{aligned}||\mathbf{U_j}^T(\underline{\mathbf{Y}} - \mathbf{U}_{\mathbf{S}^*}\tilde{\mathbf{b}}_{\lambda_{\mathbf{n}}})|| &\leq ||\mathbf{U_j}^T(\underline{\mathbf{f}}^*(\underline{\mathbf{X}}) - \mathbf{U}_{\mathbf{S}^*}\underline{\mathbf{b}}^*_{\mathbf{S}^*})|| + ||\mathbf{U_j}^T\Pi_j\underline{\epsilon}|| \\ &+ ||\mathbf{U_j}^T(\mathbf{U}_{\mathbf{S}^*}\underline{\mathbf{b}}^*_{\mathbf{S}^*} - \mathbf{U}_{\mathbf{S}^*}\hat{\mathbf{b}}_{\mathbf{S}_{\lambda_{\mathbf{n}}}})|| + ||\mathbf{U_j}^T\mathbf{U}_{\mathbf{S}^*}(\mathbf{U}_{\mathbf{S}^*}^T\mathbf{U}_{\mathbf{S}^*})^{-1}\underline{\mathbf{r}}||,\end{aligned}$$

where $\Pi_j = \mathbf{U_j}(\mathbf{U_j}^T\mathbf{U_j})^{-1}\mathbf{U_j}^T$.

We then examine the four terms of the right hand side of the above inequality (see Lemma 7 of [3] for details and justifications).

- $||\mathbf{U_j}^T(\underline{\mathbf{f}}^*(\underline{\mathbf{X}}) - \mathbf{U}_{\mathbf{S}^*}\underline{\mathbf{b}}^*_{\mathbf{S}^*})|| \leq \max_{j\notin S^*}||\mathbf{U_j}^T\mathbf{U_j}||^{1/2}||(\underline{\mathbf{f}}^*(\underline{\mathbf{X}}) - \mathbf{U}_{\mathbf{S}^*}\underline{\mathbf{b}}^*_{\mathbf{S}^*})|| \leq O(1)\sqrt{\frac{n}{m_n}}m_n^{-\nu} \leq o(\lambda_n\sqrt{m_n})$
- $||\mathbf{U_j}^T\Pi_j\underline{\epsilon}|| \leq \max_{j\notin S^*}||\mathbf{U_j}^T\mathbf{U_j}||^{1/2}||||\Pi_j\underline{\epsilon}|| \leq \sqrt{\frac{n}{m_n}}o_P(\sqrt{m_n\log(nd)}) \leq o_P(\lambda_n\sqrt{m_n})$
- $||\mathbf{U_j}^T(\mathbf{U}_{\mathbf{S}^*}\underline{\mathbf{b}}^*_{\mathbf{S}^*} - \mathbf{U}_{\mathbf{S}^*}\hat{\mathbf{b}}_{\mathbf{S}_{\lambda_{\mathbf{n}}}})|| \leq \max_{j\notin S^*}||\mathbf{U_j}^T\mathbf{U_j}||^{1/2}||\mathbf{U}_{\mathbf{S}^*}^T\mathbf{U}_{\mathbf{S}^*}||^{1/2}||\underline{\mathbf{b}}^*_{\mathbf{S}^*} - \hat{\mathbf{b}}_{\mathbf{S}_{\lambda_{\mathbf{n}}}}|| \leq o(\lambda_n\sqrt{m_n})$
- $||\mathbf{U_j}^T\mathbf{U}_{\mathbf{S}^*}(\mathbf{U}_{\mathbf{S}^*}^T\mathbf{U}_{\mathbf{S}^*})^{-1}\underline{\mathbf{r}}|| = ||\mathbf{U_j}^T\mathbf{U}_{\mathbf{S}^*}\sum_{i=1}^{s^*}T_i^T T_i(\mathbf{U}_{\mathbf{S}^*}^T\mathbf{U}_{\mathbf{S}^*})^{-1}\underline{\mathbf{r}}|| \leq \sum_{i=1}^{s^*}||\mathbf{U_j}^T\mathbf{U}_{\mathbf{S}^*}T_i^T T_i(\mathbf{U}_{\mathbf{S}^*}^T\mathbf{U}_{\mathbf{S}^*})^{-1}\underline{\mathbf{r}}|| \leq \sum_{i\in S^*}||\mathbf{U_j}^T\mathbf{U_i}||||(\mathbf{U}_{\mathbf{S}^*}^T\mathbf{U}_{\mathbf{S}^*})^{-1}||||\underline{\mathbf{r}}|| \leq o_P(n/m_n)\frac{m_n}{n}\lambda_n\sqrt{m_n} \leq o_P(\lambda_n\sqrt{m_n})$

These inequalities allow us to conclude. □

The next result we develop here allows us to control of the gap between the quadratic errors of Group LASSO and OLS estimators.

Lemma 2 $||\underline{\mathbf{Y}} - \mathbf{U}\hat{\underline{\mathbf{b}}}_{\lambda_{\mathbf{n}}}||^2 - ||\underline{\mathbf{Y}} - \mathbf{U}_{\mathbf{S}_{\lambda_{\mathbf{n}}}}\hat{\underline{\mathbf{b}}}_{\mathbf{S}_{\lambda_{\mathbf{n}}}}||^2 = o_P(m_n\sqrt{\log(m_n^2 d)\log(nd)})$

Proof of Lemma 2 We have

$$||\underline{\mathbf{Y}} - \mathbf{U}\hat{\underline{\mathbf{b}}}_{\lambda_n}||^2 - ||\underline{\mathbf{Y}} - \mathbf{U}_{\mathbf{S}_{\lambda_n}}\hat{\underline{\mathbf{b}}}_{\mathbf{S}_{\lambda_n}}||^2 = ||\mathbf{U}_{\mathbf{S}^*}(\mathbf{U}_{\mathbf{S}^*}^T\mathbf{U}_{\mathbf{S}^*})^{-1}\underline{\mathbf{r}}||^2 - 2(\underline{\mathbf{Y}} - \Pi_{S^*}\underline{\mathbf{Y}})^T(\mathbf{U}_{\mathbf{S}^*}(\mathbf{U}_{\mathbf{S}^*}^T\mathbf{U}_{\mathbf{S}^*})^{-1}\underline{\mathbf{r}}),$$

where $\Pi_{S^*} = \mathbf{U}_{\mathbf{S}^*}(\mathbf{U}_{\mathbf{S}^*}^T\mathbf{U}_{\mathbf{S}^*})^{-1}\mathbf{U}_{\mathbf{S}^*}^T$.

By using Lemma 3 of [18], we can show $||\mathbf{U}_{\mathbf{S}^*}(\mathbf{U}_{\mathbf{S}^*}^T\mathbf{U}_{\mathbf{S}^*})^{-1}\underline{\mathbf{r}}|| = O(\sqrt{m_n \log(m_n^2 d)})$. We have $(\underline{\mathbf{Y}} - \Pi_{S^*}\underline{\mathbf{Y}})^T(\mathbf{U}_{\mathbf{S}^*}(\mathbf{U}_{\mathbf{S}^*}^T\mathbf{U}_{\mathbf{S}^*})^{-1}\underline{\mathbf{r}}) = (\underline{\mathbf{f}}^*(\underline{\mathbf{X}}) - \mathbf{U}_{\mathbf{S}^*}\hat{\underline{\mathbf{b}}}_{\mathbf{S}^*})^T(\mathbf{U}_{\mathbf{S}^*}(\mathbf{U}_{\mathbf{S}^*}^T\mathbf{U}_{\mathbf{S}^*})^{-1}\underline{\mathbf{r}}) + \underline{\epsilon}^T(\Pi_{S^*}\mathbf{U}_{\mathbf{S}^*}(\mathbf{U}_{\mathbf{S}^*}^T\mathbf{U}_{\mathbf{S}^*})^{-1}\underline{\mathbf{r}})$.

Using the Cauchy–Schwartz we obtain:

- $(\underline{\mathbf{f}}^*(\underline{\mathbf{X}}) - \mathbf{U}_{\mathbf{S}^*}\hat{\underline{\mathbf{b}}}_{\mathbf{S}^*})^T(\mathbf{U}_{\mathbf{S}^*}(\mathbf{U}_{\mathbf{S}^*}^T\mathbf{U}_{\mathbf{S}^*})^{-1}\underline{\mathbf{r}}) \leq O_P(m_n\sqrt{\log(m_n^2 d)})$
- $|\underline{\epsilon}^T\Pi_{S^*}(\mathbf{U}_{\mathbf{S}^*}(\mathbf{U}_{\mathbf{S}^*}^T\mathbf{U}_{\mathbf{S}^*})^{-1}\underline{\mathbf{r}})| \leq o_P(m_n\sqrt{\log(nd)\log(m_n^2 d)})$

Hence, the conclusion. □

References

1. Antoniadis, A., Fan, J.: Regularization of wavelet approximations. J. Am. Stat. Assoc. **96**, 939–967 (2001)
2. Antoniadis, A., Gijbels, I., Verhasselt, A.: Variable selection in additive models using P-splines. Technometrics **54**(4), 425–438 (2012)
3. Antoniadis, A., Goude, Y., Poggi, J-M., Thouvenot, V.: Sélection de variables dans les modèles additifs avec des estimateurs en plusieurs étapes. Technical report, 2015. https://hal.archives-ouvertes.fr/hal-01116100
4. Bach, F.R.: Consistency of the group lasso and multiple kernel learning. J. Mach. Learn. Res. **9**, 1179–1225 (2008)
5. Belloni, A., Chernozhukov, V.: Least squares after model selection in high-dimensional sparse models. Bernoulli **19**(2), 521–547 (2013)
6. Taieb, S.B., Hyndman, R.J.: A gradient boosting approach to the kaggle load forecasting competition. Int. J. Forecast. **30**(2), 382–394 (2014)
7. Buhlmann, P., Van de Geer, S.: Statistics for High-Dimensional Data: Methods. Theory and Applications, 1st edn. Springer Publishing Company, Incorporated (2011)
8. Candes, E., Tao, T.: The Dantzig selector: statistical estimation when p is much larger than n. Ann. Stat. **35**(6), 2313–2351 (2007)
9. Cantoni, E., Mills Flemming, J., Ronchetti, E.: Variable selection in additive models by non-negative garrote. Stat. Modell. **11**(3):237–252 (2006)
10. Eilers, P.H.C., Marx, B.D.: Flexible smoothing with B-splines and penalties. Stat. Sci. **11**(2), 89–121 (1996)
11. Fan, J., Feng, Y., Song, R.: Nonparametric independence screening in sparse ultra-high-dimensional additive models. J. Am. Stat. Assoc. **106**(494), 544–557 (2011)
12. Fan, J., Jiang, J.: Generalized likelihood ratio tests for additive models. J. Am. Stat. Assoc. **100**, 890–907 (2005)
13. Fan, S., Hyndman, R.J.: Short-term load forecasting based on a semi-parametric additive model. IEEE Trans. Power Syst. **27**(1), 134–141 (2012)
14. Goude, Y., Nedellec, R., Kong, N.: Local short and middle term electricity load forecasting with semi-parametric additive models. IEEE Trans. Smart Grid **5**(1), 440–446 (2014)
15. Hastie, T.J., Tibshirani, R.J.: Generalized Additive Models. Chapman & Hall, London (1990)
16. Hong, T., Pinson, P., Fan, S.: Global energy forecasting competition 2012. Int. J. Forecast. **30**(2), 357–363 (2014)

17. Horowitz, J., Klemela, J., Mammen, E.: Optimal estimation in additive regression models. Bernoulli **12**(2), 271–298 (2006)
18. Huang, J., Horowitz, J.L., Wei, F.: Variable selection in nonparametric additive models. Ann. Stat. **38**(4), 2282–2313 (2010)
19. Kato, K.: Two-step estimation of high dimensional additive models. Technical report, July 2012. http://adsabs.harvard.edu/abs/2012arXiv1207.5313K
20. Koltchinskii, V., Yuan, M.: Sparsity in multiple kernel learning. Ann. Stat. **38**(6), 3660–3695 (2010)
21. Lin, Y., Zhang, H.H.: Component selection and smoothing in multivariate nonparametric regression. Ann. Stat. **34**(5), 2272–2297 (2006)
22. Marra, G., Wood, S.: Practical variable selection for generalized additive models. Comput. Stat. Data Anal. **55**(7), 2372–2387 (2011)
23. Nedellec, R., Cugliari, J., Goude, Y.: Gefcom 2012: electric load forecasting and backcasting with semi-parametric models. Int. J. Forecast. **30**(2), 375–381 (2014)
24. Pierrot, A., Goude, Y.: Short-term electricity load forecasting with generalized additive models. In: Proceedings of ISAP power, pp. 593–600 (2011)
25. Raskutti, G., Wainwright, M.J., Yu, B.: Minimax-optimal rates for sparse additive models over kernel classes via convex programming. J. Mach. Learn. Res. **13**(1), 389–427 (2012)
26. Ravikumar, P., Lafferty, Jo., Liu, H., Wasserman, L.: Sparse additive models. J. R. Stat. Soc.: Ser. B (Stat. Methodol.) **71**(5), 1009–1030 (2009)
27. Stone, C.J.: Additive regression and other nonparametric models. Ann. Stat. **13**(2), 689–705 (1985)
28. Suzuki, T., Tomioka, R., Sugiyama, M.: Fast convergence rate of multiple kernel learning with elastic-net regularization. arXiv:1103.0431 (2011)
29. Thouvenot, V., Pichavant, A., Goude, Y., Antoniadis, A., Poggi, J.-M.: Electricity forecasting using multi-stage estimators of nonlinear additive models. IEEE Trans. Power Syst. (2015)
30. Tibshirani, R.: Regression shrinkage and selection via the lasso. J. R. Stat. Soc. Ser. B **58**, 267–288 (1994)
31. Wang, H., Li, B., Leng, C.: Shrinkage tuning parameter selection with a diverging number of parameters. J. R. Stat. Soc. Ser. B (Stat. Methodol.) **71**(3), 671–683 (2009)
32. Wood, S.: Generalized Additive Models: An Introduction with R. Chapman and Hall/CRC (2006)
33. Yuan, M., Lin, Y.: Model selection and estimation in regression with grouped variables. J. R. Stat. Soc. Ser. B **68**, 49–67 (2006)
34. Zhou, S.: Restricted eigenvalue conditions on subgaussian random matrices. Technical report, Dec 2009. http://adsabs.harvard.edu/abs/2009arXiv0912.4045Z

Nonparametric Method for Estimating the Distribution of Time to Failure of Engineering Materials

Antonio Meneses, Salvador Naya, Ignacio López-de-Ullibarri and Javier Tarrío-Saavedra

Abstract The aim of this work is to develop and assess a new method to estimate lifetime distribution in materials subjected to mechanical fatigue efforts. This problem is addressed from a statistical semiparametric and nonparametric perspective. Taking into account that fatigue failures in materials are due to crack formation and the subsequently induced crack growth, linear mixed effects regression models with smoothing splines (based on the linearized Paris-Erdogan model) are applied to estimate crack length as a function of the number of fatigue cycles. This model allows to simultaneously estimate the dependence between crack length and number of cycles in a sample of specimens. Knowing the crack length that induces material failure, the lifetime of each specimen is the crossing point of the crack length limit and the model crack length estimate. The authors propose to estimate the lifetime distribution function by applying nonparametric kernel techniques. In order to assess the influence of factors such as material type, material heterogeneity, and also that of the parameters of the estimation procedure, a simulation study consisting of different scenarios is performed. The results are compared with those of a procedure proposed by Meeker and Escobar (Statistical Methods for Reliability Data, Wiley, 1998, [16]) based on nonlinear mixed effects regression. Functional data analysis techniques are applied to perform this task. The proposed methodology estimates lifetime distribution of materials under fatigue more accurately in a wide range of scenarios.

Keywords Fatigue crack growth · Paris-Erdogan model · Nonparametric kernel distribution function estimation · Linear mixed effects · Statistical learning · Nonlinear mixed effects

A. Meneses
Universidad Nacional de Chimborazo, Riobamba, Ecuador

Ignacio López-de-Ullibarri
Universidade da Coruña. Escola Universitaria Politécnica, Ferrol, Spain

Javier Tarrío-Saavedra · S. Naya (✉)
Universidade da Coruña. Escola Politécnica Superior, Ferrol, Spain
e-mail: salva@udc.es; salvanaya@gmail.com

R. Cao et al. (eds.), *Nonparametric Statistics*, Springer Proceedings in Mathematics & Statistics 175, DOI 10.1007/978-3-319-41582-6_15

1 Introduction

Accurate estimation of the failure time of structural elements or mechanisms is essential to ensure their proper performance, in accordance with each piece's requirements. Of course, estimates will depend on the definition of failure, which differs depending on the piece and application. Generally, failure is defined as failing to meet the design requirements. Therefore, the lifetime of a piece made from a specific material is defined as the interval of time between the material commissioning and the time when failure occurs. Fatigue efforts largely condition the design of parts and structural elements.

Material degradation and failure due to fatigue are characterized by crack initiation and its posterior propagation under alternating stress. Crack length variation due to fatigue versus time defines a convex degradation curve [16] that characterizes the degree of degradation of a material under fatigue. The final failure is defined by the fatigue limit: the largest value of fluctuating stress that will not cause failure for an infinite number of cycles. The evolution of material degradation due to fatigue is characterized by three main patterns (Fig. 1). Region 1 accounts for crack initiation without propagation. Region 2 is defined by crack propagation and by the lineal dependence between logarithm of the derivative of crack length with respect to number of cycles, da/dN, and the logarithm of stress concentration factor variation (ΔK). In Region 3 crack propagation is faster, representing a very small contribution to the life of the material. It reflects the proximity of the unstable propagation of the crack when the value of the maximum stress intensity factor, K_{max}, reaches the fracture toughness [15]. Cracks occur due to repeated stress applications that can generate localized plastic deformation at the material surface and then they evolve into sharp discontinuities [4]. Region 3 is characterized by a greater nonlinear increment of crack propagation rate; thus, its contribution to the life of the material is small [15].

To estimate fatigue failure it is necessary to know how cracks grow with alternating fatigue loads. In order to estimate this relationship, some simplifications are customarily assumed: the fatigue lifetime of a specific material depends on the evolution of

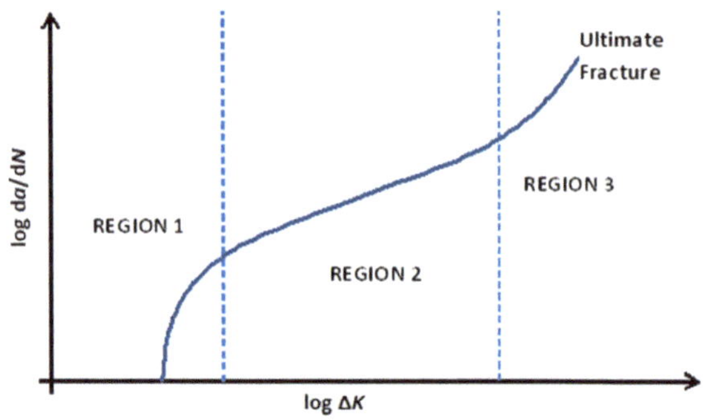

Fig. 1 Patterns of crack growth due to mechanical fatigue

the largest crack, the crack growth rate with respect to the number of cycles, da/dN, depends on the stress intensity factor range, ΔK, and the stress ratio $R = \sigma_m/\sigma_M$ (where σ_m and σ_M are the minimum and the maximum stress, respectively).

Most models or formulas used to define the crack growth in materials are based on the law of Paris or Paris-Erdogan model [19, 26]. The equation of Paris is the most used in the study of propagation of fatigue cracks because of its mathematical simplicity and the good results obtained with it [8]. The Paris law has been used to fit the crack growth corresponding to the crack propagation step, Region 2 [4, 8]. Since the main contribution to material damage occurs in Region 2, the Paris model is a widely applied and useful way to make lifetime predictions of fatigue lifetime. The Paris model is:

$$\frac{da\,(t)}{dN} = C\,[K\,(a)]^m\,, \tag{1}$$

where C and m are model parameters related to the type of material (m is between 1 and 6, $m > 4$ indicating very nonductile materials). Furthermore, it is assumed that the factor of stress concentration K is dependent on crack size, a [9, 18].

As mentioned before, Paris' law gives good approximations of the crack growth rates in Region 2, but tends to overestimate da/dN in Region 1 and to underestimates it in Region 3. Modifications of the Paris model have been proposed to improve its adequacy for those regions. For example, we mention the models by [13] for Region 3 (Eq. 2), and by [7] for Region 1 (Eq. 3).

$$\frac{da}{dN} = C\frac{[\Delta K]^m}{(1-R)\,K_{Ic} - \Delta K}, \tag{2}$$

$$\frac{da}{dN} = C\,(\Delta K - \Delta K_{th})^m\,. \tag{3}$$

Among others, FKM-Guideline [12] summarizes additional models to fit this dependence. There are other models where crack closure or elastic-plastic fracture parameters are considered (Elber 1970; Nishioka and Atluri 1982). The relationship between da/dN and ΔK can also be estimated by numerical analysis, without assuming potential laws depending on fracture mechanics parameters. That is the case of the Moving Least Squares method used by [1] and [8]. Advanced statistical methodologies have been less frequently used than numerical analysis procedures in fatigue problems. The application of statistical modeling is particularly relevant due to the inherent variability of the problem: different *a-N* curves are obtained for the same material at the same experimental conditions. Thus, the implementation of mixed effects regression models, instead of fixed effects regression modeling, is justified. These models [20] are frequently used to account for the variability between replicates and its effect on model parameters. Pinheiro and Bates [20–22] proposed an approximated maximum likelihood estimate of the parameters of the degradation model and a numeric method implemented in the `nlme` function of the R package `nlme`. Furthermore, Meeker and Escobar [16] used the numerical method of Pinheiro and Bates [21] and Monte Carlo simulation to estimate the failure time distribution in the problem of fatigue crack growth. This method is based on the simultaneous fitting of *a-N* curves corresponding to different specimens of the same material at the same conditions, using nonlinear regression mixed models with the Paris equation. This simultaneous fitting allows to estimate the joint distribution of C and m, which is assumed to be Gaussian, and thus the lifetime distribution function by maximum likelihood methods. The goal of the Meeker and Escobar method is to estimate the complete fatigue lifetime distribution under certain experimental conditions from a parametric point of view, assuming a Gaussian distribution for the parameters.

As in the procedure of Dong et al. [8], the errors caused by numerical differentiation of the noisy discrete crack length data could be attenuated by using semiparametric statistical regression techniques, such as B-splines regression [25, 30]. This method has not been sufficiently studied in the context of fatigue. Also, the lifetime distribution function is not usually studied in engineering works. The use of kernel estimates of the distribution function [24, 28] is advisable in order to obtain more information (position, variability, probability of fail) about the fatigue lifetime at operational conditions. Thus, we propose a methodology consisting of the following steps: (a) estimating with B-splines the da/dN or da/dt corresponding to a group of experimental tests that provides a number of experimental curves, (b) fitting the da/dN or da/dt data versus ΔK by applying mixed effects linear regression techniques through the linearized Paris equation, (c) obtaining the fatigue lifetimes as the crossing point of the crack length limit and the model estimates, and (d) estimating the fatigue lifetime distribution function by nonparametric kernel methods. The combination of Paris model (and also other parametric alternatives), B-splines fitting, mixed effects regression modeling, and nonparametric kernel estimation of the distribution function could improve the knowledge and estimation of damage tolerance and lifetime of many mechanism elements subjected to fatigue efforts in aircraft, automotive industry, and civil engineering. To sum up, the aim of our work is to develop an alternative nonparametric methodology to estimate the fatigue lifetime distribution function more accurately. In Sect. 2 the methodology and simulation scenarios used in our study are described in detail. Simulation results are presented in Sect. 3. We conclude in Sect. 4.

2 Simulation Study and Methodology Description

In this section, the procedure to generate simulated data is described. Crack growth has been simulated under different scenarios in order to evaluate the proposed methodology. Our procedure for estimating the lifetime distribution is also described from a computational point of view.

2.1 Simulation of Fatigue Crack Growth Data

Two assumptions related to Fracture Mechanics Theory have been made. Firstly the crack plastic deformation area is assumed to be small with respect to the crack area. Therefore, we meet the assumptions of linear fracture mechanics by defining $\Delta K\ (a)$ with the following expression [18]:

$$\Delta K = F \Delta S \sqrt{\pi a}, \tag{4}$$

where a is the crack length, $\Delta S = \Delta\sigma$ is the stress amplitude $\sigma_{max} - \sigma_{min}$, and F is a parameter depending on the crack and specimen geometry. For the simulation study, we have fixed $\Delta S = 1$, and F has been defined as:

$$F = \sqrt{\frac{2}{\mu\alpha} \tan \frac{\pi\alpha}{2}} \left[\frac{0.923 + 0.199 \left(1 - \sin \frac{\pi\alpha}{2}\right)^4}{\cos \frac{\pi\alpha}{2}} \right], \tag{5}$$

where $\alpha = a/B$ and B is the specimen width, and $\mu = 1$, assuming the opening mode of fatigue testing, plotted in Fig. 2 [9]. Secondly, we consider the study of cracks inside a big piece or sheet of material subjected to remote cyclic stresses (Fig. 2).

The simulated curves of crack length growth, a, versus time (they could be also obtained depending on N) have been obtained using the solution of Paris equation [16]:

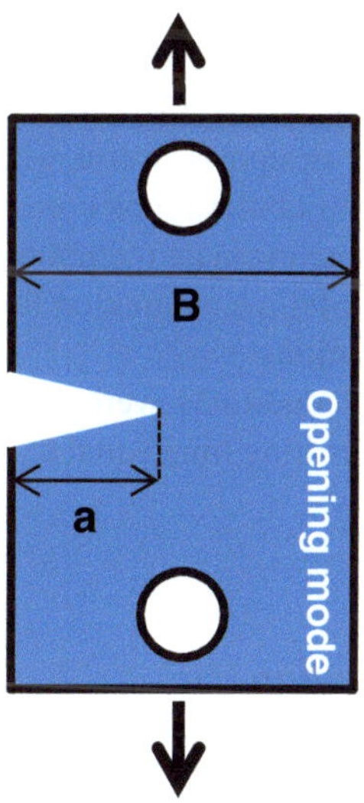

Fig. 2 Opening mode fatigue testing

$$a(t) = \left[a_0^{\left(1-\frac{m}{2}\right)} + \left(1 - \frac{m}{2}\right) C \left(FS\sqrt{\pi}\right)^m t\right]^{\frac{2}{2-m}}, \text{ for } m \neq 2. \tag{6}$$

To evaluate the proposed methodology 32 different scenarios defined by different mean values of C, m, which we denote by μ_C and μ_m, respectively, and their covariance matrix, Σ_{Cm}, were considered. The means of C and m are chosen in a representative range of real fatigue data. C and m are assumed to be jointly normally distributed. It is important to note that the values of C and m are related to the material type. Thus, we can test the methodology by simulating different types of materials and different levels of heterogeneity and dependence of the parameters. The simulation scenarios are defined by the following values:

$$\mu_C = \begin{cases} 6 \\ 5 \end{cases}, \quad \mu_m = \begin{cases} 4 \\ 3 \end{cases}, \quad \Sigma_{Cm} = \begin{pmatrix} \sigma_C^2 & \sigma_{Cm} \\ \sigma_{Cm} & \sigma_m^2 \end{pmatrix},$$

$$\sigma_C^2 = \begin{cases} 0.5 \\ 0.1 \end{cases}, \quad \sigma_m^2 = \begin{cases} 0.5 \\ 0.1 \end{cases}, \quad \sigma_{Cm} = \begin{cases} -0.02 \\ -0.09 \end{cases}.$$

We have obtained 1000 curves of a versus t in each of the $2^5 = 32$ different scenarios defined by combining the values of the parameters. These curves are evaluated in a range of time between 0 and 1, measured in arbitrary units (a.u.). Crack length is also measured in a.u.. The failure is defined at a specific crack length (critical crack length, a_c), and then the lifetime of each simulated specimen can be obtained. Thus, the empirical distribution function of lifetime is also obtained from those 1000 values for lifetime (Fig. 3), without time censoring. This distribution can be compared with the estimates obtained by the proposed methodology.

It can be observed that each factor has 2 levels, so the 32 combinations correspond to a factorial design. In this case, a 1/4 fractional design with III resolution, 2_{III}^{5-2}, has been performed [17]. This design is useful when the experimental time is a critical parameter or there are no observations corresponding to all the levels. The 2_{III}^{5-2} design is used in this work to evaluate methodology performance and to check the influence of parameters on lifetime estimation through the L_2 distance between empirical and estimated distributions.

Additionally, several initial conditions were set for the simulations: an initial crack growth equal to 0.1, a censoring time equal to 0.15 and a critical crack length $a_c = 0.5$.

2.2 *Methodology Description*

In this section, the method proposed by Meeker and Escobar [16] and our procedure for estimating the distribution function are described. Nonparametric kernel distribution function estimation and bootstrap procedures of functional data analysis (FDA) are also presented.

2.2.1 Nonlinear Mixed Effects Regression to Estimate the Fatigue Lifetime Distribution

The distributional parameters related to C and m, which define the degradation path due to fatigue, can be estimated by the parametric method introduced by Meeker and Escobar [16]. This method is based on Pinheiro and Bates [21, 22] procedure to obtain an approximated maximum likelihood estimate of the parameters of a nonlinear mixed effects regression model. The required computations are performed using the R package `nlme` [22]. The procedure (denoted by PB-nlme) consists of the following steps:

1. Several *a*-*t* curves are randomly chosen from the set of 1000 curves simulated in each scenario. In this work, 15 curves were taken, a number of samples plausible for real experimental studies.
2. Mixed effects nonlinear regression modeling is applied using the Paris solution function (6), and μ_C, μ_m and Σ_{Cm} are estimated by the likelihood method [16, 21], assuming that C and m are normally distributed.
3. Assuming that C and m have a bivariate Gaussian distribution, 1000 pairs (C, m) are drawn, substituting the population parameters by their sample estimates:

$$\hat{\mu} = \begin{pmatrix} \hat{\mu}_C \\ \hat{\mu}_m \end{pmatrix}, \quad \hat{\Sigma} = \begin{pmatrix} \hat{\sigma}_C^2 & \hat{\sigma}_{Cm} \\ \hat{\sigma}_{Cm} & \hat{\sigma}_m^2 \end{pmatrix}$$

The fatigue lifetimes are obtained by Monte Carlo simulation as the crossing points of the simulated cracks with a_c. The previous steps are repeated 100 times to estimate the mean distribution function.

2.2.2 Proposed Methodology to Estimate the Fatigue Lifetime Distribution

The proposed methodology includes the application of semiparametric models (B-splines), parametric models (linear mixed effects models using the `lme` function [22, 27]), and nonparametric models (kernel distribution function estimation using the `kde` function of the R package kerdiest [24]. Broadly, this procedure estimates the C and m parameters from the linearized Paris equation 7. The values of da/dt were estimated by fitting a basis of B-splines to the *a*-*t* curves with the R package `smooth.Pspline` [25].

$$\log\left(\frac{da}{dt}\right) = \log(C) + m \log(\Delta K(a)) \tag{7}$$

The proposed method (denoted by NP) consists of the following steps:

1. Several *a-t* curves are randomly chosen from the set of 1000 curves simulated in each scenario. In this work, 15 curves were taken a number of samples plausible for real experimental studies.
2. The 15 curves are translated to the coordinate origin and fitted using B-splines [22, 25].
3. The da/dt values are obtained from the B-spline estimation and then logarithms are applied to linearize the curves.
4. The `lme` model [21, 27] based on Eq. 7 is applied and the pairs (C, m) for each curve are estimated.
5. The (C, m) pairs are plugged in Eq. 7 and the estimated 15 *a-t* curves are obtained. They are used to predict the lifetime corresponding to the critical crack length, irrespective of censoring time.
6. Fifteen lifetimes are obtained from which the distribution function is estimated. The kernel estimation method is used as implemented in the `kerdiest` R package [24]. The expression $\hat{F}_h(x) = n^{-1}\Sigma_{j=1}^{n} H\left(\frac{x-x_j}{h}\right)$, where $H(x) = \int_{-\infty}^{x} K(t)\, dt$, K is the kernel and h the bandwidth, represents the estimated distribution function. An Epanechnikov kernel with plug-in bandwidth is used [23, 24].
7. The previous steps are repeated 100 times, thus obtaining 100 different distribution functions from which the mean is estimated.

2.3 Estimation Accuracy: FDA Bootstrap Confidence Bands and L_2 Distance

The L_2 distance is used to measure the precision of distribution estimates compared with the empirical distribution. The NP and PB-nlme methods are compared using the L_2 distance. In addition, the smooth bootstrap for functional data is used to obtain confidence bands for the mean distribution function [6]. The R package fda.usc [11] is used to construct confidence bands and to calculate L_2 distances between distribution functions.

3 Results and Discussion

In this section, the fatigue lifetime distribution for each simulated scenario is estimated by the NP method, and then it is compared to the empirical distribution of the simulated data. The results of the new procedure are also compared with the distribution estimates obtained by the PB-nlme method. The simulation scenarios were defined using a fractional factorial design of experiments, 2_{III}^{5-2} (Sect. 2.1). Thus, for simplicity, and also for reducing computing time and preventing the appearance of missing data, 8 different scenarios were analyzed. The performance of the NP and PB-nlme method is studied and the dependence of estimates with respect to material type and material heterogeneity is checked. The different scenarios replicate the fatigue performance of different type of materials.

Figure 3 shows the results of 4 scenarios. Each row shows the data and estimates corresponding to each scenario. The left panels plot the simulated crack growth data versus time with the empirical distribution. The censoring time (0.15) and critical crack length ($a_c = 0.5$) for which the fatigue failure occurs have been also highlighted on the right column. Crossing points of crack length and critical crack length are shown in red on the left plot. They indicate the lifetime of each simulated specimen. In the right panels, the empirical distribution of fatigue lifetime $\hat{F}$, and the mean distribution estimates $\hat{F}_{NP}$ and $\hat{F}_{PB-nlme}$ are plotted. For simplicity, both time and crack lengths are measured in arbitrary units (a.u.). The right panels of Fig. 3 show that the estimations corresponding to NP methodology tend to be closer to the empirical distribution $\hat{F}$ of the simulated data. Table 1 shows the L_2 distances between $\hat{F}$ and $\hat{F}_{PB-nlme}$ and $\hat{F}$ between $\hat{F}_{NP}$. The results of Table 1 support the conclusion that the NP method generally provides better estimates of empirical distribution of fatigue lifetime. Although the PB-nlme method provides good estimates of the empirical distribution in many scenarios, the proposed methodology fits better the actual empirical lifetime distribution, possibly because of the greater flexibility inherent to its nonparametric component. It is important to note that very accurate estimates of the distribution function were obtained.

An ANOVA study was implemented using the 2^{5-2}_{III} design of experiments. Taking into account the NP estimations, the effects of the means of the parameters C, m, and their structure of variance and covariance over the L_2 distance to the lifetime empirical distribution are checked. The parameters μ_C, μ_m and σ^2_C are influential over the quality of distribution estimation. Results are shown in Fig. 4. The left panel shows the half normal probability plot of the effects [2]. It is applied to compare the magnitude and to test the statistical significance of the main effects of the 2-level factorial design. The fitted line shows the position of points if the effects on the response were not significant (significant effects are labeled). The right panel of Fig. 4 shows that higher values of μ_m and σ^2_C produce higher L_2 distances to the empirical distribution. On the other hand, higher values of μ_C involve lower L_2 distances and better distribution estimates. Since C and m depend on the material, the performance of the NP method depends on the properties of the material subjected to fatigue. For example, $m = 4$ means that the corresponding material is not very ductile.

In addition, preliminary studies were also performed to evaluate the influence of factors like kernel type and values such as censoring time and critical crack length.

One of the advantages of the NP method is its flexibility due to the use of nonparametric techniques. In fact, the distribution estimates can be improved in some scenarios by changing factors such as the type of kernel used to fit the distribution function. Figure 5 shows the results of using a Gaussian kernel instead of the Epanechnikov kernel in the scenario defined by $\mu_C = 5,\ \mu_m = 3,\ \mu^2_C = 0.1,\ \sigma^2_m = 0.5$ and

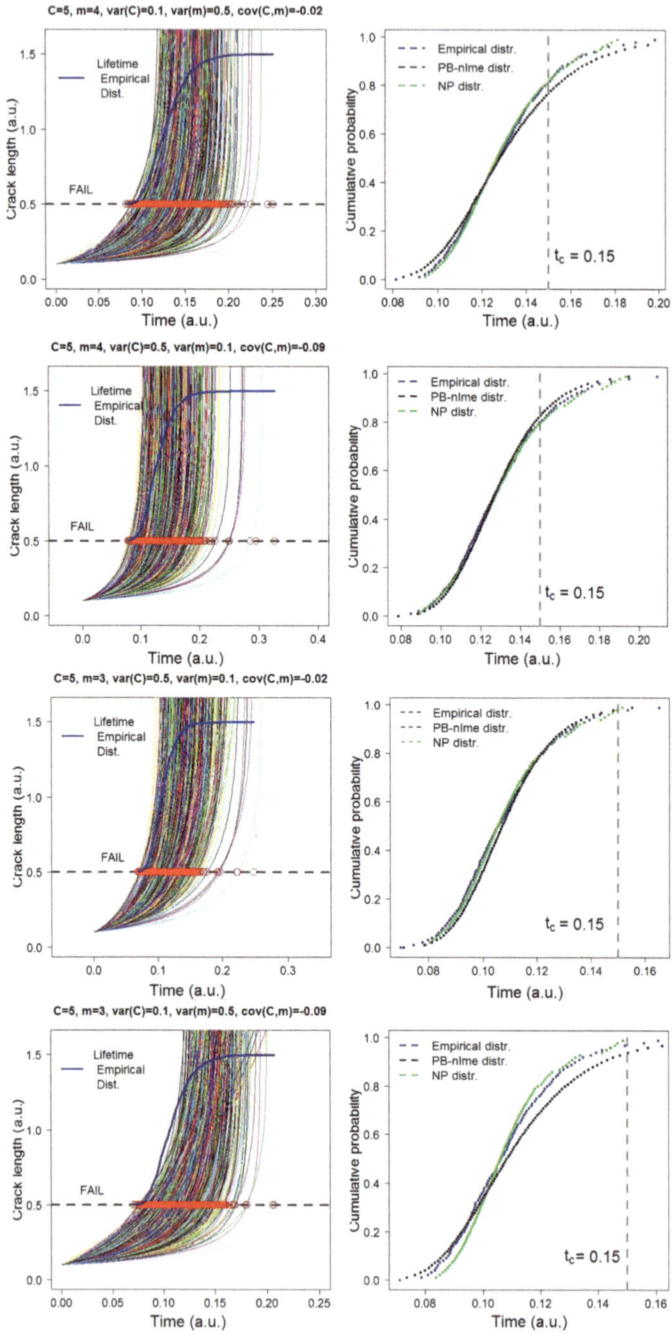

Fig. 3 *Left column*: Simulated crack growth data for 4 scenarios. *Right column*: The empirical distribution, PB-nlme and NP distribution estimates, $\hat{F}_{PB-nlme}$ and $\hat{F}_{NP}$, are shown. Censoring time (*right column*) and critical crack length (*left column*) are also shown

Table 1 L_2 distances between the empirical distribution $\hat{F}$ of simulated data and PB-nlme and NP distribution estimates, $\hat{F}_{PB-nlme}$ and $\hat{F}_{NP}$, respectively

Scenario	L_2 distance between $\hat{F}$ and $\hat{F}_{PB-nlme}$	L_2 distance between $\hat{F}$ and $\hat{F}_{NP}$
$\mu_C = 5, \mu_m = 4, \sigma_C^2 = 0.1, \sigma_m^2 = 0.5, \sigma_{Cm} = -0.02$	0.005323	0.003299
$\mu_C = 5, \mu_m = 4, \sigma_C^2 = 0.5, \sigma_m^2 = 0.1, \sigma_{Cm} = -0.09$	0.006763	0.005091
$\mu_C = 5, \mu_m = 3, \sigma_C^2 = 0.5, \sigma_m^2 = 0.1, \sigma_{Cm} = -0.02$	0.004887	0.003773
$\mu_C = 5, \mu_m = 3, \sigma_C^2 = 0.1, \sigma_m^2 = 0.5, \sigma_{Cm} = -0.09$	0.005569	0.001609
$\mu_C = 6, \mu_m = 4, \sigma_C^2 = 0.5, \sigma_m^2 = 0.5, \sigma_{Cm} = -0.02$	0.002559	0.003696
$\mu_C = 6, \mu_m = 4, \sigma_C^2 = 0.1, \sigma_m^2 = 0.1, \sigma_{Cm} = -0.09$	0.000945	0.001721
$\mu_C = 6, \mu_m = 3, \sigma_C^2 = 0.5, \sigma_m^2 = 0.5, \sigma_{Cm} = -0.09$	0.003578	0.003284
$\mu_C = 6, \mu_m = 3, \sigma_C^2 = 0.1, \sigma_m^2 = 0.1, \sigma_{Cm} = -0.02$	0.001537	0.000349

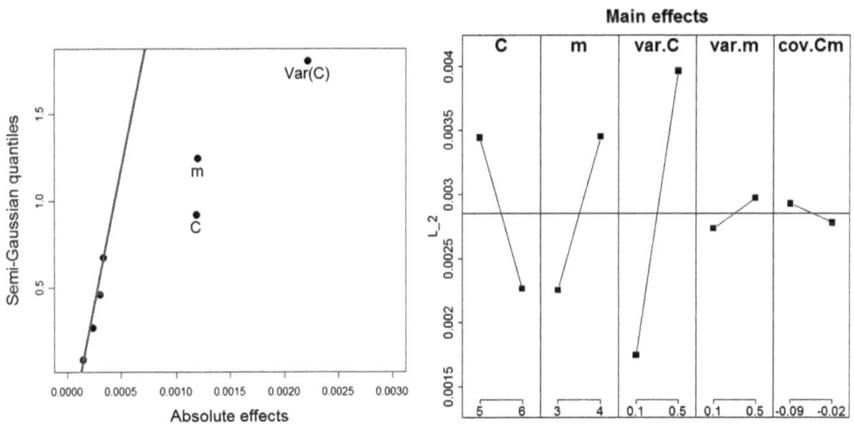

Fig. 4 Graphs related to factorial design of experiments. *Left panel*: significant effects at 95 % confidence level. *Right panel*: evaluation of the significance of principal effects

$\sigma_{Cm} = -0.09$. The estimates with the Gaussian kernel are closer to the empirical distribution than those obtained with the Epanechnikov kernel (cf. Fig. 3).

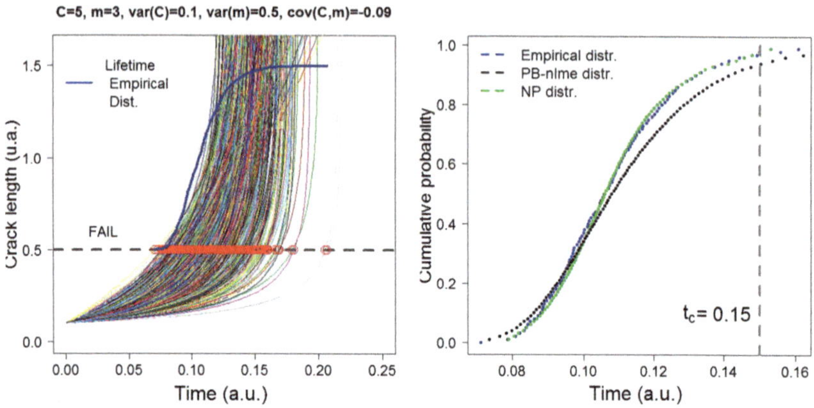

Fig. 5 Simulated data for $\mu_C = 5$, $\mu_m = 3$, $\mu_C^2 = 0.1$, $\sigma_m^2 = 0.5$ and $\sigma_{Cm} = -0.09$ scenario, empirical distribution and PB-nlme and NP distribution estimates. The mean $\hat{F}_{NP}$ estimates were obtained using a Gaussian kernel

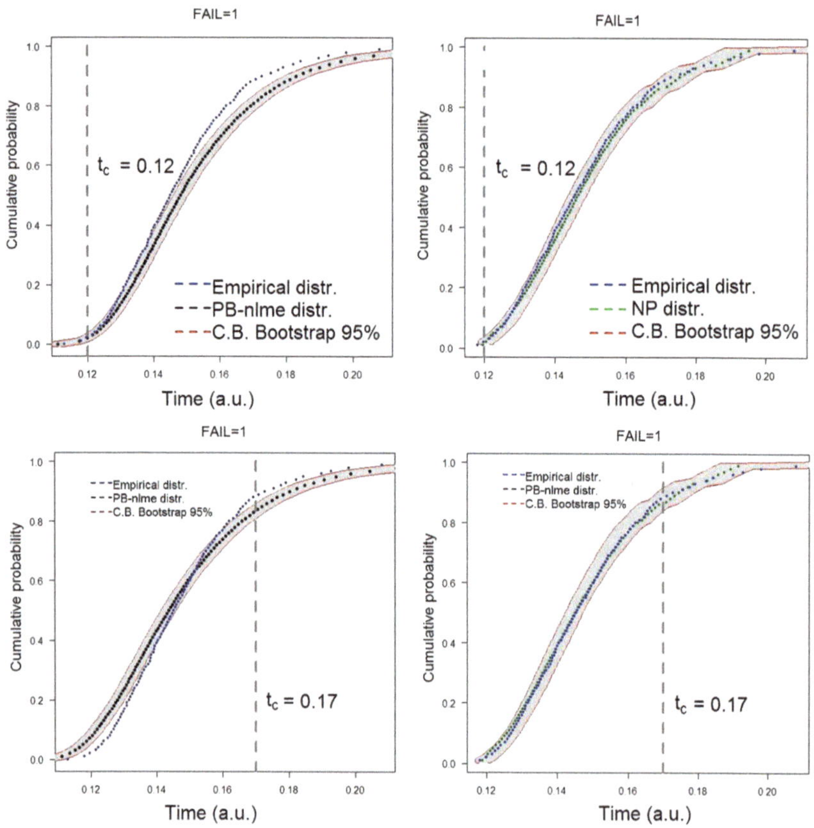

Fig. 6 FDA bootstrap confidence bands for the mean of the fatigue lifetime distribution estimated by the NP method with different censoring times, at 95 % confidence level

Regarding the influence of initial conditions, such as censoring time and critical crack length, confidence bands for the $\hat{F}_{NP}$ functional mean were obtained by smooth bootstrap. These bands contain the empirical distribution even in extreme conditions, with high censoring level (see Fig. 6). Thus, the performance of the NP method could be to some extent independent of censoring time.

4 Conclusions

A new methodology based on B-splines fitting, the application of linear mixed effects regression models (based on the linearized Paris Erdogan equation) to crack growth data, and nonparametric kernel distribution function estimation has been successfully introduced to estimate the fatigue lifetime distribution accurately.

The distribution function estimates obtained by our proposal are compared with the lifetime distribution estimated by the methodology proposed by Meeker and Escobar, based on nonlinear mixed effects regression. The proposed procedure estimates more accurately the material lifetime distribution under mechanical fatigue in a wide range of scenarios. Thus, the use of our methodology seems justified and deserves further study.

A complete simulation study was performed in order to know how parameter values affect the quality of the distribution estimates. The ANOVA study of the factorial experimental design shows that the variance of C, and the means of C and m are influential parameters over the quality of the estimates of lifetime distribution. Thus, performance depends on the properties of the material subjected to mechanical fatigue. The higher the value of σ_C^2, (i.e., the higher the material heterogeneity), the worse are the distribution estimates. Also, the higher the mean of m (i.e., the less ductile the material), the worse are the distribution estimates.

According to the bootstrap confidence bands for the mean of lifetime distribution estimates, the performance of the proposed methodology could be independent of censoring time.

Acknowledgments This research has been supported by the Spanish Ministry of Economy and Competitiveness, grant MTM2014-52876-R (ERDF included), and by the Secretariat for Higher Education, Science, Technology and Innovation of Ecuador (SENESCYT).

References

1. Atluri, S.N., Han, Z.D., Rajendran, A.M.: A new implementation of the meshless finite volume method, through the MLPG "mixed" approach. Comput. Model. Eng. Sci. **6**, 491–514 (2004)
2. Box, G.E., Hunter, J.H., Hunter, W.G.: Statistics for Experimenters: Design, Innovation, and Discovery. Wiley (2005)
3. Braz, M.H.P.: Propriedades de Fadiga de Soldas de Alta Resistência e Baixa Liga com Diferentes Composições Microestruturais. Tese de M. Sc, USP, So Carlos, SP, Brasil (1999)

4. Callister, W., Rethwisch, D.G.: Materials Science and Engineering: an Introduction, 9th edn. Wiley (2013)
5. Campbell, F.C.: Elements of Metallurgy and Engineering Alloys. ASM International (2008)
6. Cuevas, A., Febrero, M., Fraiman, R.: On the use of the bootstrap for estimating functions with functional data. Comput. Stat. Data Anal. **51**, 1063–1074 (2006)
7. Donahue, R.J., Clark, H.M., Atanmo, P., Kumble, R., McEvily, A.J.: Crack opening displacement and the rate of fatigue crack growth. Int. J. Fract.Mech. **8**, 209–219 (1972)
8. Dong, L., Haynes, R., Atluri, S.N.: On improving the celebrated Paris' power law for fatigue, by using moving least squares. CMC-Comput. Mater. Con. **45**, 1–15 (2015)
9. Dowling, N.E.: Mechanical Behavior of Materials, Engineering Methods for Deformation, Fracture and Fatigue, Pearson Prentice Hall (2012)
10. Faraway, J.J.: Extending the Linear Model with R. Generalized Linear, Mixed Effects and Nonparametric Regression Models. Chapman & Hall/CRC (2005)
11. Febrero-Bande, M., Oviedo de la Fuente, M.: Statistical computing in functional data analysis: the R package fda.usc. J. Stat. Soft. **51**, 1–28 (2012)
12. FKM-Guideline: Fracture Mechanics Proof of Strength for Engineering Components (2004)
13. Forman, R.G., Kearney, V.E., Engle, R.M.: Numerical analysis of crack propagation in cyclic-loaded structures. J. Basic Eng. **89**, 459–463 (1967)
14. Jones, R.: Fatigue crack growth and damage tolerance. Fatigue Fract. Eng. Mater. Struct. **37**, 463–483 (2014)
15. Meggiolaro, M.A., De Castro, J.T.P.: Equacionamento da curva de propagação de trincas por fadiga. III Seminário de Mecânica da Fratura/Integridade Estrutural (1997)
16. Meeker, W., Escobar, L.: Statistical Methods for Reliability Data. Wiley (1998)
17. Montgomery, D.: Design and Analysis of Experiments, 8th edn. Wiley (2013)
18. Montes, J., Cuevas, F., Cintas, J.: Ciencia e Ingeniería de los Materiales. Ediciones Paraninfo (2014)
19. Paris, P.C., Erdogan, F.: A critical analysis of crack propagation laws. J. Basic Eng. **85**, 528 (1963)
20. Pinheiro, J.C., Bates, D.M.: Approximations to the loglikelihood function in the nonlinear mixed-effects model. J. Comput. Graph. Stat. **4**, 12–35 (1995)
21. Pinheiro, J.C., Bates, D.M.: Mixed-effects Models in S and S-plus. Springer (2000)
22. Pinheiro, J., Bates, D., DebRoy, S., Sarkar, D., R Core Team.: nlme: linear and nonlinear mixed effects Models (2014). R package version 3.1-116. http://CRAN.R-project.org/package=nlme
23. Polansky, A.M., Baker, E.R.: Multistage plug-in bandwidth selection for kernel distribution function estimates. J. Stat. Comput. Simul. **65**, 63–80 (2000)
24. Quintela-del-Rio, A., Estevez-Perez, G.: Nonparametric kernel distribution function estimation with `kerdiest`: an R package for bandwidth choice and applications. J. Stat. Softw. **50**, 1–21 (2012)
25. Ramsey, J.: pspline: penalized smoothing splines (2013). R package version 1.0-16. http://CRAN.R-project.org/package=pspline
26. Shariff, A.: Simulation of Paris-Erdogan crack propagation model with power value, m = 3: The impact of applying discrete values of stress range on the behaviour of damage and lifetime of structure. Asian J. Appl. Sci. **2**, 91–95 (2009)
27. Verbeke, G., Molenberghs, G.: Linear Mixed Models for Longitudinal Data. Springer (2009)
28. Wand, M.P., Jones, M.C.: Kernel Smoothing. Chapman & Hall/CRC (1995)
29. Wang, K., Wang, F., Cui, W., Hayat, T., Ahmad, B.: Prediction of short fatigue crack growth of Ti-6Al-4V. Fatigue Fract. Eng. Mater. Struct. **37**, 1075–1086 (2014)
30. Wood, S.N.: Generalized Additive Models: An Introduction with R. Chapman & Hall/CRC (2006)

MIX
Papier aus verantwortungsvollen Quellen
Paper from responsible sources
FSC® C105338

If you have any concerns about our products,
you can contact us on
ProductSafety@springernature.com

In case Publisher is established outside the EU,
the EU authorized representative is:
Springer Nature Customer Service Center GmbH
Europaplatz 3, 69115 Heidelberg, Germany

Printed by Libri Plureos GmbH
in Hamburg, Germany